ORE DEPOSITS
OF THE USSR
Volume I

ORE DEPOSITS OF THE USSR
Volume I

Edited by V I Smirnov

Pitman Publishing

LONDON · SAN FRANCISCO · MELBOURNE

Pitman Publishing Ltd
Pitman House, 39 Parker Street, London WC2B 5PB, UK

Pitman Publishing Corporation
6 Davis Drive, Belmont, California 94002, USA

Pitman Publishing Pty Ltd
Pitman House, 158 Bouverie Street, Carlton, Victoria 3053, Australia

Pitman Publishing
Copp Clark Ltd
517 Wellington Street West, Toronto M5V 1G1, Canada

Sir Isaac Pitman Ltd
Banda Street, PO Box 46038, Nairobi, Kenya

Pitman Publishing Co SA (Pty) Ltd
Craighall Mews, Jan Smuts Avenue, Craighall Park,
Johannesburg 2001, South Africa

First published in the USSR by Nedra Press, Moscow 1974
First English edition published by Pitman Publishing Ltd 1977
Translated by D A Brown

Reproduced and printed by photolithography at
Biddles of Guildford

ISBN 0 273 01034 4

PREFACE

Deposits of all metals have been revealed in the confines of the Soviet Union, including huge accumulations that are of extremely great commercial value. We shall consider the ore deposits of all genetic types that have been formed during all metallogenic epochs, and in all the most important regional and local geological structures that are capable of controlling the origin and distribution of ores of various metals.

Ore deposits of pre-Palaeozoic, Caledonian, Hercynian, Cimmerian, and Alpine metallogenic epochs are known in the USSR. *Formations of pre-Palaeozoic age* have been concentrated in the base of the Russian and Siberian Platforms, and also in blocks of the very ancient associations of the folded regions in all the subsequent cycles of geological development. Various schemes of subdivision into geological cycles and corresponding metallogenic epochs have been proposed for them. If, without such subdivision, we speak of this period as a whole, then we may assign to it the greatest deposits of iron ores, and also quite substantial deposits of titanium, nickel, copper, zinc, lead, and gold.

The Caledonian metallogenic epoch, most completely defined along the southern margins of the Siberian Platform, in Kazakhstan and in the Urals, is represented by a limited, yet quite important range of deposits of iron, copper, zinc, lead, and gold. *The Hercynian epoch*, which is reflected in the folded regions of Siberia, the Urals, Kazakhstan, Middle Asia, and the Caucasus, is distinguished by a variety of ore deposits, which include those of iron, chromium, titanium, vanadium, nickel, cobalt, aluminium, copper, zinc, lead, antimony, mercury, tin, tungsten, molybdenum, and other rare metals, gold and platinoids. The *Cimmerian and Alpine metallogenic epochs*, most completely observed in the Far Northeast, the Soviet Far East, in Transbaikalia, and in the Caucasus, are especially remarkable for deposits of gold and silver, and also lead and zinc, tin, tungsten and molybdenum, and antimony and mercury.

In our country, ore deposits of all phases of the geosynclinal and platform stages of geological development have been identified.

In the early phase of all the geosynclinal cycles, especially during that of the Hercynian cycle, magmatic deposits of chromites and titanomagnetites associated with basaltic magmatism, skarn deposits of iron, and also pyritic ores of copper, zinc, and lead, have been formed. During the middle phase of the geosynclinal stage, corresponding to the period of the principal phases of folding and injection of batholithic masses of granitoids, pegmatite, greisen, and high-temperature hydrothermal deposits of tin, tungsten, lithium, and beryllium, were formed. The concluding phase of the geosynclinal periods, transitional to the platform regime, has been distinguished by an unusual variety of plutonogenic and volcanogenic hydrothermal deposits of non-ferrous, rare, noble, and radioactive metals.

During the phase of platform stabilization, exogenic deposits of aluminium, iron, manganese, and uranium were formed in the sediments of the platform cover. During the phase of resonance and autonomous tectonomagmatic activation of the platforms, magmatic deposits of copper-nickel sulphide ores, diamonds in kimberlites, carbonatite deposits of rare elements, and hydrothermal accumulations of non-ferrous, rare, and noble metals developed.

Ore deposits of all three genetic series (endogenic, exogenic, and metamorphogenic) are known in the Soviet Union.

Amongst the endogenic deposits, formations of the magmatic group play a significant role, being represented by deposits of the copper-nickel sulphide ores of the eastern and northwestern regions of the country, the chromites and titanomagnetites of the Urals, the apatites of the Kola Peninsula, and the diamonds of Siberia. The pegmatites, which are widely distributed in all the ore provinces of the Soviet Union, serve as a source of non-metallic mineral raw materials, but deposits of rare metals are also found amongst them. The carbonatites, which are mainly restricted to the elements of tectonomagmatic activation of the Russian and Siberian Platforms, contain rare elements. Amongst the skarn deposits, there are those of iron and copper, associated with basaltic magmatism, especially evident in the Urals, and also deposits of tungsten, molybdenum, lead, zinc, and certain other metals, controlled by granite magmatism in the folded regions of the Far East, Siberia, Kazakhstan, Middle Asia, and the Caucasus.

Deposits of the feldspar-metasomatism group, the so-called albitites, known in the ancient structures of the Ukrainian and Aldan Shields, and also in the apical zones of granite stocks in Kazakhstan and Transbaikalia, contain commercial accumulations of rare metals. Greisen deposits, on the one hand clearly associated with pegmatites, and on the other hand, with albitites, are recognized on the basis of the amount of tin and tungsten in them. Commercial resources of ores of non-ferrous, rare, noble, and radioactive metals have been concentrated in a vast group of plutonogenic and volcanogenic hydrothermal deposits, which comprise the principal value of many ore provinces of the USSR. Pyritic deposits, known in Kazakhstan, Middle Asia, in the Urals, the Caucasus, the Rudnyi Altai, and in other regions, are important sources of copper, zinc, and lead. The stratiform deposits of Kazakhstan, Middle Asia, and Siberia are even more significant for the supply of these three metals.

In the exogenic series, the most significant are the classical sedimentary deposits of iron, manganese, and aluminium, known both on the platforms and in the folded regions of the Precambrian, Palaeozoic, and Mesozoic-Cainozoic. In the group of weathering deposits, including both residual and infiltration formations, the role of deposits of iron, nickel, aluminium, and uranium is considerable. Deposits of titanium have been concentrated in present-day and fossil placers, and deposits of gold, platinoids, tungsten, and tin have also retained their importance.

In the metamorphogenic series, we recognize the metamorphosed deposits of ferruginous quartzites of the type seen at Krivoi Rog, the Kursk Magnetic Anomaly (KMA), and other regions of the country. No ore deposits have been definitely identified amongst the metamorphic rocks proper, although some investigators assign veins of rich ferruginous-quartzite ores and certain others to them.

Thus, the variation in the geological structure in the Soviet Union has permitted our geologists to discover and assess the whole range of series, groups, and classes of deposits of heavy, light, non-ferrous, rare, noble, and radioactive metals. But this does not mean that all the varieties of deposits known on our planet are included in their ambit. We must mention here the deposits of metalliferous conglomerates of the South African type, the mercury ores of the North African type, the lateritic bauxites of the type seen in Guinea and Australia, molybdenum of the Climax type in the USA, and in general we must develop new large and rich deposits of all metals. The features of the geological structure of our country, progress in the theory of ore-formation, and the attempts to prospect, explore, and assess the ore deposits, instil confidence in the reality of future discoveries.

The present three-volume work is directed towards that which we can locate and that which we still do not possess. The principal object of the book is to give some idea of the most characteristic deposits of the major metals in the Soviet Union. In all, we have selected 28 metals or their groups, to the description of which 42 geologists, out of those who know them most completely and who have made substantial contributions with respect to them, have contributed, based on long years of personal study.

The authors have agreed on an approximately common order of presentation of the material, in accordance with which there is first of all a classification of the deposits of each metal, and then depending on the grouping adopted, a description of the most representative deposits is presented. Although the commonly-known order with recognition of series, groups, and classes is used as the basis of the classification, reflected in this preface, the authors have not stuck to the rigid frameworks of such a classification. On the contrary, most of the authors have used their own variations depending on the features of the conditions of formation of the deposits of the various metals, and also the individual approaches of the authors to a genetic classification of the ore deposits and the practical assessment of the groups, classes, associations, and types adopted. The descriptions of the deposits also reflect the individual whims of the authors. The deposits of certain metals have been described with minute details of their geological location, structure, and composition; deposits of other metals have been defined in a more abbreviated fashion.

It is completely natural that the editor of the present group monograph has not considered it proper to interfere with the scientific principles of the authors, leaving to them the responsibility for the approaches to the classification of the deposits and the manner of their description. The editor's objective has been to compile the general plan and program of the work, the organization of the preparation of the manuscript and the illustrations, the elimination of clearly defined contradictions in the statements of the various authors, the scrutiny of the scientific and literature style in the way that is possible for the present multilateral work, and the difficult task of cramming all the information into the limits allotted to it. In the latter situation we have no possibility of describing all or even the most significant deposits and have been restricted to only some of them, in most cases the most typical.

We have attempted to create a work, defining a factual picture of the ore deposits of our country, without setting a goal of discussing theoretical and especially genetic problems. However, the information in the three-volume work presents a thorough basis for the most varied constructions in the field of theory of ore-formation.

For the recognition and study of the ore deposits of the Soviet Union, we are obliged to a large group of our talented geologists. The number of people who have played a particular role in the discovery, exploration, study, and assessment of individual deposits and their groups, is so substantial that their listing alone would occupy many pages; for economy's sake they have not been presented in full, with the exception of those cases when an actual reference to individual persons is made in a particular situation. The deposits of every metal have a large bibliography, which we also cannot present in its complete form, but each chapter is accompanied by a brief list of basic literature on the deposits of the particular metal.

Up till now there has been no such compilation of the group characteristics of the ore deposits of our country. However, the idea of preparing a work of such a kind aroused so much positive response in the widest geological circles, that this alone seemed sufficient justification to take the complicated work through to its completion, regarding this as a first attempt, requiring improvement in the future. In its achievement, we were inspired not only by the moral support of leading mining geologists, but also by the actual scientific-organizational help, especially from V.I. Kazansky, N.P. Laverov, and A.D. Shcheglov.

We realize the inadequate finality of a number of sections in all three volumes, but at the same time we would hope that both the production as a whole and also its particular chapters in a sense will play a positive role in the development of the mineral – raw material basis of metallurgy and in the progress of our understanding in the vast sphere of the geology of ore deposits.

Academician V.I. Smirnov.

MAP REFERENCES

Note:- 5630/10410 = Lat. 56°30'N, Long. 104°10'E

Abagas, Krasnoyarsk Krai	5330/8840
Abail, Kazakhstan	4230/7014
Abakan, Krasnoyarsk Krai *ca*	5235/8940
Akkermanovka, Ovenburg Obl.	5115/5815
Alapaevo, Sverdlovsk Obl.	5752/6142
Amangel'dy, Kazakhstan	4807/6522
Angren, Kirgizia *ca*	4020/7000
Anzas, Krasnoyarsk Krai	5214/8930
Ayat, Kazakhstan	5250/6235
Bakal, Chelyabinsk Obl.	5456/5848
Bakchar, Tomsk Obl.	5701/8205
Balbraun, Karaganda Obl.	4744/6743
Batamshinsk, Kazakhstan	5036/5816
Belinsk, Kazakhstan	5125/6312
Berdsk-Maisk, Salair *ca*	5415/8510
Berëzovka, Chitá Obl.	5115/11940
Bokson, Buryat ASSR	5158/10020
Bol'she-Tokmak, Ukraine	4714/3544
Burshtyn, Ukraine	4916/2438
Central, Krasnoyarsk Obl.	5845/9855
Chadobets, Krasnoyarsk Obl.	5840/9851
Chiatura, Georgia	4219/4318
Dashkesan, Azerbaidzhan SSR	4029/4605
Durnovo, Kemerovo Obl.	5437/8521
Dzirula, Georgia	4205/4309
Forty Years of Kaz. SSR - - Molodëzhnoe	5015/5822
Goroblagodat, Sverdlovsk Obl.	5818/5943
Goryachegorsk, Krasnoyarsk Krai	5524/8855
Grushevsk-Basan, Ukraine	4735/3440
Gusevogorsk, Sverdlovsk Obl.	5845/5923
Ibdzhibdek, Krasnoyarsk Krai *ca*	5840/9850
Iksa, Arkhangel'sk Obl.	6138/4024
Ilikta, Khabarovsk Krai *ca*	5300/10500
Indygla, Krasnoyarsk Krai *ca*	5900/9400
Ingolets, Ukraine	4743/3316
Ir-Nimiisk, Khabarovsk Krai *ca*	5420/13430
Ivanovsk, Krasnoyarsk Krai *ca*	5900/9300
Ivdel', Sverdlovsk Obl.	6042/6024
Kachar, Kazakhstan	5327/6259
Kachkanar, Sverdlovsk Obl.	5845/5923
Karazhal, Kazakhstan	4802/7049
Keiv, Murmansk Obl.	6735/3800
Kempirsai, Kazakhstan	5036/5815
Kerchen, Ukraine	4520/3600
Khoshchevat, Ukraine	4818/2957
Kirgitei, Krasnoyarsk Krai	5816/9445
Kiya-Shaltyr, Kemerovo Obl.	5502/8828
Klevakinsk, Sverdlovsk Obl.	5635/6143
Klyuchevsk, Sverdlovsk Obl.	5707/6056
Kolpashevo, Tomsk Obl.	5820/8250
Kopansk, S. Urals	5500/5925
Korshunovo, Irkutsk Obl.	5630/10410
Kovdor, Murmansk Obl.	6733/3029
Kozlovsk, Sverdlovsk Obl.	5630/6110
Krasnooktyabr'sk, Kazakhstan *ca*	5120/6110
Krivoi Rog, Ukraine	4755/3324
Kusa (Kusinsk), Chelyabinsk Obl.	5521/5928

Laba (Labinsk), Krasnodarsk Obl.	4439/4044
Lisakovo, Kazakhstan	5240/6430
Livanovka, Kazakhstan	5204/6200
Malyi Khingan (Bureya Range), Khabarovsk Obl.	ca 5000/13300
Mangyshlak, Kazakhstan	4429/5230
Matkal, S. Urals	ca 5450/5930
Medvedevsk, S. Urals	5510/3255
Mikhailovsk, Ukraine	5219/3519
Molodëzhnoe, Kazakhstan	5015/5822
Murozhnoe, Krasnoyarsk Krai	ca 5900/9300
Naurzum, Kazakhstan	ca 5230/6230
Near-Magnitogorsk, Chelyabinsk Obl.	ca 5330/5900
Nikopol', Ukraine	4734/3425
Nizhne-Angarsk, Buryat ASSR	5547/10933
Nizhneudinsk, Irkutsk Obl.	5455/9903
Novo-Berezovo, N. Urals	? 6050/6030
Obukhovo, Altaisk Krai	5410/8455
Oldakit, Khabarovsk Krai?	ca 5640/11300
Olenegorsk, Murmansk Obl.	6809/3318
Pervomaisk, Sverdlovsk Obl.	5736/6134
Pervoural'sk, Sverdlovsk Obl.	5654/5958
Peschansk, Sverdlovsk Obl.	5936/6010
Polunochnoe, Sverdlovsk Obl.	6052/6028
Porozhna, Krasnoyarsk Obl.	5810/9715
Pudozhgorsk, Karelian SSR	6215/3557
Punya, Krasnoyarsk Obl.	5923/9945
Revda (Revdinsk), Sverdlovsk Obl.	5648/5957
Rudnogorsk, Irkutsk Obl.	5715/10342
Sagan-Zaba, Baikal	? 5150/10410
Samokhvalovsk, Kurgansk Obl.	5639/6443
Saranovsk, Permsk Obl.	5830/5853
Sarbai, Kazakhstan	5258/6307
Sheregesh, Kemerovo Obl.	5257/8802
Shorzha, Armenian SSR	4030/4518
Sokolovsk, Kazakhstan	5258/6313
Sulaksha, Krasnoyarsk Obl.	ca 6000/9300
Taëzhnoe, Amursk Obl.	5153/12840
Tagar, Krasnoyarsk Krai	ca 5840/9910
Takhta-Karacha, Uzbek SSR	3916/6654
Taldy-Éspe, Kazakhstan	4712/5837
Tashtagol, Kemerovo Obl.	5247/8753
Tatar (Tatarskoe), Krasnoyarsk Krai	ca 5900/9300
Taunsor, Kazakhstan	5109/6220
Tëya, Krasnoyarsk Obl.	5257/9015
Tikhvin (Tikhva), Leningrad Obl.	5939/3331
Tsentral'noe, Krasnoyarsk Obl.	5845/9855
Twenty Years of Kaz. SSR	5015/5822
Ulutelyak, Bashkirsk ASSR	5455/5700
Usa, Kemerovo Obl.	5405/8845
Verblyuzh'egorsk, Chelyabinsk Obl.	5300/6307
Verkhoturovo, Krasnoyarsk Krai	5824/9518
Vezhayu-Vorykva, Komi ASSR	6444/4940
Vislovsk, Ukraine	ca 5040/3630
Vysokopol'e, Ukraine	4729/3331
West (Zapadnyi) Karazhal, Kazakhstan	ca 4800/7000
Yakovlevsk, Ukraine	5051/3627
Yenda, Krasnoyarsk Obl.	5835/9554
Zaglik, Azerbaizhan SSR	4032/4602

CONTENTS

DEPOSITS OF IRON

CLASSIFICATION OF IRON-ORE DEPOSITS

Iron-ore deposits have been formed under different geological conditions, which naturally also lead to variation in the mineral aspects of the iron ores. Up till now, the grouping and classification of iron-ore deposits have been presented according to the nature of the principal ore mineral, the assumed processes of formation, the associations of the country rocks, the local geological conditions, the association with regional tectonic structures, etc.

In the present monograph, the iron-ore deposits have been separated on the basis of genetic groups (magmatic, contact-metasomatic, hydrothermal, marine and continental sedimentary, weathering crust deposits, and metamorphosed types), and within each group, classes or associations are recognized on the basis of features of the mineral composition of the ores and the geological position. The assignment of actual iron deposits to a particular genetic group is determined in most cases quite clearly. However, the origin of a number of deposits, especially those of the endogenic type, is often disputable. This is explained by the fact that certain investigators of deposits, normally assigned on the basis of features of their geological position and mineralization to the magmatic, skarn (contact-metasomatic), or hydrothermal groups, have found a primary-sedimentary or volcanogenic-sedimentary origin for the corresponding ore concentrations, which in their opinion, subsequently underwent regional, contact, or hydrothermal metamorphism.

Such viewpoints have been stated, for example, about the origin of the large magnetite deposits of the Turgai province, Gornaya Shoriya (Chuguev-skaya, 1969; Derbikov & Rutkevich, 1971), the earlier exploited group of Bakal sideritic deposits in the South Urals, and the widely known magnetite-apatite deposits of Sweden (Kiruna, etc.). For a long time, the origin of the rich ores in the association of Precambrian ferruginous quartzites (in particular, the origin of the large segregations of rich ores in the Krivoi Rog iron--ore basin) has been explained in different ways: some geologists have considered them to be primary-metamorphogenic; for others, the rich ores have been formed primarily from low-grade ferruginous quartzites as a result of ancient weathering effects on them.

Such formulation of the problem on the genetic assignment of a number of iron-ore deposits is completely justified, although in each actual case a reliable basis for hypotheses on the primary-sedimentary or primary-metamorphogenic (endogenic in the broad sense) origin of ore segregations is necessary, which is not always the case.

In future, in denoting the genetic assignment of a particular deposit, the twofold origin (when ore segregations and bodies primarily have a different origin from that indicated by the observed geological position, structure, and mineral composition of the deposits), will only be accounted for in those cases when it is quite reliably and convincingly based. In such cases, one and the same deposit will appear in two different groups. The deposit may also be assigned to two groups, and if a thick zone of oxidation is developed in it, then this part of the deposit is assigned to the 'weathering crust' group.

In regard to deposits not affected by substantial metamorphic trans-formations, there is no doubt in most cases about their assignment to a partic-ular group.   In these cases, investigations and discussions are already direc-ted towards the clarification of details under conditions and processes of the formation of deposits, and not to their assignment to a particular genetic group.

Taking account of all the above, a classification of the iron-ore dep-osits of the USSR is presented (Table 1);   some deposits have been assigned to the appropriate group provisionally, because their origin remains in doubt. Thus, the magmatic origin of the Kovdor and Afrikanda deposits is not recognized by some investigators, who consider the iron ores of these deposits to be metasom-atic.   Not all the investigators accept hypotheses on the contact-metasomatic formation of the magnetite magnesian-skarn deposits of Southern Yakutia (Taezhnoe, Pionerskoe, etc.), and regard them as metamorphic.

Recently, widely distributed albite pseudomorphs after scapolite have been found in the Abakan deposit, which suggests that this deposit belongs to classes 'c' and 'd' of the contact-metasomatic group.   The Kholzun deposit was primarily formed as a volcanogenic-sedimentary type, and then in individual places underwent contact metamorphism with the formation of typical skarn magne-tite ores.

During the history of the Earth, an alternation of epochs of intense and feeble formation of iron-ore deposits has been observed.   Depending on the degree of intensity of investigations and accumulation of new data, the recog-nition of iron-ore epochs becomes more firmly based.   We shall take advantage of the latest ideas of Nikol'sky & Kaukin (1968), based on time of publication, and will consider the formation of the principal classes of iron-ore deposits during the epochs so recognized.

During the A r c h a e a n   -   E a r l y   P r o t e r o z o i c
e p o c h   (A - Pt$_1$), the largest iron-ore accumulations in the USSR, as throughout the entire world, were formed in iron-ore basins with widely developed facies of primarily chemogenic and terrigenous-chemogenic sediments, later metamorphosed to ferruginous quartzites.   They have been localized in the Precambrian shields and in the basement of the Russian and Siberian Plat-forms.   They include the huge iron-ore basins of the KMA and the Krivoi Rog, and the smaller, but still large, basins in the more easterly parts of the Ukrainian crystalline shield and in the Baltic Shield (the Olenogorsk, Kostamu-ksh deposits, etc.).   The total resources of iron in the ferruginous quartzites of the USSR comprise hundreds of billions of tonnes, far exceeding the resources of all the other classes and groups of iron-ore deposits.

During the L a t e   P r o t e r o z o i c   e p o c h   (Pt$_{2-3}$) dep-osits of ferruginous quartzites on a considerably smaller scale developed in the massifs and cores of Precambrian rocks in the regions of younger folding in Northern Kazakhstan (Karsakpai), the Sayany (Sosnovyi Baits), and in the Malo--Khingan massif in the Ussuri Basin in the Soviet Far East.   In addition, weakly metamorphosed sedimentary hematitic and sideritic deposits have been developed in miogeosynclinal sedimentary deposits: the Komarovo-Zigazinsk, Akhten, and other deposits on the western slope of the South Urals, the Lower Angarsk and other deposits of the Angara-Pit basin in Western Siberia, the Dzhetym deposit

Table 1. *Classification of Iron-Ore Deposits of the USSR*

Genetic group	Class (Association)	Deposit [1]
Magmatic	Low-titanium magnetite, in intrusives of the dunite-pyroxenite-dunite association	Kachkanar, Gusevogorsk, Pervoural'skoe (Urals); Lysansk (East Sayan)
	Titanomagnetite-ilmenite, in gabbroic and gabbro--amphibolitic intrusives	Kusinsk, Kopansk (Southern Urals)
	High-titanium titanomagnetite, in gabbroic and gabbro-diabase intrusives	Pudozhgorsk, Koikar (Karelia); Kharlovo (Altai)
	Perovskite-titanomagnetite and apatite-magnetite, in alkaline-ultramafic intrusives with carbonatites	Afrikanda, Kovdor (Kola Peninsula)
Contact-metasomatic	Magnetite calc-skarn	Magnitogorsk, Vysokogorsk, Lebyazhinsk, Goroblagodat, North Peschansk, etc. (Urals); Adaevo and other deposits of the southern half of the Turgai iron-ore province, Dashkesan (Azerbaidzhan), Atansor (Central Kazakhstan); Belorets and Kholzun (Gornyi Altai); Tashtagol', etc. (Altai-Sayan district); Chokadam-Bulak (Tadzhikistan)
	Magnetite magnesian-skarn and magnesian-calc-skarn	Teya (Kuznets Alatau); Kazsk, Sheregesh (Gornaya Shoriya); Zheleznyi Kryazh (Iron Ridge) (Eastern Transbaikalia); Taezhnoe, Pionersk (Southern Yakutia)
	Scapolite-albite and scapolite-albite-skarn magnetite	Kachar, Sarbai, Sokolovsk (Turgai province); Goroblagodat (Urals); Anzas (West Sayan)

[1] In this table, almost all the significant deposits are shown that possess proved reserves of over 100 M tonnes. Smaller deposits are indicated in those cases when significant deposits are not known for comparison.

Genetic group	Class (Association)	Deposit [1]
	Magnetite and hematite, hydrosilicate	West Sarbai (Turgai province); Abakan (Khakassia); individual sectors of deposits of preceding classes
Hydrothermal	Magnomagnetite, associated with traps	Korshunovsk, Rudnogorsk, Tagar, Neryuda, etc. (Eastern Siberia)
	Magnetite specularite, intensely metasomatic	Paladaur (Georgia); Kutimsk (western slopes of the Northern Urals)
	Iron-carbonate vein--metasomatic	Bakal (Southern Urals); Abail (Southern Kazakhstan)
Marine sedimentary (weakly metamorphosed and unmetamorphosed)	Sideritic (brown-ironstone in the zone of oxidation) layered, in marine terrigenous-carbonate sediments	Komarovo-Zigazinsk, Katav--Ivanov, and other groups (Southern Urals)
	Hematitic, in marine carbonate-terrigenous sediments	Nizhne-Angara (Eastern Siberia)
	Hematitic and magnetite--hematitic, in eruptive--sedimentary sequences	Atasu group (Central Kazakhstan); Kholzun (Gornyi Altai)
	Siderite-leptochlorite--hydrogoethite, pisolite--oolitic, in marine carbonate-terrigenous sediments	Kerchen (Crimea); Ayat (Turgai province); Bakchar (Western Siberia)
	Magnetite, partially titaniferous marine placers	Modern 'black' beach sands of the coasts of the Black, Caspian, and Japan seas; fossil beach sands in Azerbaidzhan, etc.
Continental sedimentary	Hydrogoethite, pisolite--oolite, lacustrine--paludal	Large number of small deposits on the Russian Platform and other parts of the Union
	Siderite-leptochlorite--hydrogoethite, pisolite-oolite, naturally alloyed with chromium and nickel, lacustrine--paludal, associated with weathering crust of ultramafic rocks	Orsk-Khalilovo group (Southern Urals); Serovo (Northern Urals); Malka (Northern Caucasus)

Genetic group	Class (Association)	Deposit [1]
	Siderite (brown-ironstone in zone of oxidation) hypergene-metasomatic, in littoral-lacustrine coarsely-clastic, predominantly carbonate sediments	Berezovo (Eastern Transbaikalia)
	Siderite-leptochlorite--hydrogoethite, in ancient fluvial sediments	Lisakoyo (Turgai province); Taldy-Éspe, etc. (Northern Aral region)
	Predominantly martite eluvial-deluvial (cobbly)	Vysokogorsk (Central Urals)
Weathering crusts (residual and infiltration)	Goethite-hydrogoethite (brown-ironstone) and martite-hydrogoethite zones of oxidation of deposits of sideritic and skarn-magnetite ores	Bakal, etc. (Southern Urals); Berezovo (Eastern Transbaikalia); Vysokogorsk (Urals)
	Goethite-hydrogoethite, ocherous, naturally alloyed with chromium and nickel, in weathering crust of ultramafic rocks	Yelizavetinsk (Central Urals)
	Hydrogoethite, in eluvial--deluvial sediments in karst limestones	Alapaevo (eastern slopes of Urals)
	Martite and hydrohematite, in ferruginous quartzites	Yakovlevsk, Mikhailovsk, etc. (KMA); Saksagan' group (Krivoi Rog)
Metamorphic (metamorphosed)	Precambrian ferruginous quartzites	Krivoi Rog, Kremenchug, Belozero, Mariupol' (Ukraine); Olenegorsk (Kola Peninsula); Kostamukshsk (Karelia); Karsakpai (Central Kazakhstan); Malyi Khingan, Ussuri (Soviet Far East)
	Magnetite and magnetite--specularite contact--metamorphosed sedimentary (with relicts of sedimentary iron ores)	Kholzun (Gornyi Altai)

in the Southern Tyan'-Shan', and others, and during this same epoch, some magmatic deposits were formed (the Pudozhgorsk and Koikar in Karelia, and the
Kusinsk, Kopansk, and others in the South Urals.

The e a r l y P a l a e o z o i c e p o c h (Cm - O) is relatively poorly productive in iron ores.   Only magmatic and contact-metasomatic
deposits of limited reserves are associated with this epoch in the Altai-Sayan
folded region (the Patynsk, Kul'-Taiginsk, Kondomsk, and Shamansk), and in
Northern Kazakhstan (Atansor).

The m i d d l e P a l a e o z o i c e p o c h (S - D_1) is characterized by extremely significant iron mineralization, associated with the concluding magmatic phase of Caledonian tectogenesis.   Large magmatic deposits were
formed during this epoch (the Kachkanar, Gusevogorsk, and Pervoural'sk in the
Urals, and the Teya, Irbinsk, and Krasnokamensk in the Altai-Sayan region).
Sedimentary and metamorphic deposits were not formed during this epoch.

The u p p e r P a l a e o z o i c e p o c h (D_2 - P_1) is characterized by maximum development (as compared with other epochs) of contact-metasomatic magnetite deposits, associated with Hercynian tectogenesis.   These include the huge deposits of the Turgai Province (the Kachar, Sokolovsk, Sarbai,
etc.), deposits in the Urals (the Magnitogorsk and the Serovsk group), in the
Gornyi Altai (the Insk and Belorets), and in Central Kazakhstan (the Ken'-Tyube
and Togai).   Ore formation on a moderate scale has also been revealed in the
Gornyi Altai (the Kholzun and Korgonsk), in Central Kazakhstan (the Atasu group),
and on the western slope of the Urals (Pashiisk Group).

The L a t e P e r m i a n - T r i a s s i c e p o c h (P_2 - T),
like the early Palaeozoic, is comparatively poorly productive in iron-ore.   It
includes the small contact-metasomatic deposits of the Tyan'-Shan' (Turangly and
Chokadam - Bulak) and the substantially more significant hydrothermal magnomagnetite deposits of the trap region of the Siberian Platform (Korshunovo, Rudnogorsk, Tagar, Neryunda, etc.).   Commercial sedimentary iron-ore deposits of
this epoch are unknown.

During the J u r a s s i c - E a r l y C r e t a c e o u s
e p o c h (J - Cr_1) numerous, though small, deposits almost without commercial
value at the present time, developed, which belong to three classes: sedimentary
lacustrine-swamp, brown ironstones (Tul'sk, Vyatka, Lipetsk, etc.), naturally-
alloyed lacustrine-swamp types (the Serovsk and the Orsk - Khalilovo in the Urals,
and the Malka in the Caucasus), and infiltration weathering crusts (the Alapaevo
Group in the Urals).

During the L a t e C r e t a c e o u s - P a l a e o g e n e
e p o c h (Cr_2 - Pg) vast marine and continental sedimentary fluvial iron-ore
basins and deposits were formed on the young epi-Hercynian platforms and plates
(Ayat, Lisakovo, Kirovsk, and Near-Aral in Kazakhstan, and the Bakchar deposit
in the West Siberian iron-ore basin).   Endogenic deposits of this epoch comprise
only the Dashkesan contact-metasomatic deposit in the Caucasus and a few small
skarn deposits in Primor'e (the Ol'ga deposit, etc.).

During the N e o g e n e - Q u a t e r n a r y e p o c h (N - Q),
the large Pliocene Kerchen-Taman iron-ore basin was formed with marine sedimentary deposits of pisolite-oolite ores.

We must make individual mention of the epochs of ancient weathering, which led to the formation of thick seams of rich martite and hydrogoethite--martite ores in the weathering crust of ferruginous quartzites, and also naturally-alloyed iron-ores in the weathering crust of the ultramafics.

The ancient weathering of ferruginous quartzites, so widely expressed in the KMA and Krivoi Rog, developed during those periods when the Precambrian ore-bearing complexes with ferruginous quartzites were subjected to continental conditions and were exposed by erosion; it terminated after marine transgressions which extended over the Precambrian complexes.   A different age and varying duration of the periods of ancient weathering are suggested for the individual iron-ore basins.   Thus, in the Krivoi Rog during the early Proterozoic period, rich ores of an old weathering crust of ferruginous quartzites were formed, later overlain by deposits of the upper shale group and then metamorphosed along with all the Proterozoic deposits of the Krivoi Rog geosyncline. The second, late Proterozoic period of ancient weathering of Precambrian complexes was more widely and intensely displayed, and these complexes were exposed on the Earth's surface by orogenic movements at the end of the Proterozoic. During this period the bulk of the rich martite and martite-hydrohematite--hydrogoethite ores of the Krivoi Rog were formed.   Finally, in the latest pre-Tertiary and Tertiary phases of weathering, rich brown-ironstone ores were formed from the ferruginous quartzites, being characterized by a markedly smaller distribution and low commercial value.

Similar periods of ancient weathering are also recognized for the KMA, with this difference, that the Proterozoic weathering crust here is still the object of discussion.   Iron ores have also been formed in the ancient weathering crusts of ultramafics in the Urals, in Northwestern Kazakhstan, and partly in the Northern Caucasus.   In all the other regions, the ancient weathering crust is Triassic-Jurassic and partly Early Cretaceous age.   The naturally--alloyed ores formed in the ultramafics have a small distribution, although in other parts of the world, especially in the equatorial and subequatorial regions, ores of the ancient weathering crust, commonly termed lateritic, are extremely widely distributed.   One possible reason for their low distribution in the Soviet Union is the intense erosion, which has destroyed a large portion of the lateritic deposits.   The ancient weathering crust of the Jurassic-Early Cretaceous epoch is associated with infiltration deposits of the so-called Alapaevo type, which also has no significant resources.

A generalized view of the epochs of formation of iron-ore deposits of different genetic classes in the Soviet Union is given in Figure 1, in which the genetic groups of iron-ore deposits recognized in Table 1 are arranged along the horizontal axis, and epochs of formation along the vertical axis. The circles of varying dimensions indicate the total resources of ores in the deposits of each group based on epochs of ore formation.   Figure 1 clearly shows the basic trend of changes in the genetic groups of iron-ore deposits during the geological history of the USSR.

In the oldest Archaean - Early Proterozoic and Middle - Late Proterozoic epochs, there is a marked predominance in the formation of metamorphosed marine sedimentary deposits.   The next maxima of ore-formation occurred during the middle and late Palaeozoic, in the Silurian - Early Devonian and Middle Devonian - Early Permian, when deposits, already of different groups (magmatic and contact-metasomatic), developed.   And, finally, the third maximum of iron deposition took place in Mesozoic - Cainozoic time (Jurassic - Early Cretaceous,

Middle Cretaceous - Palaeogene, and Neogene epochs), when vast marine sedimentary iron-ore deposits, unaffected by metamorphism, were formed.

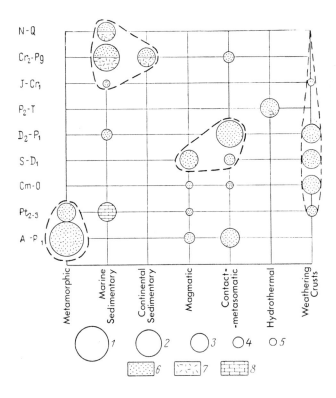

Figure 1.
Distribution diagram of iron-ore deposits of the USSR by different genetic groups based on iron-ore epochs in the Earth's history.

1-5) total reserves of iron ores: 1) hundreds of billions * of tonnes; 2) up to 100 billion tonnes; 3) up to 10 billion tonnes; 4) up to 1 billion tonnes; 5) a few hundreds of millions of tonnes; 6-8) mineral types of ores: 6) iron-oxide;
7) carbonate-silicate-oxide;
8) carbonate.

* Billion in US sense =
= 10^9.

During each of the three maxima listed, vast deposition of iron ores took place, the total resources of which comprise hundreds or tens of billions of tonnes. It is possible, with a considerable degree of probability, to advance reasons for the origin of each maximum. The conditions for intense ore-deposition during the Archaean - Proterozoic maximum, which is represented by metamorphosed marine sedimentary deposits, were, on the one hand, the formation of a thick and widely distributed Archaean - Lower Proterozoic weathering crust preceding the deposition of a substrate of ferruginous quartzites, with the substantial role of basic and ultramafic magmatic rocks or their metamorphic equivalents in the rocks which underwent weathering. On the other hand, the widespread occurrence of submarine basic volcanism contributed to the deposition of iron ores, as a result of which the waters of the marine basins were enriched in iron components. The second, middle and late Palaeozoic ($S - D_1$ and $D_2 - P_1$) maximum, in which magmatic iron-ore deposits are represented in geosynclinal regions during the evolving initial phase of the geosynclinal process, is associated with a period of especially widespread injection of basic magma from the upper mantle, as a result of which basic volcanogenic complexes of the spilite series and comagmatic differentiated plutons of ultramafics, basic rocks, and granitoids of increased basicity and in part alkalinity, developed. These plutons and their magmatic foci were the sources of iron during the formation of magmatic and contact-metasomatic deposits. Additional sources of iron in the deposits of the latter group were the surrounding rocks, from which iron was extracted by magmatogenic solutions and vadose waters, mixed with juvenile hydrotherms and discharged in the area of ore deposition.

The Mesozoic - Cainozoic maximum with marine and continental sedimentary deposits, is again associated with the preceding period of a widespread and intense ancient weathering. During the Late Triassic - Jurassic - Early Cretaceous epoch on the Siberian Platform and its margins, including the Turgai Basin and the Aral region, and during the Palaeogene in the Crimea - Caucasus region, thick weathering crusts were formed, and these were also a source of iron for the deposits of the Mesozoic - Cainozoic maximum.

MAGMATIC DEPOSITS

Low - titanium magnetite deposits in intrusions of the gabbro - pyroxenite - - dunite association are represented in the folded regions during the well-defined initial stage of geosynclinal development and the injection during the late phases of this stage of basic magmas, which formed differentiated intrusions of the gabbro-pyroxenite-dunite association. The ore concentrations have been localized mainly in the pyroxenite or hornblendite-pyroxenite and gabbroic facies of the intrusions, and less frequently in limited amounts in the peridotite and dunite facies. The ore bodies are zones of increased segregation, with schlieren and vein-lensoid segregations of low-titanium titanomagnetite in the corresponding intrusive rocks. The position of the ore zones and bodies in the intrusion is determined by the distribution of the petrographic zones in the differentiated intrusion, since the ore-bearing rocks belong to completely defined petrographic zones, recognized during a detailed petrographic mapping. Depending on the form of the intrusions and the arrangement of the petrographic zones in them, the ore-bearing zones lie either gently or steeply (up to vertical). The boundaries of the zones of ore segregation are usually not sharp, with a rapid but gradual transition into non-ore rocks, and the boundaries of the schlieren and vein-like ore accumulations are often sharp.

The principal ore mineral of the deposits of this class is titanomagnetite with a dissociation texture of a solid solution, containing a limited amount of finely banded ingrowths of ilmenite. Grains of magnetite, ilmenite, and spinel are present in subordinate amounts. The accompanying gangue minerals consist of rock-forming silicates of the surrounding basic rocks or ultramafics (olivine, pyroxene, plagioclase, and their hysterogenic products, amphibole, epidote, zoisite, chlorite, serpentine, sphene, etc.).

The ores contain tenths of a percent of vanadium pentoxide, usually hundredths of a percent of sulphur, and hundredths and thousandths of a percent of phosphorus. In some deposits disseminated platinum has been recorded. In addition to those listed in Table 1, numerous deposits of this class are known on the eastern slope of the Urals, although for the most part they have no great commercial importance. Foreign deposits of this class are not known. Examples in the USSR are the Kachkanar and Gusevogorsk deposits.

Titanomagnetite - ilmenite deposits in gabbroic and gabbro - amphibolitic intrusions are known only on the western slopes of the Southern Urals, where the ore-bearing intrusions, which have the form of gigantic dykes, cut Upper Proterozoic deposits in rises of ancient Baikalian structures in the cores of anticlinoria of the Hercynian orogeny.   The intruded complexes have a miogeosynclinal character and do not contain volcanics and rocks of the vol- canogenic-sedimentary associations.   In direct contact with the basic ore-bear- ing intrusions there are later granite massifs, the rocks of which have been metamorphosed to gneiss-granites.   The basic intrusions have a banded structure with an alternation of gabbros, subordinate pyroxenites, anorthosites, and horn- blendites.   The basic rocks have undergone intense metamorphism resulting from tectonic effects and the granitic intrusion, as a result of which various amph- ibolites, and also biotite and biotite-chlorite schists are quite widely distr- ibuted here.

The titanomagnetite ores are late-magmatic differentiates of a basic magma, enriched in varying degree in ore minerals, with the formation of uniform and segregated ores.   They have been subjected, along with the country rocks, to the metamorphism noted.

Metamorphosed ores are a characteristic feature of the Southern Urals deposits of this group.   Metamorphism exerted a specific effect on the titano- magnetites and magnetite-ilmenites, leading to selective recrystallization of ilmenite, and solution and redeposition of magnetite, and has significantly incr- eased the amount of individualized grains of ilmenite.   The primary gangue min- erals have been converted during the metamorphism into amphibole, epidote, actin- olite, chlorite, and partly into biotite.   The metamorphism was accompanied by plastic deformation of the amphibolites and vein-like ore bodies, as a result of which fold- and flexure-like curves in the latter have appeared along the dip and strike.   However, metamorphism has been manifested with differing deg- rees of intensity in the various ore-forming intrusives.

In addition to the deposits listed in Table 1, this class includes the South Kusinsk, Chernorechensk, Matkal, and other deposits.   Similar types have not been described in foreign countries.   As an example, the Kusinsk deposit will be described.

High - titanium titanomagnetite dep- osits in gabbroic and gabbro - diabase intrusions are known in Karelia.   They are localized in intrusions consisting of banded-differentiated gabbros and gabbro-diabases.   The ore- -bearing intrusions occur in the region of late Karelian folding among Proterozoic metamorphosed rocks.   The ore bodies represent a facies of gabbro-diabases, en- riched in varying degree (up to extremely significant) in titanomagnetite and ilmenite.   Such bodies are conformable with the banding of the rocks in the in- trusion, having layered- and dyke-like forms, maintained along the strike and dip, and gradually pass through low-grade segregated ores into non-ore gabbro- -diabases.

In the USSR three representatives of this type are known (see Table 1). Abroad, this class may generally include the Taberg and some other deposits in Sweden.   As an example, the Pudozhgorsk deposit will be described briefly.

Perovskite - titanomagnetite and apatite - magnetite deposits in alkaline - - ultramafic intrusions of the central t y p e are known on the Baltic Shield and on the Siberian Platform.   The origin of such ore-bearing intrusions is associated with the tectonomagmatic

activation of the shields and platforms.   The intrusions have a zoned structure, and are annular in plan, with ultramafic rocks (olivinites, pyroxenites,
and peridotites) in the central parts, and alkaline rocks (ijolites, melteigites, etc.) on the periphery.   Carbonatites are developed to a greater or lesser degree in the form of veins, lenses, and sometimes uniform fields.   In the
intrusions with a predominance of ultramafic rocks, the most central, specifically ore-bearing zone is formed by coarse-grained pyroxenites with an eruptive
breccia of olivinites, transected by abundant veins of nepheline-pyroxene pegmatites.   Calcite-amphibole-pyroxene and phlogopite metasomatites have been
developed through all these rocks.

The iron ores, in which the ore minerals are titanomagnetite and a
cerium perovskite (knopite), have been concentrated predominantly in the central
part of the intrusives.   Several types have been recognized: 1) veins of ore
nepheline-pyroxene pegmatites, phlogopitized to a significant degree; 2) titan-
omagnetite-perovskite-phlogopite veins; 3) separate nests and lenses in pyroxenites, enriched in titanomagnetite and knopite, also phlogopitized to a certain
degree; 4) small veinlet segregations or ore minerals in pyroxenites, accumulations of ore minerals at the contacts between the pyroxenites and calcite-amphibole-pyroxene metasomatites and in places in the selvedges of the olivinite
blocks.

In the intrusions, apatite-forsterite, apatite-calcite, and calcite
metasomatites after ultramafic rocks are distributed with considerable development of carbonatites.   The iron-ore bodies in such massifs are in the main
apatite-forsterite rocks with abundant segregations, veins, and veinlets of
low-titanium titanomagnetite.   They contain fine uneven segregations of pyro-
chlore and baddeleyite, which control the presence of zirconium and niobium in
the ore.

A magmatic origin with subsequent metasomatic effects is most likely
for the perovskite-titanomagnetite deposits in the intrusions with predominant
ultramafics with weak development of carbonatites.

In the case of the apatite-magnetite deposits in the intrusions with
significant development of carbonatites, a number of investigators advocate
their metasomatic formation.

In addition to the Afrikanda and Kovdor deposits listed in Table 1,
several small deposits of this class are known on the Kola Peninsula and in
Karelia, which do not have any real commercial value.   On the northern margin
of the Siberian Platform, in the huge Gula alkaline-ultramafic intrusion and in
Yakutia, in the Arbarakhta intrusion of the same type, identical titanomagnetite
deposits have been found, which are still poorly known.   Abroad, deposits of
such type include those of Africa such as Sukuku (Uganda), Lulekop (South
Africa), and Dorowa (Rhodesia).   As an example, we shall use the Kovdor deposit.

CONTACT-METASOMATIC DEPOSITS

The contact-metasomatic magnetite, in part hematitic, deposits, are
widely distributed, and are known in large numbers, including extremely large
ones with reserves of up to one billion and more tonnes.   Common features of
all the iron-ore deposits of this group are their occurrence in the zones of

contact aureoles of granitoids and their accompaniment by zones of peri-ore alterations of the country rocks, comprising skarns and other high-, medium-, and low-temperature metasomatites.   In some deposits of this group, in spite of the development of high-temperature parageneses, active intrusions have not been found in the accompanying metasomatites.   It is likely that the latter lie at a considerable depth (down to 2-3 km), which, for example, has been confirmed for the Kachar deposit by geophysical data.   G.A. Sokolov has proposed that such deposits be termed 'remote contact' types.

Investigators have shown in the last two decades that contact-metasomatic deposits exist, in which the peri-ore alterations consist in the main not of skarns, but of scapolite, albite-scapolite, and hydrosilicate metasomatites.   Amongst the skarns, two essentially different types have been recognized (calcic and magnesian).   Deposits with different types of peri-ore metasomatites have been formed under somewhat different geological conditions.   All this has also served as a basis for subdividing the contact-metasomatic iron-ore deposits into classes (Sokolov, 1967), listed in Table 1.

The   m a g n e t i t e   c a l c - s k a r n   d e p o s i t s   are localized in geosynclinal volcanogenic-sedimentary complexes with volcanics of predominantly basic and intermediate composition, which contain layers or seams of carbonate rocks (limestones or dolomites), and also calcareous tuffs and tuffites.   These complexes are typical of geosynclines of the femic type and belong to the early phases of their development.   An essential condition for the development of skarn magnetite deposits is the injection of granitoid intrusions, mostly of increased basicity and alkalinity (diorite, granodiorite, granosyenite, and syenite compositions), into a volcanogenic-sedimentary sequence.   These intrusions are the product of a gabbroic magma.   Only rare and small deposits are known in the contact aureoles of granitoid intrusions of the granite series.

Calc-skarn magnetite deposits are restricted to hypabyssal intrusions alone;   calc-skarns have not been developed in abyssal depths.

The injection of active intrusions occurred in zones of deep-seated faults, which also continued to develop after the rise of the magmatic masses, when the magnetite calc-skarn deposits were formed.   Localization of the deposits was controlled by a system of plastic and disrupting fractures, linked with deep-seated faults.   The disrupting fractures, which affected both the exocontact regions and the endocontact zones of intrusions, appeared in the form of faults or thrusts, zones of jointing and brecciation, and exposures of interlayer partings.   On these grounds, various forms of ore-bodies have been observed (irregular, vein-like, and layer-like).   The layer-like skarn-ore bodies, conformable with the stratification of the country rocks are typical, and they develop during infiltration of solutions along the opened-up interstratal partings.

The formation of iron ores in these deposits is one of the events in the general contact metasomatic process.   The ores replace the mainly earlier--formed skarns.   It has been established, however, that there is a direct replacement of the skarns of the primary surrounding rocks (limestones, dolomites, tuffs, and tuffites, in particular, calcareous) by ore parageneses along the periphery.   It is typical that the thin-bedded uniform carbonate rocks have been subjected in lesser degree to metasomatic replacements by skarns and ores, and the fine-bedded facies with alternation of limestones, volcanics, tuffs, and tuffites are more amenable to metasomatic replacements and mineralization.

Deposits, for instance, at Dashkesan, are also known in which the volcanogenic facies, devoid of layers and seams of carbonate rocks, have been subjected to skarnification and mineralization.   However, these carbonate rocks are present in the underlying horizons, where they have not been affected by skarnification.   In some cases, facies of sedimentary iron ores participate in the composition of the primary volcanogenic-sedimentary sequences, which surround the contact-metasomatic iron-ore deposits.   These facies have also been subjected to metasomatic transformations with conversion into skarn-ore bodies, amongst which there are relicts of sedimentary ores.   But such deposits are rare, and the most convincing example is the Kholzun deposit in the Gornyi Altai.

Some investigators suggest a considerably wider distribution of the primary-sedimentary contact-metamorphic deposits, arising mainly from layer--like forms of skarn-ore segregations and their conformable attitude with the country rocks.   However, in the absence of clear relicts of sedimentary ores, the form and attitude of the skarn-ore bodies cannot serve as a convincing argument in favour of the primary-sedimentary origin of the ore segregations.

The following are involved in the mineral association of the calc--skarn deposits: 1) skarn minerals proper of the pyroxene-salite type, and garnets of the andradite-grossular series; 2) hysterogenic minerals (epidote, zoisite, actinolite, chlorites, and idocrase); 3) iron-ore minerals (magnetite, muschketowite, and hematite (in the oxidation zone, martite)); 4) sulphide minerals (pyrite, pyrrhotite, chalcopyrite, sphalerite, etc.); 5) late gangue minerals (calcite and quartz).

The most typical trace-elements in the ores of the calc-skarn magnetite deposits are: cobalt (mainly in pyrites, sometimes in the form of cobalt sulphides [cobaltite, glaucodot] (Dashkesan deposit); vanadium (as an isomorphous additive in magnetites and ferruginous pyroxenes or amphiboles); copper (in chalcopyrite and other copper sulphides); and zinc (in sphalerite).
It is impossible to list here all the numerous deposits of this class in the Soviet Union; only the principal examples are given in Table 1. Calc-skarn magnetite deposits are just as numerous abroad: they are known in California, Utah, New Mexico, etc. (USA); in the Erzgebirge (Central Europe); in Italy, Romania, Japan, China, and other countries. However, there are no particularly large deposits amongst them.

As examples, we shall describe the North Peschansk, Goroblagodat, Tashtagol, and Dashkesan deposits. No new data have been obtained recently on the widely known, classical representatives of this class, the Vysokogorsk and Magnitogorsk, repeatedly described in review publications; therefore they have not been described.

M a g n e t i t e,   m a g n e s i a n - s k a r n,   a n d m a g n e s i a n - c a l c - s k a r n   d e p o s i t s were not recognized earlier, since the association of magnesian skarns has been clarified and investigated only in the last two to three decades.   In contrast to the calc--skarn magnetite deposits, they have been formed both at hypabyssal and abyssal depths.   For the deposits of hypabyssal depth, the general geological situation is similar to that of the calc-skarn deposits, although the active intrusions usually belong to the granite association proper, and not to the acid product of gabbroid magmas.   In this respect, the hypabyssal magnesian-skarn magnetite deposits may also occur in the sedimentary complexes of the middle stages of development of the geosynclines.   In addition, the presence of dolomites is mandatory in the surrounding sequences.   In regard to the deposits of abyssal depths, they are typical of the Archaean complexes of the Aldan Shield (Taezhnoe and other deposits) and belong to the purely magnesian-skarn representatives; they have been formed in zones of assumed deep-seated faults and at the contacts between fields of granitization and gneiss-dolomite complexes.   There is an hypothesis on the sedimentary-metamorphic origin of such deposits, although the absence of strict stratigraphic control and extremely clearly defined occurrences of precipitation of magnetite during the process of skarnification of the dolomites contradict such views.

The formation of the deposits of the magnesian-skarn group is favoured by complexes containing dolomites, distributed in the regions of ancient shields and Precambrian folding, and also complexes of the middle stages of the development of eugeosynclines with dolomites and polyphase granitoid magmatism.

The magnesian skarns, including those associated with magnetite ores, have been formed either during the magmatic phase up to the crystallization of active intrusions in which cases endoskarns do not develop, or during the post--magmatic phase as contact-infiltration skarns at the contacts between the dolomites and gneisses or schists.   Post-magmatic magnesian skarns are not typical of hypabyssal depths, but magnesian skarns of the magmatic phase develop both under abyssal and hypabyssal conditions.

In the deposits of the hypabyssal series the formation of calc-skarns after magnesian skarns is commonly observed, and the magnesian skarns are preserved only in the form of relicts.   Mixed magnesian-calc-skarn magnetite deposits develop in this way.

The mineral associations of the magnetite magnesian-skarn deposits are characterized by the development of magnesian silicates (forsterite, an aluminous diopside-fassaite, and spinel);   the last minerals consist of phlogopite, humite, hornblende, pargasite, scapolite, serpentine, chlorite, brucite, and borates (ludwigite, etc.).

In the mixed magnesian-calc skarns, the high-temperature minerals of the magnesian skarns are present as relicts in the aggregate of minerals of the calc-skarns;   in the association of apo-skarn minerals, late minerals of both the magnesian skarns (especially serpentines) and the calc-skarns are developed.

The principal deposits of this class in the Soviet Union are listed in Table 1.   Of the foreign occurrences, we shall note the Adirondacks, Iron Hat, Key Canyon, etc. (USA), Malko-Turnovo (Bulgaria), Okna-de fer (Romania), and a number of others.   As examples, we shall describe the Taezhnoe, Teya, and Sheregesh deposits.

The   scapolite - albite   and   scapolite - - albite - skarn   magnetite   deposits,   like the calc-skarn deposits, occur in volcanogenic-sedimentary complexes of the early phases of development of eugeosynclines and have been associated with intrusions of acid products of gabbroic magmas.   All that has been said about the tectonic control of the calc-skarn magnetite deposits applies also to the scapolite-albite deposits.

The principal feature of the deposits of this group is appearance, unusual for other groups of contact-metasomatic iron-ore deposits, of an intense sodium-chloride metasomatism, which is expressed in the wide development of albite-scapolite metasomatites with subordinate pyroxene, replacing the various aluminosilicate rocks (with feldspars) of the ore field.   The carbonate rocks have not been affected by this metasomatism.   In the deposits of mixed type, following the albite-scapolite metasomatites, skarn metasomatites are developed both independently and superposed on the preceding albite-scapolite metasomatites.   In the deposits with predominance of albite-scapolite metasomatites (Kachar), the magnetite mineralization has occurred either in the form of segregated albite-scapolite-magnetite ores, or in the form of uniform magnetite ores, replacing the carbonate rocks simultaneously with scapolitization and albitization of the adjacent aluminosilicate rocks.   In the deposits of mixed type, the ore magnetite has been formed in the calc-skarn deposits at the end of precipitation of the skarn minerals proper or immediately after them.

The largest magnetite deposits of the Turgai iron-ore province (Kachar, Sarbai, and Sokolovsk) definitely belong to this class.   Moreover, the presence among the peri-ore metasomatites of albite-scapolite types does not certainly determine the large scale of the deposits.

Investigations of recent years (Pavlov, 1971) have shown that the intense development of sodium-chloride metasomatism in deposits of this class is explained by the special geological conditions of their formation, when the juvenile magmatogenic solutions in the sphere of ore deposition or along its paths could be mixed with brine waters of vadose origin, heated intensely by the magma.   These brines were also the source of chlorine and sodium, and contolled the intense sodium-chloride metasomatism with the formation of scapolite and albite-scapolite metasomatites.   One of the features in the participation of the vadose brines is the presence, noted in some deposits, of an hydrothermal anhydrite, so typical for example, of the Kachar deposit.

The above-noted special geological conditions have not been expressed within the deposits themselves.   They result in the existence of local depressions on the surface during the period of ore-formation (marine lagoons with increased salinity and marginal faults).   The clarification of such conditions has been achieved by an analysis of palaeogeographical and palaeohydrological data for the ore region.

The mineral associations of the deposits of this class are complicated as compared with those of the calc-skarn deposits by the presence of a sodic chlor-scapolite, albite, and anhydrite (not always), and by a widespread development of zeolites.   The early co-scapolite magnetite is frequently characterized by a small amount of titanium.

Deposits of this class are so far known in limited numbers and the principal examples are listed in Table 1.   Amongst them are deposits with unique large reserves.   Such deposits are as yet unknown abroad.

The magnetite and hematite hydro-silicate deposits of the contact-metasomatic group are characterized by a geological situation similar to that of the skarn deposits - they occur in the contact aureoles of intrusions, having been injected into volcanogenic-sedimentary complexes of the early stages of development of geosynclines. They have frequently been located in ore fields in common with skarn deposits, but are situated at some distance from the contacts with active intrusions. In addition, deposits are known (Abakan, etc.), in the region of which skarn deposits have not been found.   Transitional types have been observed in which the peri-ore hydrosilicate metasomatites contain relicts of high-temperature skarn minerals. Individual deposits of this class may be regarded as 'remote contact' types.

The features of structural control and morphology of the ore bodies in the hydrosilicate and skarn deposits are quite similar.

The mineral composition of the peri-ore metasomatites and ores of the hydrosilicate deposits reflects temperatures of commencement of their formation that are lower than those of the skarn deposits. The following minerals in varying ratios are involved in the peri-ore metasomatites: epidote, actinolite, sometimes albite, garnet, chlorite, zeolites, calcite and other carbonates, and quartz. Sometimes relicts of skarn minerals (pyroxene or garnet) are observed in such metasomatites, and in one case (the Abakan deposit), relicts of albite pseudomorphs after scapolite. The principal ore mineral of the ores is usually magnetite, but in individual cases, it is hematite in the form of specularite. In the magnetite ores there are also muschketowite and hematite (specularite). The remaining minerals of the ores, accompanying magnetite or hematite, include those that form the peri-ore metasomatites, and also sulphides of iron, copper, and zinc, arsenopyrite, and occasionally, safflorite.

The deposits of this class do not always have independent importance. Many of them are sectors of the skarn-magnetite deposits. The more significant are shown in Table 1. As an example, we shall describe the Abakan deposit.

HYDROTHERMAL DEPOSITS

The three classes recognized in this group are markedly different in their geological features of formation. The magnomagnetite and iron-carbonate classes contain deposits of great commercial importance. The deposits of the iron-mica class in the Soviet Union do not possess reserves regarded as of national importance.

M a g n o m a g n e t i t e   d e p o s i t s ,   a s s o c i a t e d w i t h   t r a p s , are known only in the Soviet Union. They have all been concentrated within the trap region of the Siberian Platform, where about 30 deposits and ore-shows have been found. The deposits occur in the Palaeozoic deposits of the platform cover. Their area of distribution is the same as that of the intrusive traps on the Siberian Platform, in which the lower horizons of the platform deposits contain facies of halide deposits. The presence of the latter is a specific feature of the Siberian Platform alone. Magnetite deposits are not known in them in any other trap association in the world.

The distribution of the magnomagnetite deposits within the platform is clearly associated with the distribution of zones of fractures and at the same time, intense manifestation of trap magmatism. Such zones form two belts: 1) those with northwesterly strike, along the western marginal portion of the trap region (from the middle course of the Angara River in the southeast to the River Kureika in the north), and 2) those with northeasterly strike, along the southeastern margin of the Tungusska Syneclise. These two belts coincide in the Angara-Ilim region. The magnomagnetite ore bodies are vein-metasomatic types and have been controlled by tectonic joints in the fracture zones. They are all accompanied by intrusive traps, forming sills and dykes, and less frequently intrusive veins.

The most important deposits occur in the region where the two above-mentioned belts of fractures come together. Here, the deposits have been restricted to specific structures (explosion pipes), distributed in this part of the platform as a direct result of increased tectonic activity, which has itself been controlled by the conjunction of the two fracture belts. The explosion pipes are cone-like bodies of irregularly rounded and oval shape in cross-section, narrowing at depth, cutting the Cambrian-Silurian stratified sequence of argillites, marls, sandstones, and limestones, and filled prior to the processes of metasomatism and mineralization with trappean tuffs, tuff-breccias, and agglomerates of Permian-Triassic age, with blocks and fragments of the surrounding sedimentary rocks and traps.

Along the tectonic fractures, which cut the bodies of the pipes and the surrounding sequence, dykes and sills of intrusive traps were injected, and then solutions penetrated which caused the metasomatic changes in the rocks and the mineralization, most intensely manifested in the bodies of the pipes, which were the most permeable. The metasomatic processes controlled the development of skarn-like and lower-temperature chlorite-serpentinite-carbonate metasomatites through the rocks which filled the pipes. The skarn-like metasomatites are made up of pyroxene and garnet.[1]

The lower-temperature metasomatites contain, in addition to chlorite and serpentine, actinolite, epidote, calcite, quartz, pyrite, zeolites, opal, and chalcedony. They have been developed both after the skarn-like metasomatites, and to a considerable degree also after the primary rocks of the pipes. The ores consist of zones of segregation in the metasomatites, metasomatic vein bodies, and filling veins. Although mineralization has been manifested mostly in the pipes themselves, the ore veins in a number of cases pass beyond the limits of the pipes, and cut the rocks surrounding the pipes.

The ore-forming magnetite in these deposits always contains an isomorphous addition of magnesium and belongs to the magnomagnetite variety or, otherwise, magnesiomagnetite. It has been emphasized that with increase in depth, the formation of deposits containing magnesium in the magnetite decreases. This is explained by lowering of the oxygen potential of the ore-forming solutions, since the entry of magnesium into the magnetite molecule occurs during the oxidation of portion of the divalent iron to the trivalent form.

In addition to the deposits listed in Table 1, this class includes the poorly studied deposits of the Tunguz, Bakhta, and Ilimpei regions of the Siberian Platform, and also the smaller deposits in the Angara-Ilim, Middle Angara, and Angara-Kat iron-ore regions. As examples, we shall briefly describe the Korshunovka, Rudnogorsk, and Tagar deposits.

Magnetite and iron - mica vein - metasomatic deposits, not belonging to the contact-metasomatic group,

---

[1] These rocks are usually simply termed skarns; V. Zharikov, who has separated them into a special group of 'auto-reaction skarns', has recorded that they develop as the result of the metasomatism of ultramafic and basic rocks, involving the participation of solutions rich in calcium.

are located among the volcanogenic-sedimentary or terrigenous-carbonate geosynclinal deposits at a distance from the intrusions.   In the Soviet Union, only
individual representatives of this class are known.   They are extremely limited
in reserves and have been poorly studied.   In the Paladaur deposits (Georgia),
segregated and massive hematite ores occur among hydrothermally-altered tuffs
and tuffogenic rocks of Cretaceous age.   During hydrothermal alterations, the
surrounding rocks have been replaced by chlorite, quartz, iron sulphides, and
rarely by garnet.   The massive ores form concordant and discordant veins, and
cement the brecciated rocks.

The Kutim magnetite-muschketowite-hematite deposit occurs in Proterozoic metamorphic terrigenous sediments on the western slope of the Northern Urals.
The small number of deposits and their poor state of investigation do not permit
us to make any significant generalizations about this class.

Iron - carbonate vein - metasomatic
deposits are localized in the limestone-dolomite sediments, involved
in the miogeosynclinal terrigenous-carbonate sequences of the folded regions.
The two well-known representatives of this class with commercial significance,
are located in regions of Hercynian folding (the Abail deposit in a Silurian
limestone-shale sequence, and the Bakal, in a sequence of limestones, dolomites,
and calcareous-clay shales in the rise of an ancient Proterozoic core of an Hercynian anticlinorium).   Volcanogenic formations are absent, and intrusive formations are represented in the Bakal deposit only by diabase dykes, and are absent
from the Abail deposit in general.   The surrounding rocks have been crumpled
into folds and have been broken into tectonic blocks with restricted amplitudes
of displacement.

The ore bodies are vein-metasomatic formations in the zones of tectonic
fractures and have been accompanied by jointing.   In the Bakal deposit, the
iron-carbonate mineralization, clearly associated with diabase dykes, for
which Zavaritsky has demonstrated a pre-ore age, forms stratified veins and nests
in the layers of dolomites and limestones, sometimes beginning at the contacts
with the diabase dykes.   The Abail deposit has been formed by lensoid and columnar bodies of iron carbonates;   the latter are restricted to the intersections
of the tectonic joints.

The ore minerals consist of siderite and sideroplesite, and at the
contact with the surrounding carbonate rocks, transitional zones of ankerite
and recrystallized dolomite have frequently been observed.   The ores contain
pyrite, chalcopyrite, galena, sphalerite, hematite, etc., as minor and accessory minerals.

Quite a deep zone of oxidation has been developed in the Bakal deposit,
and in this the iron carbonates have been converted into ocherous ores.   In this
deposit, the latter are the principal commercial type of iron ore.

Abroad, this class of iron-carbonate hydrothermally-metasomatic deposits includes those at Bilbao (Spain), Erzberg (Austria), Siegerland (West
Germany), Ljubija (Yugoslavia), Gelar and Telyuk (Romania), Kremikovtsky (Bulgaria), Wenza (Algeria), Jerissa (Tunisia), and a number of others.

MARINE SEDIMENTARY DEPOSITS (WEAKLY METAMORPHOSED AND UNMETAMORPHOSED).

S i d e r i t i c (brown-ironstone in the zone of oxidation) layered deposits in marine terrigenous-carbonate sediments are similar in geological respects to the above-mentioned deposits of the Bakal group; they occur in the Proterozoic shale-carbonate sequences on the western slopes of the Southern Urals in the ancient core of an Hercynian anticlinorium.   They differ from the Bakal deposits in their somewhat different stratigraphical position in the Proterozoic sequences and in the absence of diabase dykes within the deposits.   The deposits have been traced in drill-holes mainly in the deep zones of oxidation, where brown-ironstone ores have been found.   The primary sideritic ores have been revealed in drill-holes in a few places; their relationship to the surrounding carbonate-shale rocks has been poorly investigated.   The origin of the sideritic ores on the strength of this remains unclear, and their assignment to the class of marine sedimentary deposits is quite arbitrary.   In the Komarovo- -Zigazinsk and Katav-Ivanovsk regions of the Southern Urals, a number of deposits of this class are known which are relatively small with reserves of less than ten million tonnes, and which do not at present have commercial importance.   Only the Tukan deposit is somewhat larger with reserves of more than 40 million tonnes.

H e m a t i t i c   d e p o s i t s   in marine terrigenous-carbonate sediments are known in the Angara-Pit iron-ore basin on the right bank of the lower course of the Angara River.   The ore segregations are near-shore facies of Upper Proterozoic geosynclinal marine sediments (argillaceous, sand-clay, and in part carbonate), forming a syncline complicated by folds of higher orders. The ore layers are interstratified with argillites, and ferruginous sandstones and shales with mutual facies transitions along the strike.   Hematite, hydro- hematite, goethite, and rarely siderite, magnetite, psilomelane, braunite, and pyrite are involved in the composition of the ores; the gangue minerals are chamosite, sericite, clay minerals, and quartz.   Gravelite ores from a hematitic gravel in a hematitic matrix, including clay and quartz-sand material also, predominate.

Besides the Lower Angara deposit in the Angara-Pit basin listed in Table 1, there are the Ishimba, Udorong, and smaller deposits.

H e m a t i t i c   and   m a g n e t i t e - h e m a t i t e d e p o s i t s   in eruptive-sedimentary sequences are restricted to eugeosyn- clinal associations and are located in the synclinal zones of the latter.   The known deposits are of Devonian or Early Carboniferous age.

In some of the representatives of this class (Kholzun and other deposits), the clearest connexion has been observed with the volcanogenic facies of the volcanogenic-sedimentary associations, expressed in the occurrence of ore layers amongst the volcanics, their tuffs and tuffites, with the presence of layers and lenses of volcanogenic rocks in the ore layer itself, and also in the presence of pyroclastic particles in the ore.   In others, assigned to this class of deposits, for example, the Atasu, the rocks immediately surrounding the ore layers and lenses are interbedded limestones, and chert-carbonate, jasperoid, and argillitic rocks, although the ore-bearing sequence is underlain by typic- ally volcanogenic-sedimentary associations, including volcanics, and their pyro- clasts, and in minor quantities, conglomerates, sandstones, and siltstones.   The jasperoid rocks and the jaspers of the ore-bearing sequence itself may be regard- ed as rocks which have developed in association with volcanic activity.

It is typical that the volcanic rocks of the ore-bearing sequences belong mainly to the acid series (rhyolites, porphyries, and albitophyres, and their
pyroclasts).

The ore segregations, which are layers and lenses in the sequence surrounding the ore, have been subjected to folding and disruptive deformation along
with the entire surrounding sequence, which controls their conformable occurrence
in the folded structures of the ore fields. In the Kholzun deposit, the sequence
has been cut by granitoid intrusions, causing partial contact-hydrothermal metamorphism of the ore segregations.

The composition of the ores, including their contact-metamorphosed varieties, involves hematite, and to a lesser degree, magnetite, siderite, and pyrite,
and sparsely distributed arsenopyrite, chalcopyrite, sphalerite, and galena. The
gangue minerals include chlorite, sericite, ferrostilpnomelane, quartz, chalcedony,
opal, dolomite, ankerite, and apatite; and in the weakly developed oxidation zones
martite, goethite, and hydrogoethite.

In addition to the deposits listed in Table 1, this class includes the
Korgon (Altai), and the Keregetas, Bes-Tyube, Ktai, and Dzhumart deposits in the
Atasu ore region (Central Kazakhstan). Abroad, deposits of this class are known
in Canada, where they have been recognized as the 'Algoman type', and occur in
weakly metamorphosed Precambrian associations (Michipicoten deposits and others).

Deposits   with   siderite - leptochlorite
- hydrohematite pisolitic - oolitic ores in
marine carbonate-terrigenous deposits are represented in the Soviet Union in three
Mesozoic-Cainozoic iron-ore basins (the Kerchen, Ayat, and West Siberian). The
ore-bearing associations are involved in sequences either in the marginal foredeep
of the region of Alpine folding (the Kerchen Basin), or the cover of the West Siberian epi-Palaeozoic platform (the Ayat and West Siberian basins). In the Kerchen
Basin, the ore-bearing sequences form gentle troughs with the limbs dipping
at 3 - 4°, rarely up to 6°, being the remnants of eroded gentle brachyanticlines;
in the Ayat and West Siberian basins, they lie almost horizontally. The ore layers belong to nearshore marine facies of Late Cretaceous, Palaeogene, and Neogene
age. In the Kerchen and Ayat basins, a single ore layer is present, and in the
West Siberian Basin, up to four layers with average thicknesses of 2 to 20m. The
ore layers rest on the underlying strata with some form of break, and have themselves undergone partial erosion during the deposition of the sediments of the
roof.

The ores consist mainly of oolites of various dimensions of hydrogoethitic, goethitic, leptochloritic, or sideritic composition, fragments of oolites,
and traces of clastic sand-clay material, cemented by the same minerals that form
the oolites. There are siderite lenses, and seams of the surrounding sand-clay
rocks. In the Kerchen deposits, the composition of the ores is distinguished by
certain features: in the primary unoxidized ores, ferruginous leptochlorites predominate in the oolites and the matrix, and phosphates of iron and hydroxides of
manganese, magnesian-iron-manganese carbonates, iron sulphides, and arsenic minerals are involved in small quantities. In the oxidation zone, the ores have been
converted into brown ironstones with traces of phosphorus, sulphur, and arsenic.

In the Ayat and West Siberian basins, a characteristic pattern of change
in the mineral composition of the ores appears along the direction from the former
shoreline towards the open sea, that is, in the direction of deepening of the sea
(for the Ayat deposit in an easterly direction, and for the West Siberian Basin,

in easterly and northerly directions): hydrogoethite is gradually replaced by leptochlorites. In the Ayat deposit, in addition, a later sideritization of the hydrogoethite and leptochlorite ores has been observed, which B. Krotov associates with a regression of the Cretaceous sea, and the formation of sealed lagoons.

Abroad, deposits of this class are widely distributed in the vast Lotharingian Basin of minette ores (finely oolitic) with an area of 1100 km^2, located mainly in France, and partly in West Germany, Belgium, and Luxemburg. They are widely developed in China. We shall briefly describe the Kerchen, Ayat, and Bakchar (West Siberian Basin) deposits.

M a g n e t i t e, and in part, t i t a n i f e r o u s   m a r i n e p l a c e r s are known both amongst present-day and ancient sediments. The present-day deposits consist of black beach sands with an increased content of magnetite granules in individual sectors of the coasts of certain seas (the Black Sea, the Caspian, the Sea of Japan, etc.). Fossil marine placers of such composition, but lithified, are known in Azerbaidzhan, Armenia, and in a number of foreign countries. In the Soviet Union, such formations are not regarded as a metallurgical iron-ore raw material, and they are briefly described in the section on titanium deposits.

CONTINENTAL SEDIMENTARY DEPOSITS

H y d r o g o e t h i t e   p i s o l i t i c - o o l i t i c l a c u s t r i n e - p a l u d a l   d e p o s i t s are represented by a large number of small deposits, especially on the Russian Platform. The ores consist of large or small segregations of hydrogoethite geodes and other forms of nodules with oolitic structure in clay-sand lacustrine-paludal sediments. Ores of such type of Jurassic age are known in the Tula and Lipet regions, and in the upper reaches of the Rivers Vyatka, Kama, and Sysola; in the northern part of the Russian Platform, they are Quaternary in age, and are being formed at the present time.

The ores are characterized by low amounts of iron (30 - 40%). At the present time the deposits of this group lack commercial importance.

D e p o s i t s   w i t h   s i d e r i t e - l e p t o c h l o -r i t e   o r e s,   p i s o l i t i c - o o l i t i c,   a n d   n a t u r -a l l y - a l l o y e d   w i t h   c h r o m i u m   a n d   n i c k e l, l a c u s t r i n e - p a l u d a l, and associated with a weathering crust of ultramafic rocks lie practically horizontally in lacustrine sediments, in part resting with a break on dunite-peridotite massifs with a Mesozoic weathering crust, and in part adjoining them. The ores are thus basal continental platform sediments. They consist of hydrogoethite-leptochlorite oolites with a cementing matrix of the same composition and they contain clastic material made up of the products of erosion of the weathering crust of the ultramafic rocks (ocherous, nontronitic, nickel-silicate, cherty, chrome-spinel, and magnetite particles and fragments). Portion of the ores does not contain oolites, being compact, and thin- and thick-bedded.

The principal deposits of the USSR are listed in Table 1. Abroad, deposits of this class are not recognized independently, and they are distributed as parts of widely developed lateritic iron and iron-nickel deposits in the equatorial belt. The Akkermanovsk deposit will be briefly described.

S i d e r i t i c   h y p e r g e n e - m e t a s o m a t i c   d e p - o s i t s   i n   n e a r s h o r e - l a c u s t r i n e ,   c o a r s e - - c l a s t i c   s e d i m e n t s   are characterized by quite specific geological conditions of occurrence.   They are associated with a Cretaceous tectonic depression in Palaeozoic sediments and are restricted to its marginal portion, where they consist of nearshore-lacustrine unsorted coarse-clastic sediments of landslide and talus facies, consisting of rock fragments of variable size, forming the margin of the depression, with predominant limestones.   The layers and lenses dip gently towards the centre of the depression, in which they wedge out through facies transitions into silty lacustrine sediments.

The limestone fragments of the 'conglobreccias' and their fine clastic matrix have undergone replacement by siderite, as have also in lesser degree the adjacent limestones of the margin of the depression along the tectonic joints. In the zone of oxidation, the siderite ores have been converted into brown ironstones.   It is likely that the source of the ferruginous solutions was muddy bottom sediments, which were transformed after diagenesis into silt-pelites. They contained organic substances of plant origin, which controlled the presence of ferrous iron in the solutions infiltrating from the muddy sediments into the underlying coarse-clastic sediments, and replacing their carbonate components by siderite.   The significant distribution of the predominant calcareous 'conglobreccias' controlled the huge reserves of sideritic and brown-ironstone ores.

A representative of this class is the Berezovo deposit in the Near - - Argun' depression in Eastern Transbaikalia described below.   In addition, several poorly studied deposits, restricted to this same depression, are known.

S i d e r i t e - l e p t o c h l o r i t e - h y d r o g o e t h i t e   d e p o s i t s   in ancient fluvial sediments have achieved great economic importance in the Soviet Union with the discovery of the vast Lisakovo deposit in the Turgai iron-ore province.   These deposits are associated with fluviatile, terrace, and often estuarine sediments of the Oligocene palaeo-rivers of the Turgai downwarp and the Turanian Plain, which were incised in the marine sediments of the cover of the epi-Palaeozoic platform.   The sand-clay ore-bearing sediments rest on the eroded surface of Lower Oligocene marine clays and are overlain by gravel-clay sediments of the Middle and Upper Oligocene.   The ore segregations extend for tens of kilometres along the channels of the palaeo-rivers, and consist of a principal fluvial segregation and accompanying lens-like, oval, and irregular masses of the terrace portion of lesser dimensions.   The ores are predominantly oolitic.

The mineral composition of the ores involves hydrogoethite, leptochlorites, siderite, ankerite, calcite, quartz (clastic), clay minerals, pyrite, marcasite, stilpnosiderite, gypsum, and in places, manganese hydroxide minerals.

In the Turgai downwarp, besides the Lisakovo deposit, there are the Oktyabr'sk and Shielinsk deposits of the same type.   In the Northern Aral region, besides the Taldy-Éspe deposit, there are the well-known Kutan-Bulak, Kok-Bulak, and a number of smaller deposits.   Deposits of this class have not been recorded in the foreign literature.

P r e d o m i n a n t l y   m a r t i t i c   e l u v i a l - d e l - u v i a l   (nodular)   d e p o s i t s   are known around some large magnetite deposits, being in essence sectors of them.   The ores are segregations in varying degree of rounded blocks, fragments, and fine-clastic particles of martitized indigenous ores in a sand-clay, eluvial-deluvial matrix.   Huge segregations of such ores, even now almost exhausted, were located, for example, on the slopes

around the Vysokogorsk magnetite deposit.   Flotation in washing drums has produced a high quality block-fragment ore 'concentrate'.

DEPOSITS OF THE WEATHERING CRUST (RESIDUAL AND INFILTRATION).

Goethite-hydrogoethite (brown ironstone), and martite-hydrogoethite zones
of oxidation of deposits of sideritic
and skarn-magnetite ores. The formation of the oxida-
tion zones is associated with epochs of ancient weathering, manifested after
the exposure of the primary deposits on the surface by orogenic movements. For
example, in the Urals, where a number of deposits which are now of interest are
distributed, their oxidation zones have been formed during the epoch of the
Mesozoic-Cainozoic weathering crust.

The distribution of oxidation zones at depth was aided by an increase
in the permeability of the ore zones in respect of hypergene solutions as a
result of the development of tectonic disruptions.   The mineralogy of the deposits of this group has been determined by the composition of the primary ores.
The sideritic ores pass into a mixture of minerals of the iron hydroxide group
(goethite, hydrogoethite, hydrohematite, and turgite), and also contain calcite,
and in minor amounts, psilomelane-wad and pyrolusite; rare minerals are aragonite and gypsum, and in the products of oxidation of the hydrothermal siderites,
rare marcasite, malachite, azurite, cuprite, native copper, and sometimes
scorodite.

Martite-hydrohematite ores develop at the expense of the skarn-magnetite ores, and they contain, in addition to predominant martite and hydrohematite, depending on the distribution of skarn and post-skarn silicates and sulphides in the primary ore, nontronite, halloysite, allophane, boehmite, calcite,
aragonite, cuprite, covellite, malachite, azurite, chrysocolla, psilomelane-wad,
lampadite, sphaerocobaltite, erythrite, etc.

A significant development of the oxidation zone, even now either completely or substantially worked out, is represented in the Uralian contact-metasomatic deposits (Vysokogorsk, Goroblagodat, Magnitogorsk, etc.).   They are
typical of the sideritic deposits of Bakal and Komarovo-Sigazinsk in the Southern Urals, and also many foreign sideritic deposits.

Deposits of goethite-hydrogoethite
ores, naturally alloyed with chromium and nickel, are located in the weathering crust of ultramafic rocks in the zones of fold regions with assemblages
of the early phases of development of eugeosynclines of the femic type, including intrusive ultramafic bodies.   The deposits of this group are typical of the
zones where complexes exposed on the surface have undergone ancient Mesozoic
weathering.   This class consists of a few examples in the Urals and the Northern Caucasus.   The ore segregations are the upper ocherous zone of a profile of
an ancient weathering crust of serpentinized dunite and peridotite massifs, which
is replaced below by zones of silicified, leached nontronitized and carbonatized
(magnesite) serpentinites.

The ocherous zones, distributed in the core of nickel silicate-oxide
deposits of the weathering crust of ultramafic rocks, are not regarded, because
of their limited thicknesses and areas, as a source of commercial ferruginous

ores.   Usually, in a particular iron-ore deposit, there are ores of the weathering crust and ores which are lacustrine sediments from the products of erosion
of the crust.   The latter, in contrast to the former, form layers and possess
an oolitic fabric.   Sedimentary ores are absent only in the Yelizavetinsk deposit in the Middle Urals.

The iron ores of the weathering crust of ultramafic rocks consist mainly
of hydrogoethite and traces of chalcedony, opal, nontronite, iron chlorites, magnesite, relict accessory chrome-spinels, and powdery magnetite.

In addition to the above-noted unusual Yelizavetinsk deposit, naturally-
-alloyed ocherous ores in association with sedimentary types, and also naturally-
-alloyed oolitic ores, are represented in the Novo-Kievsk, Novo-Petropavlovsk,
and other deposits in the Southern Urals, and in the Malka deposit in the Northern Caucasus.   In the Soviet Union, naturally-alloyed ocherous ores of the weathering crust have an extremely limited importance.   Abroad, large deposits of
iron ores of the weathering crust are known in the equatorial regions (in Cuba,
the Hawaiian Islands, the Philippines, Guyana, Surinam, etc.).

H y d r o g o e t h i t e   d e p o s i t s   i n   e l u v i a l -
- d e l u v i a l   s e d i m e n t s   i n   k a r s t   l i m e s t o n e s
are known in the literature as 'deposits of the Alapaevo type'.   They are typically developed where the Middle Urals passes into the West Siberian Plain,
where the Palaeozoic complexes of the Urals, affected by Mesozoic weathering,
are partly overlain by Cretaceous and younger deposits of the platform cover.
The deposits are located above belts of Palaeozoic limestones, on which varico-
oured unstratified clays rest, which are probably a residual weathering crust of
the limestones.   The deluvial-proluvial sediments (the so-called 'beliki') rest
on such clays, and in places where they have been eroded, they lie directly on
the limestones.   These beliki are rocks of variable composition from rapidly
replaced conglomerate-breccias with clay matrix, and clay sands and clays, in
which there is an uneven distribution of variously (mainly weak) rounded clasts
of cherty limestones, siliceous shales, jasperoid rocks, and quartz.   Higher up,
the clastic material of the beliki becomes finer, and stratification appears.
Tertiary gaize and Quaternary loams and sands rest on the beliki.

The ores form stratified segregations, lenses, and nests in the lower
portion of the beliki sediments, approximately following the attitude of the
karst surface of the limestones.   The floor of the ore horizon is formed by
coloured clays or limestones.   The ores consist of nodules, concretions, crusts,
and irregular masses mainly of hydrogoethite, but sometimes sideritic, composition, occurring amongst green clays, in part with leptochlorites, also containing
chert-quartz fragments, preserved during replacement of the beliki.   The relative quantity of ore segregations varies from almost complete absence of associated clays up to nodules or concretions disseminated in the clay.   Above groundwater level the clays are oxidized and become ocherous.   The infiltration-meta-
somatic formation of the ore segregations is not in doubt, and the most likely
sources of the iron are the products of the ancient weathering crust.   In the
marginal portion of the eastern slope of the Urals, a considerable number of
deposits of this class are known, but in most cases they are small and have no
serious commercial value.   The Alapaevo group of deposits will be briefly described.

M a r t i t i c   a n d   h y d r o h e m a t i t i c   d e p o s i t s
i n   f e r r u g i n o u s   q u a r t z i t e s   are distributed in all the

continents and are of extremely great economic importance.   Their ores are the
product of natural enrichment of ferruginous quartzites, as a result of leaching of the quartz and breakdown of the silicates during processes of ancient
weathering.   Two main morphological types of segregation of ores of the weathering crust have been recognized, mantle-like and so-called linear types.
The former occur on the tops of layers of ferruginous quartzites in the form
of flat ore-segregations of substantial area with a pocket-like floor; the
latter consist of wedge-stock-columnar and layer-like ore bodies in the ferruginous quartzites, extending to depth, and along strike, and are of considerable thickness.   These and others have been covered by the sediments of the
platform cover, and during erosion of the latter, they form outcrops on the
surface.

Segregations of the linear type develop in zones of faults, joints,
warping, crushing, and flexures.   The formation of portion of the segregations
is believed to be as products of a Precambrian weathering crust with subsequent
metamorphism and repeated weathering even in post-Precambrian time.   The mineral association of the ores in the deposits of this group involves typical
minerals of the weathering crust: martite and martitized magnetite; hematite,
both as a relict of the ferruginous quartzites, and as a dispersed form; hydrohematite; goethite and hydrogoethite; clay minerals; and secondary minerals
(pyrite and carbonates).   Maghemite, hypergene magnetite, and accessory apatite,
alunite, and sphene, have been recorded.   The ores are characterized by large
quantities of iron, and small quantities of phosphorus and sulphur.

Deposits of this class are widely represented in the Krivoi Rog iron-
-ore basin and in the KMA basin, in the Kremenchug iron-ore region.   Abroad,
numerous representatives are known in the regions of Lake Superior (USA), Brazil
(Minas Gerais, etc.), and in India, Liberia, and Western Australia.

We shall describe the Yakovlevsk and Mikhailovsk deposits of the KMA,
and those of Krivoi Rog.

METAMORPHIC (METAMORPHOSED) DEPOSITS

D e p o s i t s   o f   f e r r u g i n o u s   q u a r t z i t e s
occur in metamorphosed marine sedimentary complexes of the Precambrian geosynclines, comprising the crystalline shields, the folded basement of the ancient
platforms, or forming rises of ancient structures in the cores of anticlinoria
of the younger folded regions.   The ferruginous quartzites appertain to geosynclinal formations, but are inherent only to Precambrian folded regions.   They
are the ferruginous facies of the Precambrian sedimentary associations and are
associated with the latter through transitions.   The ferruginous quartzites,
being overwhelmingly marine chemogenic sediments, are quite clearly distinguished amongst the surrounding terrigenous and volcanogenic-sedimentary complexes,
providing transitional chemogenic-terrigenous, chemogenic-terrigenous-carbonate,
and chemogenic-volcanogenic varieties mainly in the lateral portions of their
strata.

The clastic-terrigenous materials are restrictively distributed in the
ferruginous quartzites, which indicates deposition of the primary sediment at
some distance from the shoreline of the ancient seas.

All the deposits with major reserves (billions and tens of billions of tonnes) of ferruginous quartzites (Krivoi Rog, Kremenchug, and the KMA) belong to the Lower Proterozoic greenschist associations. The principal minerals of the ferruginous quartzites are quartz, magnetite, hematite, cummingtonite, biotite, chlorite, and sometimes siderite, alkali amphiboles, and pyroxenes. The texture of the quartzites is predominantly very fine and fine-grained, rarely medium-grained, and the structure is bedded and plicated. The deposits of this type occur in sedimentary and partially volcanogenic-sedimentary rocks. These deposits have been termed deposits of the Krivoi Rog type.

More intensely metamorphosed deposits of the amphibolite association of Early Proterozoic age form smaller (hundreds of millions of tonnes) accumulations (the Olenogorsk, Kostamukshsk, etc.). The principal minerals of the quartzites of the amphibolite association are quartz, magnetite, hematite, hornblende, hedenbergite, and diopside. The texture of the quartzites is medium-grained, and the structure is bedded, in places vaguely bedded. The deposits of this type occur in sedimentary-volcanogenic rocks and belong to the so-called Keewatin type.

The most intensely metamorphosed deposits of the granulite association of Archaean age form accumulations with small reserves (tens and few hundreds of millions of tonnes) (the Mariupol, Taratash, etc.). The principal minerals of the quartzites of the granulite association are quartz, magnetite, hypersthene, actinolite, talc, and cummingtonite. The texture of the quartzites is coarse-grained, and the structure is bedded and vaguely bedded. The country rocks consist of metamorphosed sedimentary and volcanogenic-sedimentary deposits. The ferruginous quartzites are typified by average amounts of iron of about 30 - 41%, predominantly, 32-37%. The ferruginous quartzites are poor in phosphorus and sulphur. Beryllium, boron, cobalt, gallium, germanium, and yttrium have been recorded in a disseminated state in concentrations somewhat in excess of the clarke values.

In the Soviet Union, extremely significant resources of ferruginous quartzites have been concentrated in the Kola Peninsula and in Karelia (Olenogorsk, Kirovogorsk, Kostamukshsk, Mezhozersk, etc.), in the KMA basin (Korobkovo, Lebedinsk, Stoilensk, Saltykovo, Oskoletsk, Mikhailovsk, and a number of other deposits) in the Krivoi Rog - Kremenchug iron-ore basin (Skelevatsk, Ingulets, Novo-Krivoi Rog, Bol'shaya Gleevatka, Pervomaisk, Gorishne-Plavninsk, etc.); in Kazakhstan (the Karsakpai group); and in the Soviet Far East (Malo-Khingan and Ussuri groups of deposits). Ferruginous quartzites have been discovered but weakly studied in Tuva, and on the western slope of the Urals, the foremost of which is the Taratash deposit.

Abroad, large areas of ferruginous quartzites are known (the iron-ore belt of Labrador (Canada), a large group of deposits in the region of Lake Superior (USA), in the state of Minas Gerais (Brazil), a series of large regions in India, recently discovered vast ore regions in Western Australia, the deposits of the An-Shan group and others in the north of China, the Musan deposit (North Korea), and a number of areas in other countries.

M a g n e t i t e   a n d   m a g n e t i t e - s p e c u l a r i t e   c o n t a c t - m e t a m o r p h o s e d   s e d i m e n t a r y   d e p o s i t s may definitely be assigned to this class only in the presence of undoubted relicts of sedimentary ores. Examples of deposits of this class are so far single occurrences. Of the commercial iron-ore type, only the Kholzun deposit in the Gornyi

Altai is known.   In it there are clear relics of volcanogenic-sedimentary magnetite-hematite ores and relict textures of such ores in the contact-metamorphosed
ores themselves.   The contact metamorphism has resulted from the effects of granite intrusions.   The combination of skarn and hydrosilicate types of mineralization within the ore field is characteristic.

DEPOSITS OF THE MAGMATIC GROUP

The Kusinsk Deposit

The Kusinsk titanomagnetite deposit is located on the western slope
of the Southern Urals, 23 km north of Zlatoust.   The ore bodies occur in the
northern part of a gabbro-amphibolite intrusive of the same name, extending in
a northeasterly direction for a distance of about 15 km, with a width in the
region of the deposit of 0.3 - 0.8 km.   The dyke-like body of the Kusinsk intrusion dips steeply, in places almost vertically, to the southeast.   The rocks
surrounding the intrusive on the northwestern side are Proterozoic dolomites, and
on the southeastern side, gneiss-granites.   The latter also form the intrusive
which extends in a northeasterly direction with a width opposite the deposit of
about 2 km.   Proterozoic mica-quartz schists and quartzites are located at its
southeasterly contact.   The granitoids were injected after the gabbroids, along
the contact between the latter and the schists and quartzites.   Granite and
aplite dykes, and also quartz veins have been observed in the gabbro-amphibolites of the Kusinsk intrusive.

The Kusinsk intrusive was originally composed of gabbroids with subordinate pyroxenites, anorthosites, and gabbro-pegmatites, with ore bodies.   As
a result of tectonic effects and the injection of granites, the rocks and ores
of the intrusive were sheared in significant degree and underwent metamorphism
of amphibolite grade with the development of amphibolites, hornblendites, and
near the contact with the gneiss-granites, biotite schists.   Amongst the amphibolites there are epidosites, amphibole-chlorite and amphibole-garnet rocks,
and mica and chlorite schists.   The granitoids have also been subjected to
metamorphism, and have been converted into gneiss-granites.

Widely manifested processes of metamorphism are a specific feature of
the Kusinsk deposit; they have left an impression not only on the mineral composition and structural-textural features of the rocks, but also on the morphology of the ore bodies, and on the ratio between magnetite and ilmenite.

The Kusinsk deposit occupies a sector 2.8 km long in the central portion of the intrusion and consists of four subparallel layer- or vein-like bodies of uniform titanomagnetite ores elongated along the strike of the intrusion,
and also zones of segregated mineralization.   The bodies of uniform ore are
restricted to a member of banded metagabbroids of alternating thin pseudo-layers
of non-feldspar and garnet-bearing feldspar amphibolites.   Segregated ores have
been developed mainly in the melanocratic, probably apopyroxenitic amphibolites.

The extent of the bodies of uniform ores is from 180 to 2500 m, the
thickness is predominantly 2 - 3 m, and in places is up to 8 m.   In a number
of cases, they have been broken by post-ore thrusts with throws of 70 - 75 m,
and less frequently by overthrusts with throws of up to 25 m.

In the upper horizons of the deposit, the bodies of uniform ore lie homoclinally, steeply dipping to the southeast, and in places vertical. During exploration of the deep-seated horizons, a smooth change in the direction of dip of the ore bodies was unexpectedly encountered, with transition along the entire strike towards the west, then horizontal, and again to the northeast and gentle.

In section, the bodies are poker-shaped. Such deformations have also been identified in the gneissosity of the surrounding amphibolites. Such a change in dip is naturally associated with plastic deformations of the fold and flexure type, to which the rocks and ores have been subjected during their metamorphism. On the flanks of the deposit, smooth arc-like bends in the ore bodies have been established along the strike.

Two main types of ores have been recognized in the deposit, uniform and segregated. The uniform ores, 90 - 95% of the time, consist of the ore minerals, titanomagnetite, magnetite-ilmenite (ilmenite with ingrowths of magnetite), ilmenite, and magnetite. The last two are late generations, developing during the metamorphism of the two former minerals. The ore minerals listed also contain sporadic inclusions of högbomite, spinels, corundum, and rutile. Chlorite, biotite, garnet, hematite, pyrite, pyrrhotite, chalcopyrite, carbonates, and quartz are present in small amounts in the ores. The uniform ores are often crumpled and jointed. In the joints there are chlorites, phlogopite, biotite, garnet, sulphides, tourmaline, quartz, and calcite.

The segregated ores contain the same principal ore minerals, and amongst the remainder, amphibole predominates, with garnet, chlorite, plagioclases, sulphides, sphene, and sporadic apatite.

Detailed investigations (Myasnikov, 1959) have shown that, during the metamorphism of the ores of the Kusinsk deposit, when the primary rocks were converted to amphibolites, a significant regrouping of the magnetite and ilmenite took place in the ores. Recrystallization of the primary titanomagnetite with a textural breakdown of the solid solution, led to coarsening of the ilmenite inclusions with the transformation of its lamellae into grains; the grains of titanomagnetite were broken down into smaller types, along the boundaries of which newly-formed pure ilmenite and magnetite were precipitated. The amphibole and garnet formed during the metamorphism intensely corroded the grains of titanomagnetite, the magnetite was dissolved and redeposited, and the ilmenite ingrowths were preserved, which also led to the segregation of the ilmenite. Thus, as a result of metamorphism, the primary ores of the Kusinsk deposit were transformed to a certain degree into ilmenite-magnetite ores of a peculiar type, more valuable in commercial respects than the normal titanomagnetite ores.

At the present time, the Kusinsk deposit is close to complete exhaustion.

The Pudozhgorsk Deposit

The Pudozhgorsk deposit of titanomagnetite ores occurs 138 km southwest of the Medvezh'e railway station in the Karelian ASSR. It is restricted to the footwall of a layered gently-dipping Proterozoic diabase intrusion. Mineralization has been expressed in schlieren and a segregation of titanomagnetite of varying concentration in amphibolitized diabases and gabbro-diabases. The ore segregation of average thickness 14 - 17 m has been examined along the strike for 7 km and to a depth of 330 m. The dip of the layer is to the southwest, at angles from 8 to 50° (*Iron-Ore Basis ...*, 1957).

The principal ore mineral is titanomagnetite, and ilmenite, chalco-pyrite, bornite, covellite, pyrite, pyrrhotite, and sphalerite occur in small amount. The gangue minerals consist of plagioclases, amphibole, pyroxene, biotite, and chlorite. The ores possess a massive fabric.

The average amounts in the ore are as follows (in wt %): Fe, 28.7; TiO_2, 5.5 - 8.5; P_2O_5, 0.44; Co, 0.004 - 0.01; Cu, 0.07 - 0.15; S, 0.07 - 0.16; and P, 0.08 - 0.12.

The reserves of the deposit in categories A + B + C_1 are 248 million tonnes.

The Kachkanar Deposit

The Kachkanar deposit (the Gusevogorsk and Kachkanar proper) occurs in the Isovsk region of the Sverdlovsk district 30 km to the northwest of Nizhnyaya Tura railway station.

The ore-bearing Kachkanar gabbro-pyroxenite pluton occupies an area of about 110 km^2. It has an equant shape and apparently belongs to the lacc-olith type (Fig. 2). The rocks surrounding the pluton are plagioclase porph-yrites and eruptive diabases of Silurian age on the eastern contacts, and Ordo-vician mica and siliceous schists on the western contacts. In the northern and southern peripheral parts of the pluton, the gabbros are replaced by amph-ibolites. Pyroxenites form half of the area of the intrusion and comprise two massifs: the Gusevogorsk in the east and the Kachkanar in the west.

The Gusevogorsk pyroxenite massif, consisting partially of peridoti-tes, hornblendites, and gabbros, extends in a meridional direction for 8.5 km with a width of 1 - 3.5 km.

The Kachkanar pyroxenite massif, in which olivinites and peridotites are involved, extends in a northeasterly direction for 5.5 km, and its average width is 3.2 km. Within the Gusevogorsk deposit there are nine ore segrega-tions, and there is exploitation of the Glavnaya (Main) vein; the area of quality mineralization in the latter is 1.1 km^2. Within the boundary of commercial mineralization there are weak-ore (non-quality) and non-ore sectors, usually equant, with an area between 1000 and 2200 m^2. Mineralization is distributed to a depth of more than 500 - 600 m; drill-holes, put down to these depths, did not pass out of quality ores.

Both the ore-bearing and the gangue pyroxenites have been cut by a large number of dykes of hornblende and quartz plagioclasites up to 2m thick, with variable strike and dips at 20 - 90°.

The ore bodies have been formed by a segregation of titanomagnetite, and less frequently by schlieren segregations and veinlets of massive ores main-ly in the pyroxenites, gabbros, and hornblendites, and to a considerably lesser degree in the peridotites and olivinites. The ores of the deposit have been subdivided into five natural types: coarse- (more than 3 mm), medium- (1 - 3 mm), fine- (0.2 - 1 mm), very fine- (0.05 - 0.2 mm), and dispersed-segregated (less than 0.05 mm).

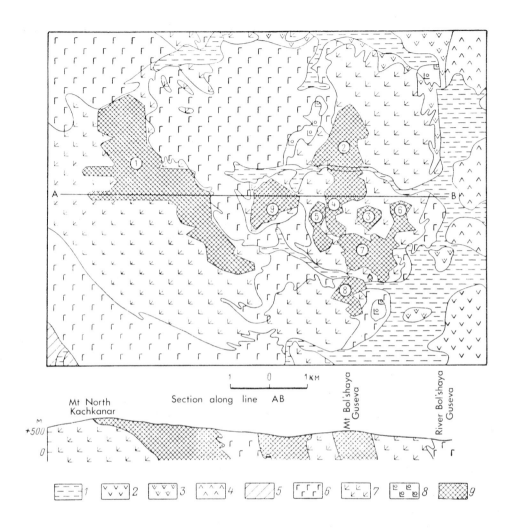

Fig. 2.   Map of the geological structure of the region of the Gusevogorsk and
Kachkanar deposits (After Z. Rupasova).

1) Modern alluvial deposits; 2-4) Silurian eruptives: 2) porphyrites,
3) plagioclase amphibolites, 4) epi-diabase; 5) albite-chlorite,
hornblende-plagioclase, micaceous, and other Ordovician schists;
6-8) intrusive rocks: 6) gabbros, 7) pyroxenites, 8) hornblendites;
9) ore segregations (figures in circles): 1 - Kachkanar deposit
proper; 2-9 - ore segregations of the Gusevogorsk deposit:
2 - Northern, 3 - Intermediate I, 4 - Intermediate II, 5 - Intermediate
III, 6 - Eastern, 7 - Main, 8 - Southern, 9 - Western.

The principal ore mineral is titanomagnetite with solid-solution textural
dissociation containing 2 - 18% of ilmenite.   The titanomagnetite contains an
isomorphous trace of vanadium.   The minor ore minerals are pyrite and pyrrhotite,

and less frequently chalcopyrite, pentlandite, and bornite, and also native platinum and platinoids.   The gangue minerals consist of pyroxenes, amphiboles (chlorite and biotite).

The ores are characterized by the low amount of economic components (in wt %): Fe, 15 - 18, with a fluctuation from 14 to 34 (quality greater than 16%); TiO_2, 0.8 - 2; V_2O_5, 0.05 - 0.31; platinum metals, in tenths of a gramme per tonne; phosphorus and sulphur are practically absent.

During the metallurgical treatment of the iron-vanadium concentrates, besides cast iron, vanadium is obtained by extraction from the convertor slags.

Reserves in categories $A + B + C_1$ in the Gusevogorsk deposit are 3.5 million tonnes, and in the Kachkanar deposit, 2.6 million tonnes, with an average iron content in the ores of 16.6%.   The Gusevogorsk deposit has been quarried, and the Kachkanar deposit has been prepared for exploitation.

The Kovdor Deposit

The Kovdor (or Yeno-Kovdor) deposit lies in the Kirovsk region of the Murmansk district.   It is confined to a massif of the same name consisting of ultramafic-alkaline rocks and carbonatites with an area of 40 km^2.   The massif is a polyphase intrusion of the central type and consists of successively injected olivinites, ijolite-melteigites, nepheline syenites, and also a complex assemblage of silicate metasomatites and carbonatites.

The magnetite ores and magnetite-bearing rocks form an ore body elongated in a submeridional direction, more than 1.3 km long and 100 - 800 m wide, occurring amongst the ijolites and pyroxenites in the southwestern portion of the massif (Fig. 3); it has been investigated to depths of 600 - 700 m.

The ore segregation is surrounded by a continuous shell of phlogopite--apatite-forsterite rocks from 20 to 120 m thick, which separates the ore bodies from the surrounding ijolites and pyroxenites.   These rocks also occur within the segregation in the form of bands, streaks, and lenses.

The ores of the deposit consist of apatite-forsterite rocks, permeated by magnetite veins and veinlets with calcite present (10 - 50%) and a predominant amount of calcite (50 - 80%).   The latter often passes into carbonates. In the deposit, ores with small amounts of calcite predominate: apatite-forsterite--magnetite, forsterite-magnetite, and phlogopite-forsterite-magnetite (*Assessment of Iron-Ore Deposits* ..., 1970).

On the basis of fabric, banded, segregated, streaky, and massive ores are recognized.   The texture of the ores is allotriomorphogranular.   The dimensions of the magnetite grains vary from 0.5 - 5 mm up to several centimetres with a predominance of grains larger than 1mm.   A characteristic feature of the magnetite is the presence in it of inclusions of olivine, apatite, calcite, and spinel.   Ilmenite, pyrrhotite, chalcopyrite, pyrite, and marcasite occur in insignificant amounts.   The distribution of sulphides is uneven.

The magnetite is distinguished by the increased amount of magnesium oxide (4.7 - 7.9%) and aluminium oxide (2 - 4.4%), from which spinel has been formed during dissociation.

All the varieties of stony ores, and also the carbonatites, contain an uneven fine segregation of baddeleyite.

The ores contain (in wt %): Fe, 20 - 55 (on average 28.8); MgO, 15 - - 17; CaO, 11 - 12; P, 2.9; S, 1.19; MnO and TiO_2, tenths of a percent.

The reserves of magnetite ores in the Kovdor deposit based on categories A + B + C_1 comprise 555 million tonnes, and in category C_2, 153 million tonnes.   In addition to the magnetite concentrate, apatite and baddeleyite concentrates may be extracted from the ores of the deposit.

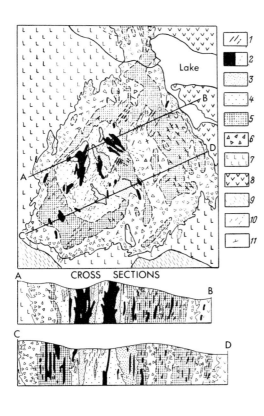

Fig. 3.
Diagrammatic geological map of the Kovdor magnetite deposit (After V. Namoyushko, A. Mikheevich & O. Rimskaya-Korsakova).
1) dolomite carbonatites; 2) calcite carbonatites (a - irregular bodies, b - vein bodies); 3) calcite-magnetite and apatite-calcite-magnetite ores; 4) magnetite ores with calcite, apatite-forsterite, and phlogopite- -forsterite ores; 5) magnetite, apatite-forsterite, forsterite, and phlogopite-apatite-forsterite ores; 6) apatite-forsterite rocks with relicts of alkaline rocks and pyroxenites, partly converted to glimmerites; 7) ijolite-urtites, ijolites, and melteigites; 8) pyroxenites; 9) fenites; 10) sectors of distribution of francolite (staffelite); 11) orientation of magnetite veins.

DEPOSITS OF THE CONTACT-METASOMATIC GROUP

The Peschansk Deposit

This deposit is located 10 km south of Krasnoturinsk in the Sverdlovsk district.   It consists of the Northern, Western, Southern, and Novo-Peschansk sectors.   The deposit is restricted to the western contact of the Peschansk diorite massif, which represents the northwestern portion of the Auerbakh intrusive complex, which also cuts the sedimentary and volcanogenic-sedimentary sequences of the Coblenzian stage of the Lower Devonian.   These rocks consist of marmorized limestones at the base, and higher up form interstratified tuff-sandstones, tuff-shales, and tuffs of andesitic porphyrites with seams of limestones, and in the upper portion of the sequence there are hornblende-plagioclase porphyrites and their tuffs with a markedly subordinate amount of tuff-sandstones. They lie gently (the dip varies from 5 to 50°) and are cut by the steep contact of the intrusion.   The surrounding volcanogenic-sedimentary rocks and the adjoining portion of the massif are cut by numerous dykes of diabasic, dioritic, and gabbro-diabasic porphyrites, spessartites, vogesites, etc. (Fig. 4).

The surrounding rocks are disrupted by tectonic fractures into blocks, some of which have been uplifted, and others depressed.   The ore segregations, especially the large ones, are flanked by wide aureoles of metasomatically altered country rocks.   Intense skarnification has affected these rocks over a radius of from 30 to 120 m;   sometimes, skarnification and bleaching of the rocks has been recorded at distances up to 200 – 500 m from the ore segregations. Albitization has been developed in the volcanogenic rocks, accompanied by epidotization, chloritization, and carbonatization.   The formation of epidosites is usually restricted to the fault zone and is accompanied by small segregations of copper-sulphide ores of the veinlet-segregation type.

Seventeen ore segregations have been investigated in the deposit, six of which embody the main reserves, and the remainder are excluded owing to their small dimensions.   The principal segregations are several hundred metres in length and width, and have a thickness of 40 – 80 m.   All the segregations do not crop out on the surface, but have been revealed as the result of geophysical investigations.

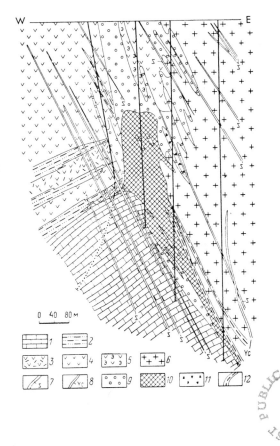

Fig. 4.
Geological section through the Northern Peschansk sector of the Peschansk deposit (After A. Usenko).

1) marmorized limestones;   2) layered tuffites and tuff-sandstones;
3) hornblende-plagioclase porphyrite tuffs;   4) hornblende-plagioclase porphyrites;   5) epidotized and diopsidized tuffs and porphyrites;
6) diorites;   7) diabase-porphyrite dykes;   8) spessartite dykes;   9) garnet and pyroxene-garnet skarns;   10) magnetite ore;   11) skarn-chalcopyrite ore (segregation and veinlets of chalcopyrite in pyroxene-garnet skarn);
12) chlorite-sericite-quartz-carbonate rocks

Magnetite, sulphide-magnetite, and skarn-magnetite ores have been recognized in the deposit.   The average amount of iron in them varies from 48.5 to 54%, sulphur from 2 to 3%, and phosphorus 0.037%.   The Western Peschansk ore segregation is distinguished by the increased content of chalcopyrite;   the average amount of copper in it is 0.61%.

The reserves of magnetite ores in the Peschansk deposit on the basis of categories $A + B + C_1$ are 173 million tonnes, and in category C_2, 24 million tonnes.   The Northern Peschansk sector is worked out.

The Goroblagodat Deposit

This deposit is located on the northeastern margin of Kushva in the Sverdlovsk District in the zone of the discordant contact between a diorite-syenite massif and volcanic and volcanogenic-sedimentary rocks of the Tura Group (Fig. 5).   The rocks surrounding the ores consist of a complex of volcanic (porphyrites) and volcanogenic-sedimentary (tuff-conglomerates, tuff-sandstones, and tuff-siltstones) rocks of the Goroblagodat sequence (upper Ludlovian), 250 - 540 m thick.   In the southern part of the deposit, these rocks have been cut and partly assimilated by a diorite-syenite intrusion, which forms with them a discordant contact of complex morphology and sublatitudinal strike, steeply plunging in depth.   Near the intrusion, the surrounding rocks have been subjected to contact-metasomatic metamorphism, as a result of which garnet and pyroxene-garnet skarns, magnetite skarns, 'variolitic' ores, and scapolite rocks have been formed.

The volcanogenic-sedimentary rocks of the Tura Group form the western limb of the Tura-Tagil synclinal structure and dip eastward at 20 - 30°.   The strike of the rocks in the northern part is 300 - 340°, and in the southern, 010 - 030°.   The general homoclinal attitude of the rocks has been complicated by second-order folding (Fig. 6).   The second-order folds are characterized by northeasterly orientation of the axes and asymmetry of construction, controlled by the steeper dip of the southern limbs.   Pre-ore and post-ore disruptions have been recorded in the deposit.   The orientation of the pre-ore fractures coincides with the axes of the second-order folds, which is evidence of their similar time of formation.   The post-ore fractures have two principal directions: northwesterly and southeasterly.   Both directions have been determined by a system of subparallel fractures and displacements, forming step faults.

The ore metasomatism has had maximum manifestation in the exocontact zone and in the weakened zones with northwesterly strike, parallel or coinciding with the principal pre-ore fracture.   The largest ore segregations occur here.   At a distance from the contact of the intrusion, the thickness of the ore bodies gradually decreases in a northerly direction (Ovchinnikov, 1960).

Fifteen ore bodies have been recognized in the deposit, spatially and genetically clearly associated with zones of skarn formation or scapolitization. Three ore horizons have been recognized on the basis of stratigraphical position. Two of them (the lower and middle) occur in the rocks of the Goroblagodat sequence, and the upper, in the zone of pyroxene-scapolite rocks of the overlying Tura-Kolyasnikovsk sequence (also upper Ludlovian).   The dimensions of the ore bodies vary in length from 200 to 930 m, and in thickness from 2 to 84 m.   Their length along the dip varies from 530 to 1600 m.

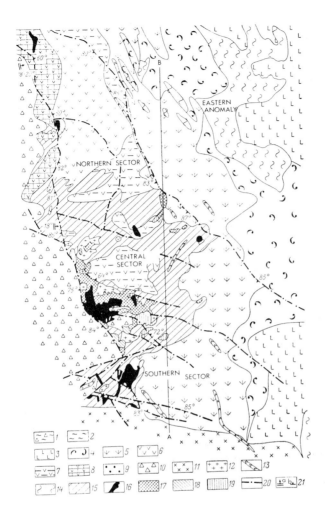

Fig. 5.
Diagrammatic geological map of the Goroblagodat deposit (After B. Aleshin, Yu. Glazov, Ye, Klevtsov, and A. Purkin).
1) Quaternary deposits and dumps; 2) Mesozoic weathering crust; 3-4) upper part of Tura Group: *3* - plagioclase
and pyroxene-plagioclase trachyandesite porphyrites, *4* - coarse clastic crystal-litho-vitroclastic tuffs of trachyandesite porphyrites; 5-10) lower part of Tura Group: *5* - eruptive plagioclase and pyroxene-plagioclase trachybasalt porphyrites, *6* - eruptive plagioclase and pyroxene-plagioclase porphyrites, *7* - fine- and
coarse-clastic tuffs of plagioclase and pyroxene-plagioclase porphyrites, *8* - basalt-limestone conglomerates with
seams of sandstones and siltstones, *9* - tuff-sandstones and siltstones, *10* - amygdaloidal pyroxene basalt porphyrites (Mysovsk sequence); 11) medium-grained biotite-pyroxene-hornblende syenites; 12-14) vein formations: *12* -
- subalkaline syenite-porphyries and microsyenites, *13* - pyroxene-plagioclase trachybasalt porphyrites, *14* - pyroxene-scapolite metasomatites, 15) garnet, pyroxene-garnet, and epidote-garnet skarns, and skarnified rocks;
16-19) ores: *16* - magnetite, *17* - garnet-magnetite, *18* - orthoclase-magnetite ('variolitic'), *19* - magnetite-
-garnet skarns; 20) tectonic disturbances; 21) attitude of stratification (*a*), and faults (*b*).

The shape of the ore bodies is layer-like, less frequently lensoid, and they rest conformably with the country rocks.   The manifestation of post--ore disruptions and vertical displacements along the steeply dipping surface considerably complicates the morphology of the ore segregations.   The deposit is characterized by a median position of the ore bodies within the skarn zone, which corresponds to one of the widespread forms of metasomatic zonation.   The volume of the skarn zone in the deposit greatly exceeds that of the ore bodies. In the northern part of the deposit, the skarn zones are small or are absent.

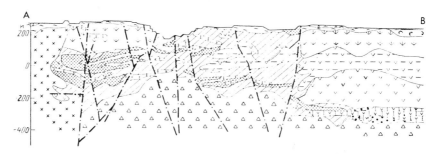

Fig. 6.   Longitudinal geological section through Goroblagodat deposit (After B. Aleshin, Yu. Glazov, E. Klevtsov, and A. Purkin. Symbols as for Figure 5.

Two types of ores have been recognized in the deposit on the basis of mineral and chemical composition: skarn and 'variolitic'.   The skarn ores are characterized by a paragenesis of garnet-magnetite or garnet-epidote-magnetite, and by the presence in the magnetite of a trace of manganese; the 'variolitic' ores have a paragenesis of orthoclase-pyroxene-scapolite-magnetite, and the presence in the magnetite of a trace of vanadium and titanium.   The 'variolitic' ores have a subordinate distribution.   They are clearly associated with dykes of microsyenites and syenite-porphyries, and are formed during the process of their replacement.

In the magnetite ores the most widely distributed fabrics are the massive, streaky, segregated, and banded types.   The principal minerals of the ores are magnetite garnet, pyroxene, orthoclase, and in places, scapolite; the minor minerals (from 1 to 10%) are pyrite, chalcopyrite, calcite, epidote, chlorite, albite, prehnite, zeolite, and there is a rare occurrence (less than 1%) of sphalerite, hematite, pyrrhotite, galena, bornite, marcasite, muschketowite, apatite, sphene, quartz, and fluorite.

The skarn ores, on the basis of mineral composition, textural-structural features, and content of iron, have been subdivided into uniform magnetite, garnet--magnetite, and magnetite-garnet ores, and magnetite-garnet skarns.   The average composition of the ores in the deposit is (in wt %): Fe, 35.5; SiO_2, 18.77; TiO_2, 0.60; Al_2O_3, 8.29; MnO, 0.95; CaO, 11.48; MgO, 1.86; V_2O_5, 0.05; S, 0.68; P, 0.053; Co, 0.022; Cu, 0.13; Zn, 0.078; calcination loss, 3.60; H_2O, 0.32.

The reserves of iron ores on the basis of categories A + B + C_1 comprise 141.2 million tonnes, and in category C_2, 16 million tonnes.   The Goroblagodat deposit has been worked by open and subsurface methods.   The main output has

been from the Central quarry.   The deep-seated horizons of the deposit have
been worked through the South shaft.

The Dashkesan Deposit

The Dashkesan magnetite deposit is located on the northeastern slope
of the Little Caucasus, in the Dashkesan region of the Azerbaidzhan SSR, 40 km
southwest of Kirovabad.   The region of the deposit belongs to the Somkhito-
-Karabakh structural-facies zone of the Little Caucasus, and consists of volcanogenic-sedimentary rocks of Middle and Late Jurassic age, forming the gentle
Dashkesan syncline of normal Caucasian strike (Fig. 7).

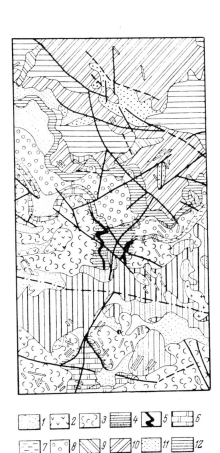

Fig. 7.
Diagrammatic geological map of the region
of the Dashkesan iron-ore deposit (After
A. Kashkai).
1) Quaternary deposits;   2) Kimmeridgian
extrusive and eruptive diabases;
3) Kimmeridgian tuffs and tuffites;
4) hornfelses after Oxfordian-Kimmeridgian
and, in part, Middle Jurassic rocks;
5) skarn-ore segregations;   6) Oxfordian-
-Kimmeridgian marmorized limestones
(Luzitansk sequence);   7) Callovian
argillites and sandstones with seams of
marls;   8) Bathonian titaniferous magnetite
sandstones and tuff-sandstones;   9) lower
Bathonian diabases and diabasic porphyrites;
10) lower Bathonian agglomerate tuffs with
seams of fine-clastic tuffs and tuff-sandstones;   11) tuff-siltstones and quartz
tuff-sandstones of the upper Bajocian;
12) upper Bajocian quartz porphyries and
their tuffs;   13-16) intrusive rocks (Lower
Cretaceous, Neocomian):   13) first intrusive phase, gabbroids (gabbros, norites,
gabbro-diorites, and diorites),   14) second
intrusive phase, granitoids (adamellite-gran-
odiorites, tonalites, and quartz diorites),
15) third intrusive phase (granite-aplites
and alaskites),   16) fourth intrusive phase
(dykes of vein rocks of basic composition);
17) tectonic fractures (a - established,
b - assumed);   18) axis of Dashkesan syncline.

The Middle Jurassic sediments consist of a sequential upward succession of quartz porphyries and their tuffs, tuff-siltstones, and quartz tuff-sandstones, agglomerate tuffs with seams of psammitic tuffs, porphyrites, bedded
yellow tuffites, magnetite tuff-sandstones and sandstones, blocky tuffites and

tuff-conglomerates, and fine-clastic tuffs and tuffites.   The Upper Jurassic sediments begin with interstratified argillites, sandstones, and marls, on which rest agglomerate tuffs, calcareous tuffites and tuff-breccias, limestones (up to 250 m thick), tuffs, and tuffites.   All these sediments have been cut by the huge Dashkesan polyphase gabbroid-granitoid intrusion of Early Cretaceous (Neocomian) age, on which Upper Cretaceous sandstones and marls have been superimposed

Four phases have been recognized in the formation of the intrusion: during the first phases, various gabbros, gabbro-syenites, quartz diorites, and syenite-diorites were injected, during the second, adamellites, granodiorites, and syenite-diorites, in the third phase, dykes and larger bodies of alaskites and granite-aplites, and during the concluding phase, numerous dykes of diabases, diabase porphyrites, and in part, lamprophyres.

The western part of the latitudinally extended Dashkesan intrusion has been cut by the valley of the River Koshkarchai, which has exposed the northern and southern contacts of the intrusion, where four sectors of the Dashkesan deposit (the Northwestern, Northeastern, Southwestern, and Southeastern) are located. The southern sectors lie 5 - 6 km from the northern ones.   The eastern sectors are separated from the western sectors by the narrow valley of the River Koshkarchai (Fig. 8).

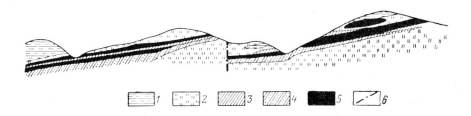

Fig. 8.   Geological section through the Northwestern sector of the Dashkesan deposit (After M.-A. Kashkai)

1) tuff-breccias and tuffs; 2) hornfelses; 3) epidote-pyroxene and pyroxene-epidote skarns; 4) garnet skarns with magnetite; 5) magnetite ore; 6) tectonic disturbance.

In the Dashkesan deposit, the layer-like skarn-magnetite segregations have been concentrated in the Upper Jurassic sediments with somewhat varying stratigraphical position within the Callovian-Oxfordian-Kimmeridgian stages (Kashkai, 1965).

The peri-ore changes of the country rocks have been expressed in hornfelsing, skarnification, and lower-temperature new-formations.   The skarns in the Dashkesan deposit have been formed mainly at the expense of silicate rocks, and limestones have been skarnified much more weakly.   Among the ore skarns, we may distinguish garnet-pyroxene skarns with magnetite, garnet skarns with magnetite, garnet skarns with hematite and magnetite, and dashkesanite skarns with magnetite.   In them, there is a widespread development of post-skarn changes and new formations, and the mineral composition of the skarns has been complicated by hydrosilicate minerals, quartz, and calcite.

The most widely distributed are the garnet skarns with epidote and calcite.   The presence in the Northeastern sector of dashkesanite cobalt--bearing skarns is a specific feature of the deposit.

The structure of the deposit varies somewhat in its four sectors. In the Northwestern sector, the skarn-ore segregation rests on marmorized limestones only in its western part.   To the east, the limestones rapidly thin out, and the thickness of the skarn-ore segregation increases, reaching 70 m as it rests on hornfelsed tuffs.   The extent of the segregation along the strike is 1.7 km, and down the dip, 1.9 km, along a slope of 10 - 12° southwestwards.

In the Northeastern sector, the skarn-ore body rests conformably between hornfelses and hornfelsed tuffites of Callovian-Oxfordian age and a Kimmeridgian volcanogenic sequence.   It is distinguished from the segregation in the Northeastern sector by its smaller dimensions and the presence in the roof of an ore body in the form of thin conformable seams of dashkesanite skarns.

In the Southeastern sector, the main layer-like skarn-ore segregation rests on limestones, dipping northnorthwestward at 10 - 12°.   Along the strike it can be traced for 2.5 km, and down dip, for 1.3 km.   A number of other layer-like bodies, partly merging with the main segregation, have been revealed below this segregation in limestones in association with seams of tuffites.

In the Southwestern sector the layer-like skarn-ore segregation has been traced in a latitudinal direction for 4 km.   It consists of a number of ore lenses, separated by non-ore skarns.

About 120 minerals have been identified in the Dashkesan deposit, of which 42 belong to the hypergene category in the oxidation zone.   Besides magnetite, the hypogene minerals include sulphides of iron, nickel, cobalt, molybdenum, zinc, lead, and copper, cobaltite, glaucodot, arsenopyrite, safflorite, native gold and electrum, maghemite, muschketowite, ilmenite, quartz, rutile, spinel, garnets, pyroxenes, amphiboles, including dashkesanite, ilvaite, epidote, allanite, talc, sericite, pyrophyllite, kaolinite, and dickite, plagioclases, chlorites, zeolites, carbonates, alunite, apatite, and fluorite.   Most of the minerals listed are weakly distributed and rare.

Among the magnetite ores of the deposit, we can distinguish uniform and segregated ores, and ore skarns.   The uniform ores form persistent bodies up to 25 - 30 m thick and separate lenses 10 - 20 m thick, and contain up to 90% of magnetite.   There are varieties, enriched in sulphides (up to 20%). The amount of iron is 45 - 60%.   The segregated ores form layer-like bodies and lenses, gradually passing into the uniform ores.   The amount of magnetite in them varies between 40 and 70%.   The ore skarns are distinguished from the segregated ores by the small amount of magnetite (30 - 40%) and the large quantity of garnet (60%).

The reserves of iron ores in all four sectors of the Dashkesan deposit in categories A + B + C_1 are 187 million tonnes and in category C_2, 7.2 million tonnes, with an average iron content of 35.1% in the Northwestern sector, 49.5% in the Northeastern, 49.4% in the Southwestern, and 42.1% in the Southeastern. The deposit is worked in a quarry.

The Tashtagol Deposit

The Tashtagol deposit occurs near Tashtagol railway station 200 km southeast of Novokuznetsk in the Kemerovo district.   The deposit occurs in a Middle Cambrian folded metamorphosed eruptive-sedimentary sequence, in the contact zone with an intrusion of syenites (Fig. 9).   In upward succession in the Cambrian sequence, we may recognize a sub-ore red-bed member of hematite-bearing sandstones, siltstones, and tuff-sandstones with seams of conglomerates and limestones, and a supra-ore member, greenish in colour, and consisting of lime-chlorit and chlorite-sericite schists, conglomeratic sandstones, and tuff-sandstones with lenses and seams of limestones.   Higher up comes a member of tuff-conglomerates, tuff-breccias, and shales with seams of marls and limestones.   The albitophyres, and pyroxene and pyroxene-hornblende porphyrites have been converted in considerable degree into porphyroids and chlorite-sericite schists.   The eruptive-sedimentary sequence dips 80° eastwards.

The syenite intrusion consists of coarse-grained quartz syenites in the centre and fine-grained varieties of syenites and syenite-porphyrites on the periphery.   The syenites are associated with small microsyenite dykes.   In a number of sectors, segregations of magnetite have been revealed amongst the syenites (*Principal Iron-Ore Deposits of Siberia*, 1970).

The ore segregation has been traced along strike on the surface for 700 m, and at a depth of 280 m, for 1000 m.   Along the dip, it has been investigated down to a depth of 1 km without evidence of tapering out.   The principal ore bodies have a layer-like or lens-like shape and in places they are dismembered into adjacent lenses joined by connecting links.   The dimensions of the ore bodies along strike are 300 - 760 m, and along the dip 500 - 1000 m and more with average thickness of 40 - 70 m.   The ore field is broken by a system of joint zones with submeridional, northwesterly, and northeasterly strike.   Two large submeridional faults divide the ore segregation into blocks descending step-like to the east.   The ore bodies are accompanied by thin shells of garnet and epidote-garnet skarns.

The ores of the deposit are magnetite, massive, and less frequently streaky and banded.   In addition to magnetite, hematite, calcite, epidote, chlorite, and quartz are present, and less frequently garnet, apatite, pyrite, pyrrhotite, chalcopyrite, sphalerite, galena, fahlore, arsenopyrite, fluorite, rhodochrosite, and secondary minerals of the oxidation zone.

In the deposit, rich magnetite ores with more than 45% iron, rich segregated skarn ores with 30 - 45% iron, and lean segregated skarn ores with 20 - 30% iron are recognized.   The average amount of iron in the ores is 44.7%, sulphur 0.11%, and phosphorus 0.1%.

The reserves of ores in categories $A + B + C_1$, comprise 241 million tonnes and in category C_2, 97 million tonnes.

The deposit has been exploited by the shaft method at a depth of 450 - 500 m from the surface.

Fig. 9.
Geological plan and sections of the
Southeastern sector of the Tashtagol
deposit (After V. Bondarets and
N. Monkevich).
1) sandstones, siltstones, and red
tuff-sandstones with disseminated
hematite, and seams of conglomerates and limestones;   2) greenish
shales and tuff-sandstones with
seams of limestones;   3) tuffites
with seams of shales and limestones;
4) porphyrites and porphyritoids;
5) syenites;   6) skarns;
7) magnetite ores;   8) tectonic
breaks;   9) worked-out area.
Ore sectors (figures on map):
1 - Main,   2 - Southeastern.

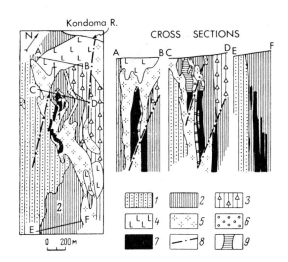

The Tёya Deposit

The Tёya deposit lies 183 km west of Abakan in the Khakas autonomous
district of the Krasnoyarsk region.   The deposit is restricted to a system of
small faults in the branching of the Deep Fault, and is located amongst dolomites and dolomitized limestones of the Upper Proterozoic - Lower Cambrian
within a pipe-like structure, consisting of tuffogenic rocks and fragments of
dolomites, limestones, amphibolites and granites.

Near the deposit there are intrusive bodies of gabbro-diorites and
diorites of Middle-Late Cambrian age, cut by granites of Late Cambrian - Ordovician age.   On the northern flank in the footwall and in the deep-seated
horizons of the deposit, there are granosyenites and syenites.   The granosyenites are accompanied by fields of albitization, quartzification, K-feldspath-
ization, skarnification, and sericitization.   In the ore field there is widespread development of pre-ore dykes of diorites, diabasic porphyrites, ortho-
phyres, felsites, porphyrites, and dolerites (Fig. 10).

The length of the ore zone with commercial mineralization is more
than 1500 m, with a maximum thickness of 300 m and a depth of mineralization
of up to 1200 m.   The dip of the ore zone is westerly at from 40 to 60°.   In
the ore zone, 12 lensoid ore bodies are involved and also skarns and metasomatites of various shapes.   The largest ore bodies have dimensions along strike
of 350 - 1150 m and down dip of 250 - 1300 m, and average thicknesses of 40 -
- 140 m.   In the deposit, almost latitudinal, post-ore fractures have been
developed with dips to the southwest and south at 58 - 85° and a horizonatal
displacement of 40 - 60 m.

The skarns are divided on the basis of composition into magnesian and
calc types.   The magnesian skarns consist of olivine, forsterite, chondrodite,
clinohumite, diopside, and spinel.   The calc skarns have been subdivided into
garnet, pyroxene-garnet, and pyroxene types.   There is a wide development of
metasomatites, consisting of amphibole, phlogopite, chlorite, serpentine, epidote, magnesite, and talc.   They contain scapolite, cordierite, tremolite,
isocrase, sphene, apatite, fluorite, tourmaline, quartz, and calcite.   Amongst
the metasomatites there are albitized pseudo-brecciated materials and quartz
K-feldspar and quartz - K-feldspar-epidote rocks.   These rocks have been formed
as a result of metasomatic replacement of eruptive breccias.

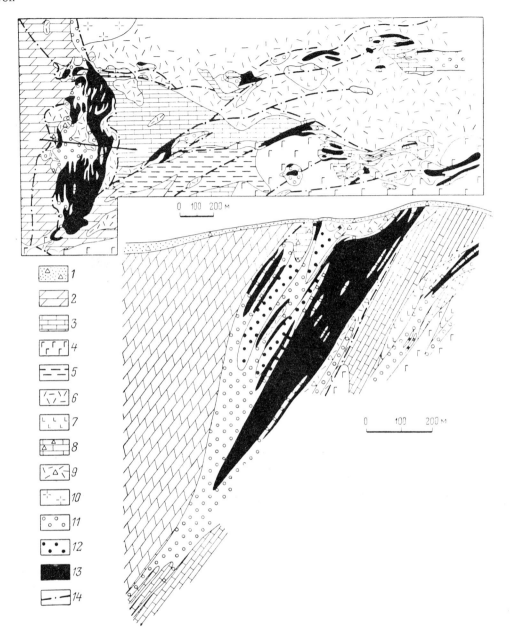

Fig. 10.
Geological plan and section of Tëya and Abagas deposits.   (After V. Semënov and
P. Zav'yalov).
1) deluvium; 2) dolomites and dolomitic marbles; 3) limestones and marbles;
4) gabbro, gabbro-diorites, and hornblendites; 5) orthoamphibolites; 6) felsites,
keratophyres, and their tuffs; 7) diabases, and diabasic and dioritic porphyrites;
8) brecciated marbles; 9) tuff-breccias of acid eruptives; 10) syenites, granosyenites, and syenite-porphyries; 11) calc and magnesian skarns and metasomatites;
12) the same, ore types; 13) magnetite ore with hematite, magnomagnetite, etc.;
14) tectonic disturbances.

In mineral composition, the ores have been subdivided into magnetite, serpentine-magnetite, carbonate-magnetite, carbonate-serpentine-phlogopite--magnetite, and hematite-magnetite types.   They are associated with magnesian skarns and their late products.   Serpentine-magnetite ores predominate.   The ores of the deposit have a streaky, breccia-like, brecciated, collomorphic, rhythmically-banded, and rhythmically-spherical fabric.   The principal ore mineral is magnetite of very fine-grained texture with grainsize of 0.005 - - 0.007 mm, rarely up to 0.5 mm.   In addition to magnetite, there are hematite, magnomagnetite, pyrite, pyrrhotite, arsenopyrite, chalcopyrite, niccolite, safflorite, and sphalerite.

In general, the ores, on the basis of iron content, belong to the iron-poor, high alumina and magnesian type with a small amount of phosphorus and sulphur.   The reserves of ores in categories $A + B + C_1$ are 144 million tonnes, with an average iron content of 32.9%.   The deposit is worked by the open-cut method.   The Abagas deposit, located 1 km from the Tëya deposit, has reserves of magnetite ores, based on categories $A + B + C_1$, of 73 million tonnes.

The Sheregesh Deposit

This deposit lies 30 km northnortheast of Tashtagol railway station, which is associated with the railway and the highway.

Middle Cambrian volcanogenic-sedimentary rocks, and Ordovician terrigenous formations, cut by gabbroids, syenites, and granites of the Mustag-Sarlyk pluton (Fig. 11) participate in the geologic structure of the deposit.

The deposit is restricted to a flexure-like curve in the Middle Cambrian rocks, consisting in the basal parts of the sequence, of andesite porphyrites and tuffs of intermediate composition, interstratified with marmorized sandstones and silty limestones, bedded limy siltstones, lithocrystalline tuffs of trachytic composition, and shales.   The middle part of the sequence consists mainly of metamorphosed limestones and dolomites.   In the upper parts of the sequence, trachyte porphyries and andesite porphyrites, and tuffs of acid and intermediate composition, are developed.

The western part of the ore field consists of Ordovician terrigenous rocks, unconformably overlying a Middle Cambrian sequence and containing basal conglomerate-breccias with fragments of all the rocks and ores surrounding the ores proper.

The deposit is restricted to the northeastern limb of the syncline. The ore-surrounding sequences form a homocline with dips of 30 - 60°.   Within the deposit there are numerous sublatitudinal faults, accompanied by joint zones, some of which have an ore-controlling character.   In the ore zone, magnesian skarns, calc skarns, and hydrosilicate peri-ore metasomatites are developed.   The formation of calc skarns and part of the magnetite ores is associated with the formation of the syenite massif.   Granites cut the rocks of the gabbro-syenite complex and the associated skarn-ore formations.

The skarn-ore zones of the deposit have layer- and lens-like shapes and a gentle dip conformable with the surrounding sequence.   The predominant part of the ores are restricted to the middle member of the Middle Cambrian rocks, and in lesser degree to the bodies of andesitic porphyrites and ancient pre-ore tectonic faults.

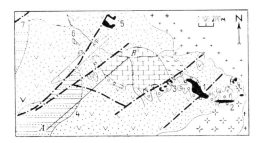

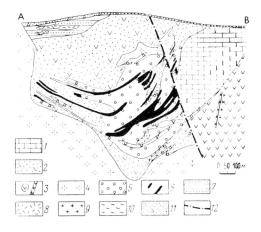

Fig. 11.
Geological plan and section of the Sheregesh deposit
(After A. Lyakhnitsky, M. Danilova, and N. Mikhailova)

1) marmorized limestones; 2) tuffs and lavas of andesitic porphyries and keratophyres; 3) pyroxene and amphibole porphyrites; 4) syenites; 5) skarns; 6) magnetite ores; 7) quartzitic sandstones and silt-stones; 8) quartz porphyries; 9) granites; 10) karst deposits; 11) deluvium; 12) tectonic breaks.

Ore sectors (figures on map):
1 - Main, *2* - Eastern, *3* - Bolotnyi, *4* - New Sheregesh, *5* - Rudnyi II, *6* - Podruslovyi, *7* - New Promplosh-chadka.

Magnesian skarns from 3 to 35 m thick form a shell round the gabbroids, and at their contact with the dolomites. In the structure of the magnesian skarns, L. Shabynin and V. Prikhod'ko have investigated the metasomatic zonation with a uniform sequence of zones; dolomite, forsterite calciphyre, spinel-forsterite skarn, spinel-fassaite skarn, pyroxene-skarn, and gabbro.

The magnesian skarns, on the one hand, have been replaced by calc pyroxene-garnet skarns, and on the other, have undergone a late change with the formation of pargasite, phlogopite, clinohumite, chlorite, and serpentine.

The ores and skarns often inherit the structures of the replaced rocks (hornfelses, porphyrites, and pre-ore breccias).

The principal mineral in the ores is magnetite, and minor minerals are pyrite, pyrrhotite, chalcopyrite, sphalerite, and galena, with rare occurrences of arsenopyrite, calcite, dolomite, and rhodocrosite. In the skarns there is a predominance of garnet, epidote, and amphibole, along with chlorite, brucite, forsterite, spinel, phlogopite, clinohumite, tremolite, serpentine, fassaite, xanthophyllite, diopside, actinolite, apatite, and biotite.

The ore magnetite in the Sheregesh deposit is related to three paragenetic associations: 1) magnetite of the magmatic phase in an association with minerals of the magnesian skarns; 2) post-magmatic magnetite in an association with minerals of the apo-skarns; 3) post-magmatic magnetite in an association with minerals of the calc skarns. The magnetites of these three paragenetic associations are distinguished on the basis of average amounts of trace-elements (Table 2).   The main part of the magnetite was formed along with magnesian skarns, and an insignificant part of the magnetite is associated with the apo- -skarns and calc skarns.

Table 2.   *Amounts of Trace Elements in Magnetites of the Sheregesh Deposit (in %).*

Association	MnO	MgO	Al_2O_3	TiO_2
Magnesian skarns	0.66	0.07	0.21	0.22
Apo-skarns	2.0	2.34	0.34	0.23
Calc skarns	0.55	0.28	0.06	0.09

In the deposit, we recognize rich (uniform) magnetite ores with an iron content of more than 45%, rich segregated skarn ores, containing 30 - 45% iron, and lean segregated skarn ores, containing 20 - 30% iron.   Reserves of ores containing more than 0.4% of zinc have been estimated separately.   The reserves of ores in categories $A + B + C_1$ are 234 million tonnes, with an average iron contant of 35%.   The deposit is being worked by the shaft method.

The Taezhnoe Deposit

The Taezhnoe magnetite deposit is located to the south of Aldan township, to the east of the Never-Aldan administrative district.   It is the largest and most representative of magnesian-skarn iron-ore deposits in Southern Yakutia.   It is located in the zone of an assumed large deep-seated fracture in the southern margin of the Central Aldan anticlinorium, in the area of direct contact between the productive horizons and the fields of granitization of the rocks of the gneiss complex of the Fëdorov Group in the Iengra Series of the Aldan Shield.

There are three horizons in the sequence of the deposit: the sub- -ore, the productive, and the supra-ore.   The sub-ore horizon, more than 400 m thick, consists of basic (pyroxene, amphibole, and biotite-amphibole) schists and gneisses.   The productive horizon, about 300 m thick, consists of magnesian skarns and peri-skarn rocks with relicts of dolomitic marbles and intercalations of magmatized schists and gneisses, altered in varying degree by the processes of skarnification.   Amongst these gneisses, and along with the basic varieties, there are also aluminous types, typical of the supra-ore horizon.   In the latter, there are biotite, cordierite, and sillimanite schists, gneisses, and gneiss-quartzites.   Various gneiss-granites are distributed in the area of the deposit (Pervago, 1966).

The skarn-ore bodies are located in various sectors of the stratig- raphical sequence.   The ore has been concentrated in the apo-dolomitic magnes- ian skarns of the magmatic phase of spinel-forsterite and diopside composition, transformed in varying degree by post-magmatic processes into clinohumite and phlogopite varieties and into serpentine rocks.   A significant portion of the mineralization (up to 25% of the total reserves) is also located in the pyroxene and phlogopite skarns, which in part have replaced the gneissic rocks of the hanging wall of the skarn bodies.   The precipitation of the bulk of the magnetite appertains to the end of the skarn phase of the post-magmatic stage, when the minerals of the original magnesian skarns were replaced by phlogopite, clinohumite, and amphiboles.   The serpentinization of the skarn silicates is post--ore.   The absolute age of the mineralization, based on phlogopite, is 1950 ± ± 50 m.y.

The deposit is located in the hinge portion of a synclinal structure (Fig. 12), in the limbs of which, on their extension southeastwards, there are a number of deposits of similar dimensions in the Legliersk ore field.   The core portion of the structure consists of gneisses and schists of the supra-ore member, which along with the rocks of the underlying horizons in the northern limb of the fold have been strongly granitized.   The overall synclinal structure of the deposit has been complicated by pre-ore disturbances (of the granitization period), transitional from flexural to fracture types.   The post-ore disruptions are few in number.   The appearance of three systems of post-Jurassic dykes of syenite-porphyries has not been accompanied by displacement of the blocks.

The shape of the ore bodies in the magnesian skarns after dolomites is layer- and lens-like, with separation along the strike and dip.   In the lower ore member, the dimensions of the bodies are up to 700 - 1200 x 80 - 120 m, passing to depths of up to 600 - 750 m without marked changes in thickness and composition of the ores.   In the upper ore member, the dimensions of the bodies are smaller.   In the skarns after the gneisses, the ore bodies have irregular shapes and consist of combinations of discordant vein and nest-like forms.   In the construction of the skarn-ore bodies of the first group (after the dolomites), metasomatic zonation is clearly defined.   In transverse section, the following alternation of rocks (from marble to gneiss) predominates: 1) dolomitic marble, 2) spinel-forsterite calciphyre, or replacing it, phlogopite-clinohumite skarn (or a serpentine rock in its place), 3) spinel-fassaite (less frequently enstatite) skarn, or replacing it, a phlogopite or pargasite skarn, 4) plagioclase--fassaite peri-skarn rock (or replacing it, a phlogopite-hornblende skarn, or tourmaline skarn), 5) pyroxene-feldspar peri-skarn rock after gneiss (or replacing it, phlogopite and hornblende skarns), 6) a near-skarn altered migmatitic gneiss, and 7) a migmatitic gneiss.

The thicknesses of the individual zones vary and for the outer portion of the column they reach many tens of metres.   On the strength of the presence of a large number of small seams of gneisses in the original bodies of dolomitic marbles, the overall section through the skarn-ore body usually represents different combinations of repeated metasomatic columns, the rocks of the individual zones in which have been altered in varying degree.   The magnetite mineralization is predominantly embodied in the forsterite zone.   In the skarns after gneisses, magnetite is usually developed on the borders of the skarn veins, although it also frequently penetrates into relict sectors of the gneisses.

On the basis of mineral composition of the skarn silicates, the ores of the Taezhnoe deposit are divided into orthosilicate (in the forsterite and clinohumite skarns, in most cases most strongly serpentinized) and metasilicate (in the skarns of pyroxene and hornblende composition) types.   The arrangement of these and others in space reflects the features of metasomatic zonation of the skarn-ore bodies, in which the zones, approximating to aluminosilicate rocks, consist of pyroxene skarns.

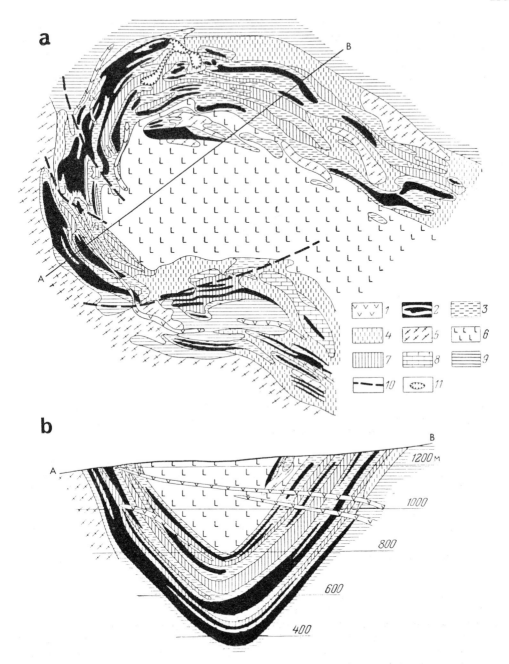

Fig. 12.
Diagrammatic geological map (*a*) and section (*b*) of the Taezhnoe deposit.
(After A. Pukharev).
1) syenite-porphyries; 2) iron ores; 3) magnesian skarns; 4) skarnified·
aluminosilicate rocks; 5) migmatites and injection gneisses; 6) biotite and
sillimanite gneisses (supra-ore); 7) biotite-amphibole gneisses and schists
(productive horizon); 8) calciphyres and dolomitic marbles; 9) pyroxene-
-amphibole gneisses and schists (sub-ore horizon); 10) flexures; 11) quarry
and dumps.

The principal ore mineral is magnetite; the zone of oxidation and hematitization is practically absent. The sulphide content (mainly pyrrhotite) comprises 3 - 8% of the volume of the ore mass. In some of the largest ore bodies, a ludwigite mineralization is manifested. In the gneiss members of the hanging wall, intense tourmalinization has been recorded, which has no stratigraphical restriction.

The average chemical composition of the ores of the deposit based on the bodies of predominantly orthosilicate composition is as follows (in wt %): Fe, 46.69; S, 2.09; P, 0.05; SiO_2, 12.79; Al_2O_3, 3.08; CaO, 1.78; MgO, 13.41. In the metasilicate varieties, the amount of CaO increases to 3 - 5% and more, and MgO falls to 5 - 3%. In the ores, cobalt, tungsten, vanadium, molybdenum, zinc, and copper have been identified in hundredths and thousandths of a percent.

The reserves of magnetite ores based on categories $A + B + C_1$ is 707 million tonnes, and in category C_2, 580 million tonnes, with an average iron content of 45.2%.

THE MAGNETITE DEPOSITS OF THE TURGAI IRON-ORE PROVINCE

Of the many magnetite deposits of this province, we shall describe the three principal examples, the Sarbai, the Sokolovsk, and the Kachar. These deposits, like the remainder, are associated with the Valerianovsk structural--facies subzone of the Turgai downwarp in the Turanian Plain of the West Siberian Platform. Three submeridional structural-facies zones (from west to east) have been recognized in the structure of the Palaeozoic basement of the Turgai downwarp: the eastern portion of the Trans-Ural anticlinorium, the tectonically depressed Tyumen-Kustanai synclinal zone, and the Tobol-Kushmurun (or Ubagan) uplift, involved in the Kazakhstan folded region.

The Tyumen-Kustanai zone is divided into two structural-facies subzones, the western, Valerianovsk, of eugeosynclinal aspect, and the eastern, Kustanai or Borovsk, of miogeosynclinal character. The Valerianovsk subzone is bounded on the west by the Livanovsk, and on the east, by the Anapovsk faults. It consists of Lower Carboniferous (Tournaisian - Namurian), predominantly volcanogenic--sedimentary deposits, crumpled into folds, in part compressed, with steep dips on the limbs, and in part gentle, of the brachyfold type. These rocks have been cut by Hercynian intrusions of gabbro-diorite-granodiorite, and in part granite composition.

On the whole, the Valerianovsk structural-facies zone belongs to the eastern margin of the Uralian folded region. The Palaeozoic folded basement within it has been covered by Mesozoic-Cainozoic sediments, lying almost horizontally, with a thickness in the region of the Sokolovsk and Sarbagai deposits of 40 - 100 m, and near the Kachar deposit, up to 180 m. The well-known Ayat sedimentary iron-ore deposit is restricted to the Upper Cretaceous sediments of the cover, and the Lisakovsk iron-ore deposit is confined to the continental fluviatile Palaeogene sediments.

The Sarbai Deposit

The Sarbai magnetite deposit was discovered in 1948 from the air on the basis of a marked deflexion of the magnetic needle; it is located in the

Trans-Ural steppe at Rudnyi and at 45 km southwest of Kustanai. The deposit consists of three ore segregations, the Eastern, Southeastern, and Western (Fig. 13). It is located in the western exocontact zone of the Sarbai diorite intrusion. The latter consists of pyroxene and quartz diorites and diorite porphyrites, and is accompanied by a series of dykes (pre-ore diorite porphyrites, and post-ore quartz-pyroxene porphyrites, and granite porphyries (Dymkin, 1966).

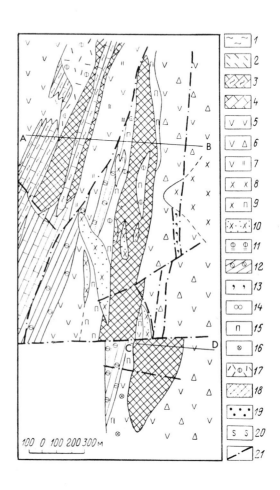

Fig. 13.
Plan of 80 m - horizon of the Sarbai deposit (After I. Kochergin).

1) Mesozoic-Cainozoic sediments;
2) clays from ancient weathering crust of Palaeozoic rocks; 3) silty and pelitic tuffites; 4) bituminous limestones with seams of tuffogenic material; 5) tuffs of intermediate composition with seams of andesite porphyrites; 6) tuffs and tuff--breccias of intermediate composition; 7) albitized tuff-breccias; 8) diorites and diorite porphyrites; 9) vein, pre-ore, diorite porphyrites; 10) vein, post-ore, quartz--pyroxene diorite porphyrites; 11) pyroxene-plagioclase hornfelses; 12) hornfelsed tuffites; 13) biotite metasomatites; 14) scapolite metasomatites; 15) pyroxene skarns and skarned rocks; 16) garnet skarns and skarned rocks; 17) epidote--actinolite rocks; 18) magnetite ores; 19) segregations of magnetite ores; 20) zones of mylonitization of rocks and ores; 21) tectonic disturbances.

The ore bodies occur amongst various metasomatic formations, which have developed after the volcanogenic-sedimentary deposits of the Valerianovsk subzone. Near the deposit, the assemblage of these deposits includes interstratified Lower Carboniferous andesite porphyrites, and their tuffs, tuff--breccias, tuffites, limestones, sandstones, tuff-sandstones, and also provisional Middle and Upper Carboniferous hematitized tuffs and tuff-lavas of basaltic composition, tuffites, and argillites. The rocks of this assemblage form the Sokolovsk-Sarbai anticline with submeridional strike. The ore bodies occur in the western limb of this anticline.

In the area of the deposit there is a widespread development of pre--ore, intra-ore, and post-ore fractures. Diorite-porphyrite dykes have been

injected along the pre-ore faults with submeridional strike and westerly dips
of 65 - 70°.   The intra-ore fractures are defined by the vein-like skarn-magnetite partings which cut the bedding.   A system of post-ore faults has controlled the block displacements of the ore bodies, and the crushing of the ores
and the surrounding metasomatites.   Dykes of quartz diorite-porphyrites and
granite-porphyries have been injected along some of the post-ore fractures.

Changes in the primary rocks in the area of the deposit consist of
pre-ore hornfelsing, the formation of biotite - K-feldspar and albite metasomatites, and the development of ore metasomatites (pyroxene-scapolite, pyroxene,
garnet, scapolite-pyroxene-garnet, epidote-actinolite) and also post-ore types
(chlorite-prehnite, calcite-quartz, and zeolite).   The widespread development
of chlorine-bearing sodium scapolite (marialite) is a marked feature of the
Sarbai and other large magnetite deposits of the Turgai province.   This indicates the powerful manifestation of sodium-chloride metasomatism and has served
as the basis for the precipitation of the scapolite and scapolite-skarn types
of iron-ore contact-metasomatic deposits.

The hornfelses have been developed mainly after tuffites and tuffs,
and the intensity of the hornfelsing decreases with distance from the intrusive
contacts.   The scapolite metasomatites have been formed only after feldspar rocks,
but the skarns and hydrosilicate metasomatites are formed after all the types of
surrounding rocks, hornfelses, and preceding metasomatites;   the post-ore metasomatites are, in particular, typical of the late tectonic disturbances.

The replacement of the various types of metasomatites is quite regular.
In an east-west direction they form the following zones, beginning from the intrusive contact: 1) biotite-albite-scapolite metasomatites up to 100 - 150 m thick;
2) garnet and pyroxene-garnet skarns, 3 - 20 m thick in the footwall of the Eastern and Southeastern segregations;   3) skarn-ore with alternation of the ores,
skarn and scapolite metasomatites, 50 - 185 m;   4) skarns and scapolite-pyrox-
ene metasomatites, 3 - 20 m thick in the hanging wall of the ore segregations;
5) pyroxene skarns with relicts of hornfelses, 10 - 30 m thick;   6) pyroxene-
-plagioclase hornfelses, to some degree scapolitized and skarned, up to 40 m
thick;   and 7) hornfelsed and albitized tuffs and tuffites, up to 160 m thick.

Around the Western ore segregation, the following sequence of rocks
(from east to west) has been identified: albitized, prehnitized, and zeolitized tuffs and tuffites, replaced by actinolitized and chloritized pyroclastics,
and farther on an ore zone, consisting of ores and mineralized epidote-actinolite metasomatites, then the same, but gangue metasomatites, and finally, a zone
of tuffs and tuffites epidotized, actinolitized, prehnitized, and zeolitized in
varying degree.

Based on data from a number of investigations, the ore bodies and
associated metasomatites of the Sarbai deposit were formed at various horizons
of the upper Viséan limestone-tuff-tuffite sequence of the Valerianovsk Group,
with a thickness of 350 - 400 m.   Originally it consisted of dark-grey bituminous limestones, clearly interstratified with calcareous tuffs and tuffites.
Recently, Chuguevskaya (1969) has suggested that the ores of the deposit are
restricted to a zone of facies transitions of tuffites into limestones, that
they occur mainly in the tuffites, and that they have a primary volcanogenic-
-sedimentary origin;   they then underwent various metasomatic changes.   However,
observations indicate the formation of a significant portion of the ores at the
expense of metasomatites after carbonate rocks and after non-ore tuffogenic
rocks, the relicts of which occur in the ore bodies.

The ore bodies of the Sarbai deposit consist of layer-like segregations, conformable along the strike and dip with the bedding of the country rocks.   They have the shape of elongated lenses of quite well-defined thickness before wedging-out occurs.   Only in the Eastern segregation, beginning at a depth of 500 m from the surface, is a gradual diminution in thickness observed as a result of the fact that in this part, the ore body is restricted to a pre-ore diorite-porphyrite dyke, cutting the stratification of the country rocks at a sharp angle.   Along the strike and dip, the ore segregations either thin out as a result of the gradual passage of the ore body into non-ore metasomatites, or, as in the case of the southern termination of the Eastern segregation and the northern boundary of the Southeastern segregation, they have been broken by large-amplitude post-ore faults.

The lensoid shape of the ore bodies is complicated by the injection of quite thick post-ore dykes of quartz diorite-porphyrites, in places by stepped forms of the surface of the ore bodies, dependent on local pre- and post-ore fractures, and narrow wedge-like 'apophyses' of the ore bodies, almost conformable with the latter (Fig. 14).

The dimensions of the ore bodies are extremely substantial: in the Eastern segregation, the length is 1700 m, and thickness up to 185 m, and it has been traced along the dip for more than 1000 m;   in the Southeastern segregation, the length is 1000 m, the observed thickness is 170 m, and along the dip up to 800 m;   in the Western segregation, the length is 1400 m, the observed thickness is 185 m, and it has been traced to a depth of 1700 - 1800 m in the central part.   The western segregation consists of two ore members, separated by a layer of rocks, which have been converted to epidote-actinolite metasomatites.

A characteristic feature of the location of the ore bodies of the Sarbai deposit is the absence of direct contacts between the ore segregations and the diorite-porphyrite intrusion;   between them there is a zone of non-ore metasomatites, 25 - 150 m thick.   The principal paths of penetration of the ore-forming solutions were not the contact surface of the intrusion, but the pre-ore faults, which had opened up under the influence of tectonic disturbances.

The ore segregations consist of alternating layers of uniform ores, segregated ores (ore skarns), and non-ore metasomatites.   The ores of the deposit consist approximately by half of uniform varieties with an iron content of more than 50% and half, of segregated ores, containing from 20 to 50% of iron.   The oxidized varieties comprise in all 0.9% of the total reserves.

All the ores are magnetite types.   In subordinate amount there are sulphide ores (pyrite, pyrrhotite-pyrite with a small amount of magnetite, chalcopyrite, and sphalerite.   The sulphide ores form layer-like segregations in the footwalls of the magnetite veins, and have no commercial importance.

The uniform magnetite ores possess banded and coarse-banded fabrics, inherited from the fabrics of the primary calcareous tuffites and limestones. In addition to the dominant magnetite, the following are present in the uniform ores (depending on the type of replaced rocks): pyroxene, scapolite, garnet, wollastonite, albite, epidote, actinolite, apatite, pyrite, calcite, quartz, and accessory idocrase and sphene.   Pyrrhotite, arsenopyrite, sphalerite, galena, and chalcocite have been rarely recorded;   an iron-micaceous hematite and muschketowite have been identified in small amounts.

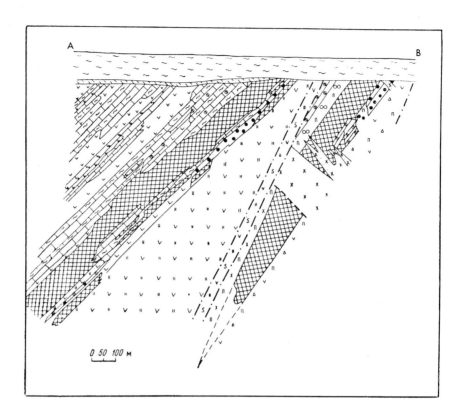

Fig. 14.   Geological section of Sarbai deposit (After I. Kochergin).
Symbols as in Figure 13.

The segregated magnetite ores are to a certain degree mineralized
metasomatites.   Like the uniform ores, they also possess banded fabrics.   In
accordance with the types of metasomatites, types of segregated ores are also
recognized: skarn (pyroxene-magnetite and garnet-magnetite), scapolite-skarn
(scapolite-pyroxene-garnet-magnetite), hydrosilicate (epidote-actinolite-magn-
etite).   The last two types of ores form the Western segregation, and the
remainder, the Eastern and Southeastern segregations.   In the skarn and scapolite-skarn ores there are present to some degree co-skarn wollastonite and
apatite, and also post-skarn minerals such as albite, epidote, actinolite,
chlorite, sulphides, calcite, quartz, and zeolites;   co-skarn sphene and idocrase have been recorded in accessory amounts.

The chemical composition of the magnetite ores is characterized by
the following average amounts (in wt %): SiO_2, 14.9 - 19.3; Al_2O_3, 0.25 - 0.36;
CaO, 5.35 - 11.15; TiO_2, 0.25 - 0.36; V_2O_5, 0.02 - 0.1;   MnO, 0.14 - 0.2; Cu,
0.03 - 0.047;   Zn, 0.03 - 0.08;   Pb, 0.015 - 0.019.

The quality ores in the deposits are regarded as those with an iron
content of more than 30%, and ores with an iron content of 30 - 20% have been

assigned to the non-quality types.   The proved quality reserves of the deposit in categories A + B + C$_1$ comprise 725 million tonnes with an average amount of iron, 45.6; sulphur, 4.05; and phosphorus, 0.13%.   There is a possible increase in the total reserves up to 1.5 billion tonnes from the deep-seated horizons of the Eastern and Western segregations.   The deposit is being worked in a quarry.

The Sokolovsk Deposit

The Sokolovsk magnetite deposit, which belongs to some of the largest in the Turgai iron-ore deposit, was discovered in 1949.   It is located 45 km southwest of Kustanai, near the Sarbai magnetite deposit.

The surrounding Lower Carboniferous sediments (Viséan-Namurian) of the Valerianovsk structural-facies subzone have been divided into three groups (in upward succession): 1) the Sarbai Group consisting of volcanic breccias and tuffs of andesitic porphyrites, up to 1500 m thick; 2) the Sokolovsk, consisting of limestones, basaltic microporphyrites and their tuffs, and calcareous tuffites, with a total thickness of up to 800 m; 3) the Kurzhunkul, consisting of andesitic-basaltic porphyrites, and their tuffs and tuff-breccias, up to 1500 m thick.   Higher up, with erosion, are upper Palaeozoic red-beds, up to 400 m thick (*Distribution Patterns* ... 1968).

The Palaeozoic complexes are overlain by Mesozoic-Cainozoic sediments of the cover of the platform, the thickness of which in the vicinity of the deposit varies from 50 to 120 m.   A Palaeozoic complex forms the Sokolovsk- -Sarbai anticline, complicated by folds of the second order and by fractures. The deposit is located on the eastern limb of the anticline (the Sarbai deposit is located in the western limb).

Intrusive rocks are widely developed near the deposit.   They comprise the Sokolovsk gabbro diorite-granodiorite intrusion, abutting the deposit on the southeast.   The intrusion is elongated in a northeasterly direction by 15 km with a width of 3.5 km; it has been cut by a large sublatitudinal fault, within which there are a number of vein bodies of albitized plagiogranites and diabase microporphyrites.   Within the deposit itself, on the south side, there are bodies of pre-ore subvolcanic diabase porphyrites, plagiogranites, and diorite porphyrites, conformable and in part discordant with the volcanogenic-sedimentary sequence, and on the other side, abundant vein bodies of plagiogranite- -porphyries and microdiabases, in part pre-ore and in part post-ore (Fig. 15).

All the primary rocks of the deposit have been subjected to intense metasomatic changes.   The assemblage of metasomatites includes: 1) alkaline metasomatites (plagioclase-biotite, pyroxene - K-feldspar, albite-quartz - - K-feldspar, pyroxene-albite, and epidote-actinolite-albite), developed after various intrusive and volcanogenic rocks; 2) pyroxene-scapolite[1] metasomatites, which most commonly occur in the hanging wall of the deposit, have been developed after aluminosilicate rocks; 3) pyroxene and garnet-pyroxene skarns have been formed after all the rocks, and the skarns after limestones and calcareous tuffites possess the most ferruginous garnet, with 80% andradite and 20% grossular; 4) hydrosilicate metasomatites (albite-actinolite, epidote, prehnite,

---

[1]   Scapolite-chlorine-bearing, with 10 - 28% of the meionite molecule.

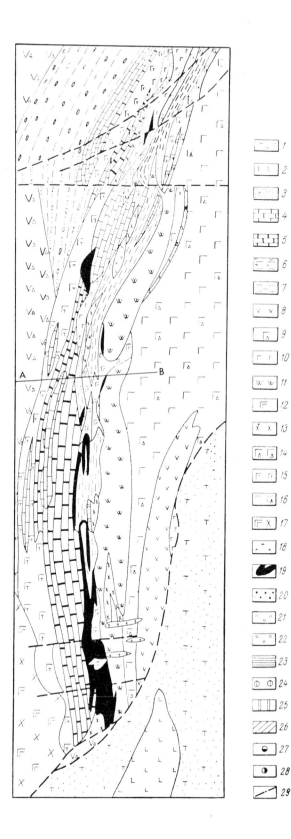

Fig. 15.
Diagrammatic geological map of the
Sokolovsk deposit (After V. Pyatun-
kin, Kh. Shangireev *et al.*).

1) Mesozoic-Cainozoic sediments;
2) upper Palaeozoic (?) basaltic,
less frequently andesitic porphyrites, their tuff-breccias and tuffs,
and tuff-conglomerates; 3) upper
Palaeozoic (?) tuff-siltstones,
tuff-sandstones, and tuff-conglomerates; 4-6) Lower Carboniferous
rocks: 4) limestones, 5) strongly
marmorized limestones and marbles,
6) interstratified tuff-sandstones,
tuff-conglomerates, and tuff-siltstones; 7) tuffites with seams of
tuffs and limestones; 8) andesitic
porphyrites; 9) tuff-breccias of
andesitic porphyrites; 10) basaltic microporphyrites; 11) subvolcanic
diabasic porphyrites; 12) subvolcanic and eruptive basaltic porphyrites; 13) dioritic porphyrites;
14) tuff-breccias of basic composition; 15) tuffs of basic composition; 16) basaltic porphyrites,
tuff-breccias and tuffs of basic
composition; 17) gabbro-diorites
and quartz diorites; 18) pre-ore
plagiogranite-porphyries; 19)
magnetite ores; 20) segregated
magnetite ores; 21) albite and
K-feldspar metasomatites; 22) bio-
titization; 23) scapolite rocks;
24) scapolitization; 25) pyroxene
skarns; 26) garnet skarns; 27)
skarned rocks with pyroxene;
28) skarned rocks with garnet;
29) fractures.

calcite-chlorite, and zeolite).   The hydrosilicate metasomatites have been
formed both after the skarns and directly after the primary rocks.

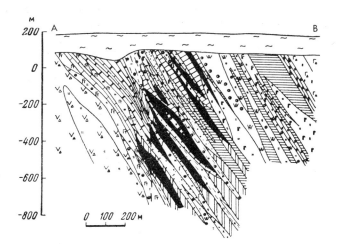

Fig. 16.
Geological section through the
Sokolovsk deposit (After Kh.
Shangireev).

Symbols as in Figure 15.

The structure of the deposit has been determined by a combination of:
1) a homoclinally bedded, facies-variable volcanogenic-sedimentary sequence;
2) the Sokolovsk deep-seated, long-developing fault along the strike of the
deposit, traversed by subvolcanic intrusive bodies; 3) a system of submeridional northeasterly and sublatitudinal fractures, coincident with this fault
and sublatitudinal fractures, controlling the zones and nodes of crushing,
jointing, shearing, opening of the interstratal partings, and also block displacements; 4) a complex system of pre-ore and post-ore dykes.

As a result, the skarn-ore zone of the deposit consists of clearly
alternating layer-like, partly discordant ore bodies and metasomatites of different composition with a predominant easterly dip (besides the southern portion
of the deposit, where the dip changes to westerly).   The thicknesses of the
ore bodies and their dimensions vary intensely throughout the length in the
various sectors of the deposit: the thickest bodies are typical of the northern and eastern parts of the deposit.   With a predominance of conformity
between the ore bodies and the stratification of the surrounding metasomatically
altered sequence, local discordant relationships are normal (Fig. 16).   Mineralized crush breccias are also frequent.

The ore bodies rarely consist of a uniform ore throughout the entire
sequence.   Usually the central portions of the ore bodies consist of rich
massive ores, which towards the periphery gradually pass into segregated ores.
The ores are characterized by relict structures of the stratification and shearing.   Statistically, the rich ores have most commonly been formed by replacement of skarns, which had themselves been formed after limestones or calcareous
tuffites.

The mineral composition of the ores reflects their formation under
conditions of falling temperatures from high post-magmatic to low, which correspond to the relatively low-temperature hydrothermal mineral associations (*Geology and Genetic Features* ..., 1969).   In the ores, in addition to magnetite,
there are: scapolite, albite, pyroxene (diopside), garnet (andradite-grossular),

actinolite, epidote, idocrase, hematite, chlorite, apatite, prehnite, calcite, pyrite, pyrrhotite, marcasite, chalcopyrite, sphalerite, zeolites, and martite.

On the basis of structural-textural features we may distinguish massive, segregated, banded, brecciated, and streaky ores. On the basis of mineral composition, we may recognize scapolite-magnetite, pyroxene-magnetite, garnet-magnetite, calcite-magnetite, and amphibole-magnetite (chlorite-epidote-actinolite-magnetite) ores, and also streaky coarse-grained, epidote-garnet-pyroxene-titanomagnetite-magnetite ores of no commercial value.

The scapolite-magnetite ores contain on the average (in wt %): S, 0.33; P, 0.07; TiO_2, 0.84; V_2O_5, 0.07; MnO, 0.42; in the calcite-magnetite ores, we have TiO_2, 0.21; V_2O_5, 0.03; and MnO, 0.87.

On the whole, the rich ores, with an average iron content of 55.6%, contain sulphur, 2.9%, and phosphorus, 0.07%, whereas the poor ores have 39.2% of iron, 2.49% of sulphur, and 0.11% of phosphorus. The greatest amount of sulphur (on average 3.3%) has been identified in the ores from the central portion of the deposit.

The quality reserves of the deposit comprise, according to categories A + B + C_1, 967 million tonnes with an average iron content of 41%. The deposit is being quarried.

The Kachar Deposit

The Kachar deposit is the largest magnetite deposit in the Turgai iron-ore province. It lies 65 km to the northwest of Kustanai and was discovered in 1943 during an aeromagnetic survey.

Like the other magnetite deposits of Turgai, it is located in the Valerianovsk structural-facies subzone of the Turgai downwarp, and the thickness of the overlying unconsolidated sediments of the cover of the platform reaches 160 - - 180 m in this region.

Two groups have been recognized in the surrounding Palaeozoic sequence: a lower, Lower Carboniferous Valerianovsk Group, and an upper, Kachar Group, provisionally of Middle - Late Carboniferous age. The Valerianovsk Group, with a thickness of more than 1000 m, consists (in upward succession) of andesitic porphyrites and their pyroclastics with seams of tuffites and limestones, clay-carbonate rocks containing seams and lenses of anhydrite near the ore bodies, basalt and dacite lava flows, plagioclase and pyroxene-plagioclase porphyrites and their tuffs. The Kachar Group (up to 800 m thick) consists of polymict tuffogenic sandstones, conglomerates, argillites, andesite-porphyrite tuffs, basalt and andesite flows, and plagioclase and pyroxene-plagioclase porphyrites and their tuffs.

The surrounding volcanogenic-sedimentary sequence has been cut by granite porphyry bodies of stock-like and irregular shape and by rare dykes of diabase porphyrites. The granite-porphyries in places have been albitized and scapolitized, that is, they are pre-ore in origin. However, investigations into the Kachar deposit negate the active role of the granite-porphyries in the formation of the magnetite mineralization based on the following considerations: 1) for the overwhelming number of the Turgai magnetite deposits, a genetic connexion has been

established with intrusions of the gabbro-diorite-granodiorite association;
2) the dimensions of the granite-porphyry bodies in the Kachar deposit are too
small to guarantee the necessary heating of the region for the formation of
the thick ore-metasomatic zones of the deposit, which is reflected in the feeble
development of hornfelses in the contact aureole of the granite-porphyries.   In
addition, according to geophysical data, a large intrusion of assumed gabbro-
-diorite composition occurs at a depth of 2 - 2.5 km from the ore bodies.

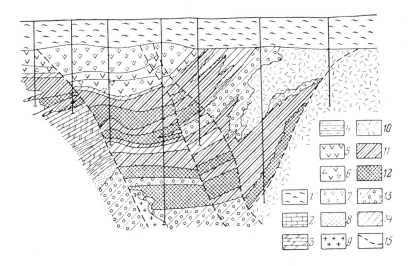

Fig. 17.   Section through the Kachar deposit.   (After N. Belyashov).

1) Mesozoic-Cainozoic sediments;   2) limestones;   3) anhydrites;
4) tuffites;   5) andesitic porphyrites;   6) tuffs of intermediate
composition;   7) andesite-basalt porphyrites;   8) sandstones and
tuff-sandstones;   9) granite-porphyries;   10) quartz porphyries;
11) segregated ores;   12) massive ores;   13) pyroxene-scapolite
metasomatites;   14) scapolite metasomatites;   15) tectonic faults.

The volcanogenic-sedimentary sequence has been crumpled into three
brachyfolds, complicated by second-order folds, and broken by fault-thrusts of
sublatitudinal, submeridional, and northeastern directions into a series of
blocks.   The largest displacements have been identified on the steeply-dipping
faults (Fig. 17).

A specific feature of the structure of the Kachar deposit, distinguishing it from most of the other magnetite deposits of the Turgai province, is
the absence of active intrusions at the depths examined, accompanied mainly by
the high-temperature nature of the strongly developed ore-metasomatic formations.

Also extremely specific is the nature of the peri-ore alterations of
the country rocks;   here widespread development of scapolitization is not common,
and the thicknesses of the zones of scapolite metasomatites reach several hundreds of metres;   in addition, there is quite widespread development of metasomatic sulphides, anhydrite, found within the magnetite deposit, and alunite in
the volcanogenic-sedimentary rocks adjoining on the east (Sledzyuk *et al.*, 1958).

Scapolite metasomatites, in which pyroxene is also involved, consist of a chlor-scapolite of the marialite type and were formed after the feldspar intrusive and volcanogenic rocks as a result of strong sodium-chloride metasomatism. Scapolitization has been most intensely manifested in the upper part of the ore zone, markedly decreasing with depth.   The mineral association of the scapolite metasomatites, in addition to the predominant scapolite and pyroxene, involves hysterogenic minerals: actinolite, tourmaline, apatite, chlorite, albite, zeolites, and calcite.

Pyroxene-albite metasomatites are also widely represented amongst the peri-ore altered rocks.   Pyroxene-garnet and garnet skarns are distributed in subordinate amounts, and replace the scapolite metasomatites;   the later actinolite-chlorite metasomatites have a limited distribution.

Anhydrite segregations have been found, on the one hand, in the form of seams and lenses amongst the limestone bodies adjoining the ore bodies, and on the other, in the shape of the metasomatic segregations, pseudomorphs after feldspars and quartz, and also nests in quartz porphyries and granite-porphyries; anhydrite, as a relict mineral, is frequently present in the magnetite ores.

According to Belyashov & Plekhova (1965), the layers and lenses of anhydrite in the limestones are syngenetic sedimentary formations, which during the subsequent metasomatic processes have partly undergone solution and redeposition.   However, as shown by D. Pavlov (Sokolov & Pavlov, 1970), the anhydrite cannot be formed in the sediments of the Valerianovsk Group on the basis of the palaeogeographical conditions;   it has been distributed only in a local association with the magnetite deposits and is not known outside their boundaries in the deposits of the Valerianovsk Group.   On the basis of origin, the anhydrite is totally hydrothermal and its presence amongst the metasomatic peri-ore formations indicates the formation of the Kachar (and a number of other deposits) with the involvement of mixed, juvenile-vadose hydrothermal solutions.

In the Kachar deposit, three sectors have been recognized (Northern, Northeastern, and Southern), corresponding to large tectonic blocks, formed even during the separation of the Valerianovsk structural-facies subzone and having undergone uplift or sinking of various amplitudes.   These sectors are distinguished by the thicknesses of the volcanogenic-sedimentary sediments, the morphology and attitude of the folds, the intensity of the fractures, and the shape and attitude of the ore bodies.

In the Northern sector, three layer-like bodies of rich ores are known, resting one upon another, separated by segregated scapolite-magnetite ores and non-ore scapolite metasomatites.   The upper body has been formed after scapolitized redbed conglomerates, sandstones, and tuffites, the middle, after limestones, and the lower, after skarns and in part after anhydritized volcanics, tuffs, and tuffites.   On the whole the mineralization of the Northern sector can be traced for 3.5 km along the strike and up to 1200 m along the dip.

In the Southern sector, there is a single layer-like body, crescentic in plan, conformably lying in the brachyanticlinal fold of the enclosing rocks in direct contact with granite-porphyries;   in the east, the ore body has been cut by a fault.   The length of the body is 600 m, and it is up to 60 m thick. The massive magnetite ores pass in the flanks of the segregation into segregated scapolite-magnetite ores.   The ore body has been formed in part directly after limestones, and in part after scapolite and pyroxene-albite metasomatites, which have replaced quartz porphyries, porphyrites, and their tuffs.

The Northeastern sector possesses ore bodies of significantly smaller dimensions.

On the whole, the mineral association of the ores of the Kachar deposit, in addition to magnetite and martite (in the zone of oxidation), includes: more common - scapolite and albite; average or poorly distributed - pyroxene, garnet, epidote, actinolite, chlorites, anhydrite, apatite, sphene, zoisite, prehnite, sericite, calcite, and sulphides (pyrite, pyrrhotite, sphalerite, chalcopyrite, galena, bornite, and chalcocite); sporadic minerals are apophyllite, acmite, tourmaline, fluorite, zunyite, and gypsum.

Amongst the ores of the Kachar deposit, on the basis of mineral composition, we may recognize: 1) uniform magnetite; 2) martite (in the zone of oxidation); 3) scapolite-magnetite; 4) skarn (pyroxene-garnet-magnetite); 5) albite-magnetite; and 6) hydrosilicate (actinolite-chlorite-zeolite-calcite-magnetite). All transitions have been observed between these types.

On the basis of structural features, we recognize massive, evenly and unevenly segregated, banded, and streaky ores, and in the zone of oxidation, powdery and honeycomb types. The ores are predominantly fine-grained, and less frequently, coarse-grained.

In the Northern sector, the quality ores contain (in wt %): Fe, 45.5; S, 0.52 (with increase to 3% in the lower ore body); P, 0.15; Zn, 0.03; V_2O_5, 0.14; and in the Southern sector, Fe, 51.18; S, 0.33; P, 0.31; Zn, 0.02; and V_2O_5, 0.12.

Manganese is present in the ores in an amount of 0.1 - 0.2%. The amount of copper varies from traces up to 0.1%. The average amount of titanium varies from 0.32 - 0.53%, and in the quality ores, up to 0.8%.

In general through the deposit, the quality reserves comprise, on the basis of categories A + B + C_1, more than 1 billion tonnes, and in category C_2, 394 million tonnes. The average iron content in the deposit is 44.9%.

The Anzas Deposit

The Anzas deposit is located on the northwestern slopes of West Sayan at heights of 1050 - 1200 m, approximately 100 km to the southwest of the Abakan mine. Administratively, it lies in the Tashtyp region of the Khakas autonomous district.

Magnetite mineralization has been localized mainly in an acmolith-like massif of metasomatically altered trachytoid gabbros, occurring in metamorphosed sedimentary-volcanogenic rocks of the lower Monok Group (Cm_1). The ore-bearing magmatic complex, termed 'gabbro-albitite', also includes dykes of albite-porphyries and post-ore ancillary intrusions of olivine gabbro. The volcanogenic-sedimentary sequence involves quartz-albite schists, amphibolites, and quartzites. The schists include lenses of marmorized limestones, diabases, and porphyrites (Fig. 18). Both the trachytoid gabbros and the sedimentary-volcanogenic rocks of the ore zone have been albitized, up to the stage of the formation of monomineralic albitites (Pavlov, 1964).   In addition, especially in the sectors distant from the contact with the surrounding sequence, the gabbros have been scapolitized, as a result of which scapolite-actinolite rocks have been formed.   Near the ores, the scapolite is often replaced by clay minerals of the montmorillonite group.   Skarn mineralization proper (pyroxene and garnet) is almost absent.

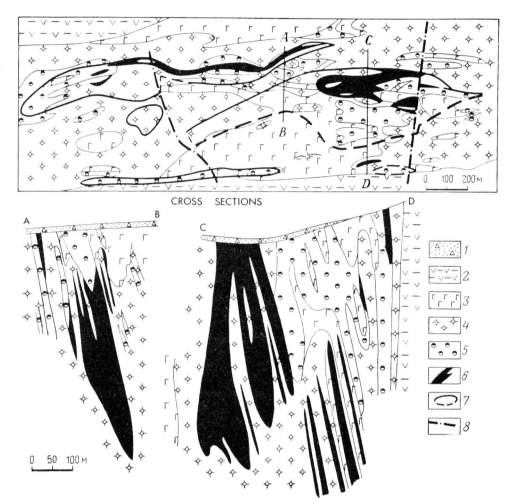

Fig. 18.   Geological plan and sections through exploration profiles VIII and XII of the Anzas deposit.   (After Sh. Kurtseraite).

1) deluvium;   2) diabases, spilites, and siliceous and argillaceous shales;   3) gabbros;   4) metasomatic albitites;   5) scapolite-amphibole metasomatites;   6) magnetite ores;   7) assumed projection of ore bodies on surface;   8) tectonic faults.

The geological structure of the deposit is determined by a zone of steeply-dipping *en échelon* fractures, feathering out from the Dzhebash deep-seated fault.   The overall width of the crush zone reaches 1000 m.   The most

valuable ore bodies are restricted to sectors of pre-ore jointing within the
gabbroid massif; in the rocks of the exocontact, a series of close-spaced thin,
rapidly wedging ore lenses has been recorded.   The age of the mineralization
is probably Ordovician, and possibly Silurian (*Principal Iron-Ore Deposits of
Siberia*, 1970).

In all, 10 lensoid ore bodies have been examined in the deposit.   The
largest measure 400 - 1700 m along the strike, and 50 - 600 m along the dip,
with an average thickness of 20 - 70 m.

In the ores, besides magnetite, there are albite, amphibole (actinolite, and much less frequently, dashkesanite), biotite, scapolite, carbonates,
clay minerals, apatite, and sulphides (pyrite (in early generations cobalt-
-bearing), pyrrhotite, chalcopyrite, and sphalerite).   The principal types of
ores are albite-magnetite and amphibole-magnetite with transitional varieties
present; a subordinate type is the magnetite-biotite.   The zone of oxidation
is not widely developed.

The quality ores on the basis of categories A + B + C_1 comprise 151
million tonnes, and in category C_2, 16 million tonnes.   The average iron content in the quality ores is 38.2%, sulphur, 0.97 - 2.66%, and phosphorus, 0.09 -
- 0.18%.

The Abakan Deposit

This deposit lies 176 km to the southwest of Abakan in the Khakas
autonomous district of the Krasnoyarsk region.   The deposit is located in the
area where West Sayan and the Minusinsk basin join and has been restricted to
the southeastern limb of an anticline, consisting of Lower Cambrian volcanogenic-sedimentary rocks, represented mainly by tuffs of augite and andesitic
porphyrites with fragments of keratophyres, diorites, shales, sand-clay shales,
tuff-shales, and limestones.   Polymict sandstones have been recorded as intercalations in them, and less frequently porphyrite bodies.   The thickness of
the sequence is 300 - 400 m.   These sediments have been cut by a massif of
albitites, probably of Devonian age (Bogatsky & Kurtseraite, 1966).   The massif
is located 3 km from the deposit.

Small Cambrian intrusions of syenite-diorites are also known in the
region of the deposit, and post-ore dykes of albitite- and diorite-porphyries,
up to 15 m thick.   Near the ore bodies there are low-temperature metasomatites,
consisting of chlorite, sericite, epidote, actinolite, calcite, quartz, and
magnetite.   Albitization has been observed, and rarely, actinolitization.   The
length of the ore zone is 1.3 km, with a width of 0.3 - 0.4 km.   Within it,
there are two steeply-dipping ore bodies, conformable *en échelon* with the
stratification of the country rocks and restricted to the largest lens of
limestones and a member of sand-clay shales.   The maximum extent of the ore
bodies along the strike is respectively 1130 and 350 m, and along the dip, 620
and 430 m, with thicknesses of 50 and 17 m.   The transitions between the ores
and the surrounding rocks are quite sharp (Fig. 19).

The ore bodies and the rocks surrounding them have been traversed by
numerous post-ore displacements with northeasterly and northwesterly strike,
and throw of up to 50 m.   Along the northeasterly joints, epidotization, actinolitization, albitization, and sulphidization have developed.   The northwesterly

joints, which are younger than those of the northeasterly direction, are associated with post-ore dykes of albitophyres and quartz porphyries, and also a superimposed ankerite, actinolite-magnetite, quartz-hematite, and sulphide-arsenide
mineralization (Sledzyuk & Sokolov, 1959).

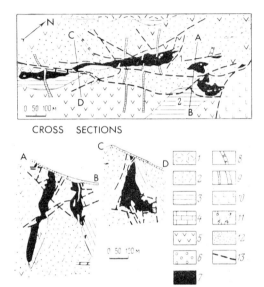

CROSS SECTIONS

Fig. 19.
Geological plan and sections of the
Abakan deposit. (After A. Trigubovich
and Ya. Bartel').

1) keratophyres and their tuffs;
2) tuffs of porphyrites; 3) sand-
-clay shales; 4) limestones; 5)
porphyrites; 6) metasomatites;
7) magnetite ores with ankerite,
siderite, etc.; 8) albitites;
9) diorite porphyrites; 10) labradorite porphyrites; 11) tectonic
breccias; 12) deluvium; 13) tectonic faults.
Ore sectors (figures on plan):
1 - Main, 2 -Third ore body.

The amount of iron in the ores increases from the flanks towards the
centre of the ore body and decreases with depth, and the amount of sulphides in
the ore increases with depth.

The ores are magnetite types.   Constant satellites of the magnetite
are actinolite, chlorite, calcite, siderite, and cobalt-bearing pyrite, and
there is a less frequent occurrence of muschketowite, hematite, albite, actinolite, epidote, zoisite, titanite, apatite, chalcopyrite, pyrrhotite, safflorite,
arsenopyrite, sphalerite, and quartz.   The predominant fabrics of the ores are
streaky, and less frequently, massive, banded, concentric, brecciated, breccia-
-like, and wavy.   In the zone of oxidation, limonite, martite, and halloysite
occur near the surface and in the sectors of development of fractures down to a
depth of 150 - 200 m.

In the quality ores, the average amount of iron is 45.3%, sulphur, 2.39%, and phosphorus, 0.19%.   The deposit is being worked in a quarry and by shafts.

DEPOSITS OF THE HYDROTHERMAL GROUP

The Korshunovsk Deposit

The Korshunovsk deposit occurs in the vicinity of Zheleznogorsk in the Irkutsk district on the Taishet-Lena railway.   It is the largest deposit in the Angara-Ilim iron-ore region, being located in the southwestern portion of the margin of the Siberian Platform.

The deposit is located in the sediments of the platform cover, which consists of argillites, limestones, marls, siltstones, sandstones, and clays of the upper Lena (Upper Cambrian), Ust'kut, Mamyr, and Bratsk (Ordovician) groups. The Upper Cambrian and Ordovician sedimentary rocks have been cut by the so-called 'explosion pipes', which have been filled with tuff-breccias and fragments of the country rocks, and subjected to considerable metasomatic alterations.   The igneous rocks of the region consist of traps, formed of steeply-dipping dykes with northeasterly, and less frequently, easterly strike, and also layered bodies, 30 m thick and more, consisting of gabbro-dolerites, dolerites, and dolerite porphyrites (Fig. 20).

An explosion pipe cuts all the lower Palaeozoic rock groups (Antipov *et al.*, 1960).   The walls of the pipe dip steeply at 65 – 75° into the field of distribution of the tuffogenic rocks.   The main ore body extends from southwest to northeast for 2.5 km with a width of 400 – 600 m.   The shape of the second ore body is close to equant with a diameter of about 500 m.   At depth, the ore bodies narrow down and can be traced to 700 m.

The morphology of the ore bodies is complex.   Layer-like bodies of metasomatic magnetite ores have been recognized, lying conformably with the lower Palaeozoic country rocks; and there are stock-like, lens-like, layered, and columnar metasomatic ore bodies in the metasomatically transformed pyroclastic rocks and steeply-dipping veins of uniform magnetite.   In combination, all these ore bodies form a large single ore segregation, the morphology of which is determined by the shape of the explosion pipe.

The peri-ore metasomatites have developed as the result of hydrothermally-metasomatic transformation of sedimentary, tuffogenic, and igneous rocks. Features of replacement are clearly defined, mainly in the rocks that fill the pipes, and less so in the rocks cut by the pipe (dolomitized limestones, marls, calcareous argillites, and sandstones with calcareous and lime-clay matrix). In the argillaceous and quartzose sandstones with clay matrix, metamorphism is quite insignificantly manifested.

Brecciated and segretated ores are the most widely developed in the deposit.   They are associated with peri-ore metasomatites through gradual transitions.   Massive and banded ores occur less frequently.   The principal ore mineral in the deposit is magnomagnetite, containing up to 6% MgO.   The minor ore minerals are represented by hematite.   In the peri-ore metasomatites and ores, there are diopside, garnet, epidote, apatite, chlorite, calcite, actinolite, phlogopite, hornblende, talc, zeolite, montmorillonite, and aragonite.   The

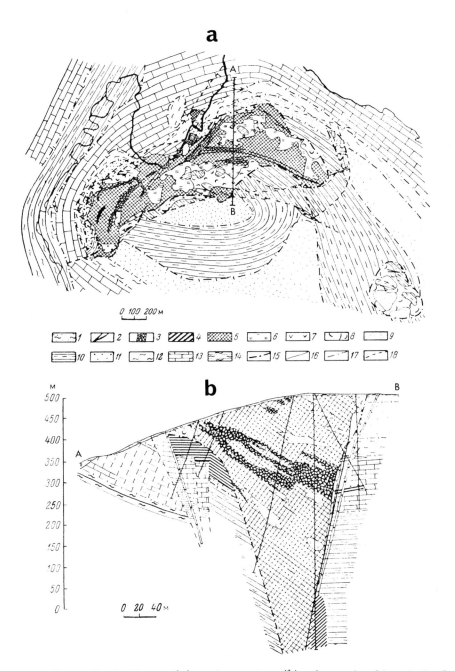

Fig. 20. Diagrammatic geological map (*a*) and section (*b*) along the line I-I of the Korshunovsk deposit (After M. Ivashchenko and V. Karabel'nikova).

1) Quaternary sediments; 2) vein magnetite ores; 3) metasomatic, massive, almost uniform magnetite ores; 4) banded magnetite ores; 5) magnetite ores with segregated and breccia-like structure; 6) peri--ore metasomatites; 7) traps; 8) tuffogenic rocks; 9) rocks of the middle Bratsk sub group; 10) rocks of the lower Bratsk subgroup; 11) rocks of the upper Mamyr subgroup; 12) rocks of the lower Mamyr subgroup; 13) rocks of the Ust'kut group; 14) rocks of the upper Lena subgroup; 15) eruptive contact – outlines of explosion pipes; 16) diffusion contact; 17) stratigraphical contact; 18) tectonic faults.

ores contain (in wt %):   Fe, 34.4 (with fluctuations from 15 to 63); SiO_2, 7.5 - 24.5;   CaO, 1.6 - 11.4;   MgO, 5.4 - 10.4; Al_2O_3, 2.1 - 5.2; TiO_2, 0.17 - - 0.46;   MnO, 0.04 - 0.11;   P, 0.19 - 0.21;   S, 0.02 - 0.03.

The reserves of the deposits based on categories A + B + C_1 are 338 million tonnes, and in category C_2, 95 million tonnes.   The deposit is being worked in a quarry.

The Rudnogorsk Deposit

The Rudnogorsk deposit is situated on the right bank of the River Ilim, 35 km northeast of Nizhne-Ilimsk, and 6 km from the Khrebtovaya-Ust'-Ilim branch of the Lena railway.   The deposit is enclosed in the sediments of the cover of the Siberian Platform.   Its main portion is located in rocks that fill a volcanic explosion pipe.

The area of the deposit is formed of horizontal Lower Silurian sediments of the Bratsk (calcareous clays, marls, calcareous and micaceous sandstones, and dolomites) and Kezhem (sandstones) groups, on which, following a substantial stratigraphical break, rest Permian-Triassic tuffs and tuff-breccias of trap composition belonging to the Tungusska Series.   Besides lying on Silurian deposits, the pyroclasts of the Tungusska Series and rare fragments of Silurian rocks fill a large and two small volcanic explosion pipes, extending almost vertically in depth.   The main pipe, or more precisely, column, has a roughly oval shape in plan with diameters of 600 and 1500 m on the surface and 300 m at a depth of 300 m.   On the southern margin of the pipe there is an east-west trap dyke (olivine diorites), 35 m thick.   A thin trap dyke has also been identified in the body of the pipe.

The ore bodies consist of several close, frequently merging, steeply- -dipping latitudinal veins of rich continuous magnomagnetite ores, cutting the body of the pipe and extending westwards and eastwards beyond its boundaries. The total length of the complex vein is more than 2 km with an overall thickness of up to 45 m.   The ore veins are associated with irregular accumulations of segregated ores, developed along a network of fine joints in the rocks of the pipe, and also on the surface of fragments of breccias and in their fine- -clastic matrix.   The tuffs and breccias, which fill the pipe, have undergone an intense peri-ore metasomatic change, and have been converted into pyroxene- -garnet, garnet-chlorite, and chlorite-serpentine-calcite metasomatites, in varying degree mineralized, with relict textures of the replaced rocks.   Chlorite-serpentine-calcite metasomatites predominate.   The principal ore mineral is magnomagnetite, containing more than 7% MgO, which isomorphously replaces FeO, and also a small trace of Al_2O_3.   The constant presence of noteworthy amounts of acicular apatite in the ores is typical.   In addition, the ores contain hematite, chlorite, serpentine, calcite, sphene, and sparsely-distributed quartz, pyrite, and iron hydroxide minerals.

A specific feature of the vein ores is the significant development of metacolloidal textures (reniform, spherically-conchoidal, endogenic oolitic, etc.).   The vein ores in general are thinly- and thickly-banded, and in places crust-like.

Two kinds of ores have been recognized: 1) rich (uniform thinly- and coarsely-banded);   2) segregated, and finely-disseminated (rich and low-grade).

The average amounts in the rich ores (in wt %) are: Fe, 53.1; P, 0.33; S, 0.05. In the segregated ores, the average iron content is 39.8 wt %.

The reserves of iron ores in categories A + B + C_1 are 209 million tonnes (including 66 million tonnes of rich ores, and the remainder segregated), and in category C_2, 60 million tonnes of segregated ores.

The Tagar Deposit

This deposit is located 110 km eastnortheast of Boguchany station in the Krasnoyarsk region, and 15 km north of the River Angara, in the southwestern part of the margin of the Siberian Platform.

The deposit is located in Lower Cambrian carbonate rocks, Middle and Upper Cambrian and Ordovician carbonate-terrigenous sediments, and Carboniferous terrigenous sediments, forming the platform cover. Layered, and less frequently discordant bodies of troctolitic olivine dolerites, and even less common gabbro--dolerites and pegmatoid dolerites are developed in the Ordovician and Carboniferous sediments (Fig. 21).

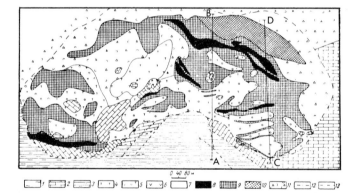

Fig. 21.
Diagrammatic geological map of the Tagar deposit. (After A. Kapinos and B. Khadikov).

1) Quaternary sediments; 2) marmorized limestones; 3) argillites, siltstones, and sandstones; 4) tuffs; 5) partly metasomatically--altered breccias and metasomatites;   6) dolerites; 7) mineralized metasomatites (10-20% iron);   8) magnetite ores with more than 50% iron;   9) magnetite ores with 20-50% iron; 10) limonite ores;   11) structural clays of the weathering crust;   12) zone of eruptive contact – outlines of explosion pipe; 13) tectonic faults.

The ore-bearing pipe-like structure (Fig. 22) consists of explosion breccias, trap agglomerates, and large blocks of the country rocks. The explosion breccias have been converted into metasomatites of differing composition. Amongst the metasomatites, there is a predominance of chlorite-serpentine-calcite and calcite types, and less frequently skarn-like metasomatites of garnet and pyroxene composition are found.

The ore-bearing structure in plan has an ellipsoidal shape with dimensions of 2000 X 1000 m. In vertical section (downwards), the columnar shape passes into a pipe-like form with an overall steep plunge of the axis towards the southeast.   The ore bodies have been recognized on the basis of a limiting iron content of 20%, and they have a complex stock- and lens-like shape with gradual transitions into the metasomatites.   The ores can be traced to a depth of 900 m in the eastern part of the deposit, and in the western part, they wedge out at a depth of 250 – 300 m.   The ore-bearing area on the surface comprises 625,000 m^2 and decreases with depth.

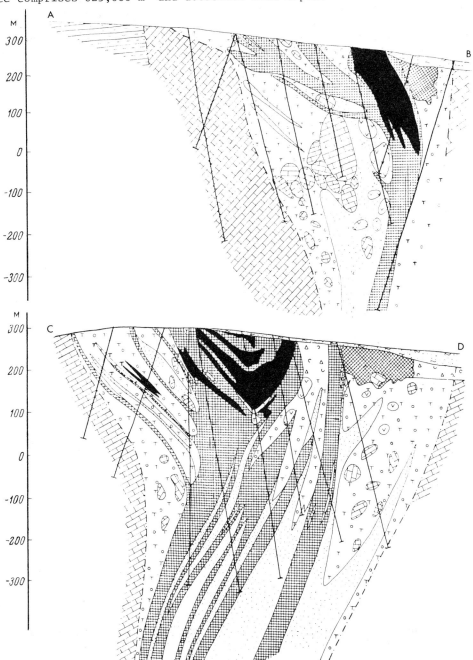

Fig. 22.   Geological sections through the Tagar deposit (After A. Kapinos and B. Khadikov).   Symbols as for Figure 21.

Among the commercial types of ores, brecciated, segregated, and mass-ive magnetite types have been recognized, and in the weathering crust, there are argillaceous and friable martite-magnetite and hematite-hydrogoethite kinds.

In the magnetite ores, the principal ore mineral is magnomagnetite, containing up to 3% MgO, and the minor minerals consist of serpentine, calcite, and chlorite, with less frequent diopside, garnets, phlogopite, epidote, chal-copyrite, and pyrrhotite.   The average amounts (in wt %) are: Fe in the comp-act magnetite ores, 28.9;   S, 1.56;   and P, 0.14;   in the friable magnetite ores, Fe, 39.5;   S, 0.04;   and P, 0.3.   The average amount of iron in the goethite--hydrogoethite ores, 45 million tonnes.

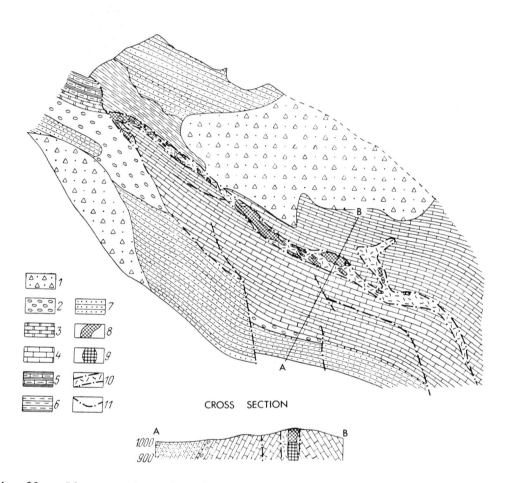

CROSS   SECTION

Fig. 23.   Diagrammatic geological map and section of the Abail deposit (After V. Gar'kovets).

1) drift;   2) coarse-pebbly conglomerates;   3) black siliceous schists; 4) limestones;   5) lime-shale horizon;   6) siliceous schists;   7) silt-stones;   8) hydrogoethite ores;   9) siderite ores;   10) fault zones; 11) tectonic faults.

The Abail Deposit

This deposit lies in the Tyul'kubas region of South Kazakhstan, 8 km from Abail station on the Turkestan-Siberian railway.

In the vicinity of the deposit, there are Silurian limestones and shales, unconformably overlain by Upper Devonian sandstone-conglomerate sequences.   The rocks have been thrown into large folds with northwesterly strike, the chief of which is the Sartur anticline, traceable for more than 20 km. Along the strike, it cuts the zone of the Sartur Fault, which in the region controls the known hydrothermal occurrences.   In the fault zone, there is a group of sideritic deposits (Abail, Kulan, Akhylbek, and Neiman) and a series of quartz veins, containing copper sulphides and native gold (Gar'kovets, 1946).

The ore bodies of the Abail deposit, totalling 32 in number, stretch for 2.4 km along the Vedushchii Fault, which is a branch of the Sartur Fault.

The ore bodies are restricted to structures that intersect and merge with the Vedushchii Fault with its accompanying joints, they are lens-like in plan, and have a columnar shape at depth and reach several hundreds of metres in length, with a thickness of a few tens of metres and an extent of more than 200 m in depth (Fig. 23).

Three types of ores have been recognized in the deposit:

1) primary sideritic, forming the lower horizons of the deposit, and consisting of sideroplesite, ankerite, subordinate quartz, and pyrite, containing 0.1 - 0.8% arsenic;   the pyrite is distributed in the siderite in the form of veinlets, nests, and segregations;

2) secondary brown-ironstone ores, consisting mainly of hydrogoethite and hydrohematite, which form the upper horizons of the deposit and are the product of oxidation of the primary ores;

3) mixed oxide-carbonate ores, occurring in the transition zone between the primary and secondary ores.   The depth of the zone of oxidation varies from 10 to 150 m.   The zone of mixed ores reaches 25 m vertically. The country rocks, and, in places, the ore bodies, are cut by a considerable number of quartz-ankerite and barytes veins and veinlets, sometimes containing chalcopyrite, fahlore, and native gold.

The ore bodies are characterized by a zoned structure.   Along the circulation channels for the hydrothermal solutions, the ores consist of siderites (in the zone of oxidation, brown ironstones).   Along the periphery, the siderite bodies are surrounded by selvedges of ankerites, which are replaced by dolomites.   In the zone of oxidation, lean brown-ironstone ores are developed after the ankerites, with substandard amount of iron (10 - 30%).   Within the ore bodies in places, there are relicts of dolomitized limestones replaced by siderite, which have not undergone dolomitization, prior to the ore formation.

The average composition of the primary ores (in wt %) is: Fe, 35.6; CO_2, 25.2; MgO, 6.57; S, 2.25; P, 0.016; As, 0.06; and of the oxidized ores: Fe, 48.4; CO_2, 7.24; MgO, 0.94; S, 0.03; P, 0.02; and As, 0.04.

The reserves of the siderite ores based on categories $A + B + C_1 + C_2$ comprise 15 million tonnes, and of the brown ironstones, 13 million tonnes.

The Bakal Deposit

The Bakal Group of deposits of siderites and brown ironstones occurs in the Satka region of the Chelyabinsk district. Here, over an area of 150 km², 24 iron-ore deposits have been counted (Fig. 24).

In the geological structure of the Bakal ore field (Yanitsky & Sergeev, 1962), the sedimentary-metamorphic rocks of the Burzyansk and Yurmatinsk series of the Upper Proterozoic are involved. Within the Burzyansk Series, we may recognize the Satka and Bakal Groups, and in the Yurmatinsk Series, the Zigal'ga, Zigazin-Komarovo, and Avzyansk Groups.

The Satka Group consists of dolomites, marls, lime-clay shales, lime-stones, and dolomitic limestones. The rocks of the Bakal Group, which rest conformably on the Satka Group, have been divided into two subgroups. The lower subgroup consists of chlorite-sericite-quartz argillaceous phyllitized slates and sandstones. The upper (ore-bearing) subgroup consists of alternating lime-stones, dolomites, and shaly rocks. It has been subdivided into 10 members. The carbonate members contain seams of siderites and brown ironstones.

On the sediments of the Bakal Group, with angular and stratigraphical unconformity, lie the rocks of the Zigal'ga Group, which consist of 'rewash shales', conglomerates with cobbles and pebbles of quartzites, and phyllites and quartz, and in the upper part, quartzites and sandstones, sometimes with ripple

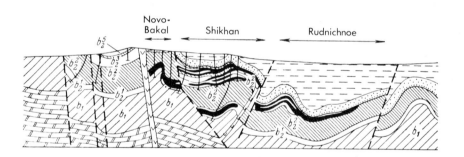

Fig. 24. Geological map, stratigraphical column, and section of Bakal ore field (After Yanitsky & Sergeev, 1962).

1) shales of the Zigazin-Komarovo Group; 2) quartzites of the Zigal'ga Group; 3) shales of upper subgroup of Bakal Group (b_2^2, b_2^4, b_2^6, b_2^8, b_2^{10}) 4) limestones and dolomites of upper subgroup of Bakal Group (Berezovo member, b_2^1, Shuida, b_2^3, Gaevo, b_2^5, Shikhan, b_2^7, Verkhnebakal, b_2^9); 5) shales of lower subgroup of Bakal Group (Makarovo member); 6) lime-stones and dolomites of Satka Group; 7) diabases; 8) iron ores; 9) outlines of 'blind' ore segregations; 10) lines of transgressive overlap; 11) lines of tectonic displacements.
Iron deposits (figures on map); *1)* Shuida II, *2)* Shuida I, *3)* Kholodnyi Klyuch, *4)* im. OGPU, *5)* Novobakal, *6)* Shikhan; *7)* Ob"edinennoe, *8)* Lenin, *9)* Kanatnaya Dorozhka, *10)* Bulandikha, *11)* Malobulandikha, *12)* Vostochnoe, *13)* Zapadnoe, *14)* Rudnichnoe, *15)* Gaevo, *16)* Northwest Irkuskaya, *17)* Aleksandrovsk, *18)* Ivanovsk, *19)* Novoalexandrovsk, *20)* Okhryanoe, *21)* Yel'nichnoe, *22)* Yel'nichnoe Bol'shoe, *23)* Nizhnebaka *24)* Petla.

Burzyansk					Yurmatinsk	Series
Satka Pt$_2$S	Bakal Pt$_2$B			Zigal'ga	Zigazinsk-Komarovo Pt$_2$zk	Group
	Lower B_1		Upper B_2		Pt$_2$zg	Subgroup
	B_1	B_2^1	B_2^2	B_2^3 B_2^4 B_2^5 B_2^6 B_2^7 B_2^8 B_2^9 B_2^{10}		Member
						Geological column
1100	450-500	0-250	120-360	120 23 60 70 100 60 120	60 150 900	Thickness, m

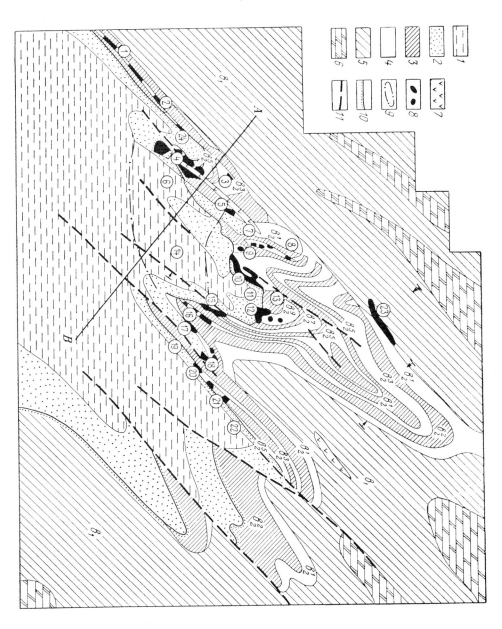

marks and desiccation cracks.   Conformably above come the sediments of the Zig-azin-Komarovo Group, consisting of quartz-chlorite-sericite, carbonaceous-seri-cite-clay, phyllitic shales, sandstones, and marly dolomites.   Thin seams of siderite (up to 0.2 m) occur in the shales.

The magmatic rocks in the ore field consist of numerous dykes and intr-usive veins of diabases and gabbro-diabases.

The interstratification of carbonate members, favourable to ore form-ation, with shales has determined the multi-phase arrangement of the ore bodies. Through tectonic movements, the ore body has been broken into a series of folded large and small blocks, displaced relatively to each other along faults for hundreds of metres.   Former erosion of the Bakal Group and the angular uncon-formity between the sediments of the Burzyansk and Yurmatinsk series have even more complicated the structure of the ore field.

The boundaries between the deposits have been plotted along tectonic faults, which separate adjacent structural blocks, along small non-ore sectors, or simply along arbitrary lines.   In the Bakal group of deposits, more than 200 individual ore bodies are known in the form of layer-like, lens-like, and nest--like segregations, and ore veins.   The largest layer-like segregations occupy an area of 1.5 - 2 km^2, with a thickness of 80 m.

The contacts between siderite and dolomite are, as a rule, sharp. The amount of iron at the contact varies from 27 - 30% in the siderite up to 2 - 5% in the dolomite.   No single case of direct contact between siderites and the limestones has been recorded in the deposits, and they are always separated by a zone of metasomatic dolomites, the width of which may vary from a few metres up to hundreds of metres.

The internal structure of the layer-like segregations is complicated. The intercalations of shales in the members surrounding the ore, the lenses of quartzite, the dykes and layered segregations of diabase, and the sand-clay material disseminated through the carbonate rocks, have not been replaced by ore, and therefore all the primary discontinuities in the construction of the carbon-ate members are preserved in the structure of the ore segregations.   These have also inherited the pre-ore plication and disjunctive tectonic structures of the members surrounding the ore and the forms of the pre-Zigal'ga karst surface. Dolomitic 'residuals' (sectors of a layer surrounding the ore, but replaced by ore) are preserved in places in the ore segregations.   As a result, along with blocks of quite simple construction, consisting of uniform ores, there are sec-tors, extremely complex and variable in form and internal construction.   Zavar-itsky (1939) has associated the diabase intrusions with faults and thrusts, which occurred mainly after the folding.   Zavaritsky, who studied the relationships between the dykes and the siderite mineralization in detail, has shown that the dykes are older than the siderite ores.

On the basis of composition, we may recognize in the deposit siderites, semi-oxidized siderites, and oxidized ores.   The last consist of compact hydro-goethites, brown ironstones and hydrohematite (turgite) ores, powdery brown iron-stones (ochres and chërnotals), and brown-ochre, cavernous-sintery, and argilla-ceous brown ironstones.

The primary fabrics of the siderites are: massive, concentrically--conchoidal (stromatolitic), bedded, and 'primary-vermiform';   the secondary

fabrics are: veinlet, banded, streaky, drusy, 'secondarily-vermiform', brecciated, and granulitic, which developed after the primary fabrics as a result of
a post-ore metamorphism and the effects of tectonic disturbances.

The principal ore minerals in the Bakal deposits are sideroplesite and
pistomesite, in which, besides Fe (25 - 40%) and MgO (7.5 - 19%), there are
CaO (up to 1.5 - 3%) and MnO (up to 2%).   Sideroplesite and pistomesite form
80 - 95% of the ore mass.   The remaining portion of the ore has been formed of
dolomite, ankerite, and barytes.   Pyrite, chalcopyrite, hematite, galena, and
sphalerite occur in small amounts in the ore.   In the joints near the ancient
surface, opal, chalcedony, and aragonite have been recorded.   The depth of
the zone of oxidation varies from 3 up to 110 m.   The boundary between the
siderites and the oxidized ores is uneven.

The oxidized ores consist of hydrogoethite (brown ironstones) and
hydrohematite (turgite).   Goethite is sparsely distributed, and forms sintery
formations.   The gangue minerals consist of quartz and clay minerals.   Dolomite, magnesite, ankerite, barytes, albite, aragonite, apatite, wad, pyrolusite,
pyrite, chalcopyrite, azurite, hematite, galena, sphalerite, magnetite and
vivianite are present in small amounts.

The average chemical composition of the siderite ores (in wt %) is:
Fe, 28.0 - 37.5; MgO, 7.5 - 13.8; CaO, 1 - 3.3; MnO, 1 - 2; SiO_2, 1.45 -
- 6.83; Al_2O_3, 0.8 - 3.43; P, 0.007 - 0.026; S, 0.058 - 0.824; calcination
loss, 30.79 - 39.47.

The average chemical composition of the brown-ironstone ores (in wt
%) is: Fe, 37.8 - 57.7; MgO, 0.16 - 1.8; CaO, 0.21 - 1.26; MnO, 0.68 - 2.28;
SiO_2, 7.5 - 18.3; Al_2O_3, 2.31 - 6.63; P, 0.015 - 0.144; S, 0.006 - 0.086;
calcination loss, 10.08 - 12.52.

The reserves in the deposits of the Bakal group on the basis of categories A + B + C_1 are: siderite ores, 560 million tonnes, and brown-ironstone
ores, 39 million tonnes.

DEPOSITS OF THE MARINE SEDIMENTARY GROUP

The Nizhne-Angarsk Deposit

The Nizhne-Angarsk [Lower Angara] deposit is located in the southern
part of the Angara-Pit iron-ore basin, 30 km from the village of South Yeniseisk
in the Krasnoyarsk region.

In structural respects, the Angara-Pit Basin is a large synclinal
fold with northwesterly strike, complicated by folds of second and higher orders.
The structure of this syncline involves Upper Proterozoic and, to a lesser degree, Lower Cambrian sediments.

All the Proterozoic sediments of the basin have been divided stratigraphically into nine groups (in upward succession): Penchenga, Uderei, Pogoryui,
Kartochki, Alad'insk, Potoskui, Kirgitei, Nizhneangarsk (ore-bearing), and
Dashka (Yudin, 1968).   The groups consist of clay and sand-clay, and to a
lesser degree, carbonate sediments.

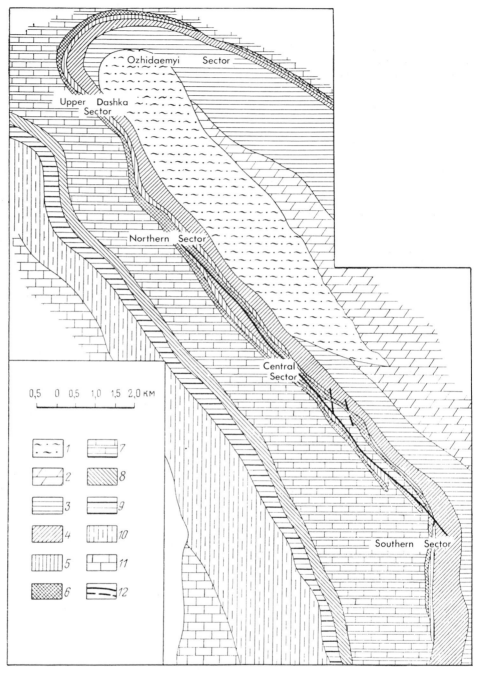

Fig. 25. Diagrammatic geological map of the Nizhne-Angarsk deposit (After
          A. Lesgaft).

          1) thick and unconsolidated Tertiary loamy sediments with clastic
          material; 2-11) Upper Proterozoic sediments: 2) dark-grey, marly
          limestones of the Dashka Group, 3) dark-grey marly shales,
          4) violet sand-clay shales, 5) grey and dark-grey sand-clay shales,
          6) ore horizon, 7) shale-carbonate horizon with limestone seams,
          8) violet shales, 9) finely-banded silty shales with chloritoid,
          10) horizon of black shales with pyrite, 11) black unstratified
          limestones of Potoskui Group; 12) tectonic faults.

The Nizhne-Angarsk deposit is restricted to the western limb of a brachysyncline. The ore-bearing group in the deposit consists of layers of hematite ores, argillites, sandstones, and shales with an abundance of ferruginous chlorite. The extent of the entire deposit is 20 km, and of the individual layers, up to 10 - 15 km, down dip they can be traced for 600 m, with dips of 45 - 60°. In the central part of the deposit (Fig. 25) for a distance of about 12 km, the ore horizon has been duplicated along the plane of the Main Fault with a throw of 150 - 400 m. Several transverse and diagonal disjunctive fractures with small throw have also been revealed. The thickest and richest ore layers in respect to iron (numbering up to 10) occur at the base of the ore horizon in the Central sector. The length of this zone is 5 km, the average thickness of the individual layers is 5 - 8 m, and the overall thickness is 50 m.

In the Northern sector, 6 km long, the number and overall thickness of the ore layers increase, but the thickness of the individual layers decreases.

On the flanks of the deposit (Southern and, in part, the Verkhne-Dashka sectors), the ore layers wedge out both along the strike and along the dip. The ore layers, as a rule, have sharp boundaries with the sand-clay rocks surrounding them, but along the strike, they often gradually pass into hematitized sandstones.

The ores of the deposit have been subdivided into hematitic (10%), sandy hematitic (60%), and clay-chlorite hematitic gravelites (30%). All these types of ores are transitional to each other. The principal ore-forming minerals are hydrohematite, hematite, and goethite, and siderite is occasionally present, with very rare magnetite and pyrite. Of the gangue minerals, we may mention quartz, leptochlorite, clay minerals, sericite, and chlorides.

The most typical textures of the hematitic ores are cryptocrystalline, flaky, tabular, collomorphic, and cemented.

The average chemical composition of the ores (in wt %) is: Fe, 40.4; SiO_2, 24.9; Al_2O_3, 7.4; CaO, 0.22; MgO, 0.22; TiO_2, 0.51; MnO, 0.05; S, 0.03; P, 0.08; Pb, Zn, Co, Ni, and V, traces.

The reserves of ores in the deposit on the basis of categories A + + B + C_1 + C_2 are 1.2 billion tonnes.

The West Karazhal Deposit

The West Karazhal deposit is located 110 km to the southwest of Zhana-Arka station in the Karaganda district of the Kazakh SSR, in the Atasu iron-ore region.

The geological structure of the deposit involves a thick group (up to 1.5 km) of eruptive and tuffogenic rocks of Early and Middle Devonian age, and a similar thickness of sedimentary rocks of the Upper Devonian-Lower Carboniferous of the Dzhail'ma trough.

The Lower and Middle Devonian volcanogenic-sedimentary rocks, distributed to the southwest of the ore segregation, consist of lavas of trachytic and rhyolitic composition and their accompanying sediments. They are overlain by polymict sandstones, siltstones, shales, and chert-clay rocks. The group of Upper Devonian - Lower Carboniferous sedimentary rocks consists in the lower part of carbonaceous-siliceous and carbonaceous-clay-siliceous lime-

a

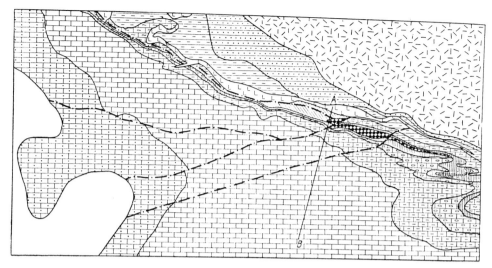

b

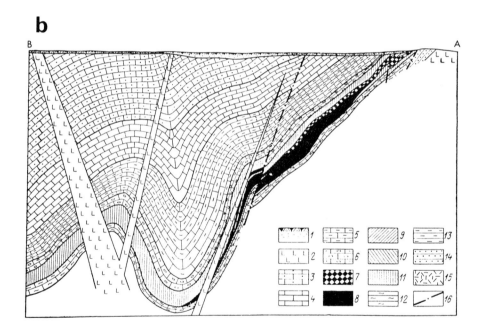

Fig. 26.   Diagrammatic geological map (a) and section (b) of the West Karazhal
deposit.   (After G. Momdzhi, V. Kavun, and S. Chaikin).

1) Quaternary loams; 2) diorite porphyries; 3) cherty limestones
with rare carbonate concretions; 4) cherty, massive-bedded limestones; 5) carbonaceous limestones with seams of argillite;
6) limestones with seams of hornfelses; 7) magnetite ores; 8)
hematite ores; 9) manganese ores; 10) low-grade barytes iron ores;
11) low-grade iron-manganese ores; 12) siliceous-carbonate rocks
with seams of hornfelses; 13) carbonaceous argillites, siltstones,
and siliceous limestones; 14) polymict sandstones and siltstones;
15) quartz porphyries, albitophyres, porphyrites, and their pyroclastics; 16) tectonic faults.

stones with seams and lenses of siliceous jasperoid rocks, eruptives of spilitic type, their tuffs, and lava-breccias, and also layers and lenses of magnetite, hematite, and manganese ores. In the middle part of the sequence there are siliceous-argillaceous-carbonate and jasperoid rocks, tuffites, and ash tuffs with subordinate seams of limestones. The upper part of the sequence consists of limestones with seams of argillites in the eastern part, and argillites and sandstones in the western part of the deposit. Down to depths of 500 - 600 m the rocks dip at 45 - 50° northwestwards, and farther on the limb of the fold dips more steeply (70 - 80°) to depths of 1000 m.

The igneous rocks consist of dykes of diorite and diorite porphyrite, which cut Lower Carboniferous sediments.

The ore segregation forms a layered body, conformable with the country rocks; it can be traced along the strike for 6.5 km and down the dip for up to 800 m. The thickness of the segregation varies from 30 - 40 m on the eastern flank down to 20 - 25 m in the central part of the deposit; on the western flank, both along the dip and along the strike, it decreases until it wedges out completely (Fig. 26). An ore layer occurs between the carbonaceous-siliceous limestones with seams of jaspers in the footwall, and limestones with seams of jasperoid rocks in the hanging wall. In the lower part of the ore segregation, hematite ores are developed, in the middle part, predominantly magnetite, and in the upper part, low-grade hematitic manganese ores.

At the base of the ore layer, there is a very thin manganese-ore layer. Such layers and lenses of manganese ore also occur within the hematitic layers (Sapozhnikov, 1963). In the upper parts of the deposit, there is a zone of baritized iron ores.

Three commercial types of ores have been recognized in the deposit: magnetite, magnetite-hematite, and hematite. The magnetite and magnetite-hematite ores are distinguished by their increased content of germanium (Grigor'ev, 1971).

The principal ore minerals are: hematite, and magnetite; minor minerals are siderite, barytes, and pyrite; arsenopyrite, chalcopyrite, sphalerite, and galena occur in small amounts. In the zone of oxidation, which extends for several tens of metres, martite and iron-hydroxide minerals are distributed. The fabrics of the ores are banded and massive, and the textures are granular and porphyroblastic.

In the high-grade iron ores, the average amounts (in wt %) are: Fe, 55.6; MnO, 0.46; SiO_2, 12.4; S, 0.6; and P, 0.03.

Reserves of iron ores based on categories A + B + C_1 are 310 million tonnes.

THE KERCHEN IRON-ORE BASIN

Within the Kerchen Peninsula there are two types of deposits of marine sedimentary oolitic iron ores of Kimmeridgian age. The first type is restricted to large tectonic brachysynclinal structures (troughs), and the second is associated with pseudotectonic structures (compensation downwarps in the zone of development of mud volcanism, the so-called 'depressed synclines' (Shnyukov, 1965).

The first type includes the Ak-Manai, Chegene-Salyn, Katerlez, Kamysh-Burun, Él'tigen-Ortel, and Yanysh-Takyl deposits, and the second, the Baksinsk, Kezen, Osovinsk, Novoselovsk, Rep'evsk, and Uzunlar deposits (Fig. 27).   The principal reserves of high-grade iron ores (with more than 30% of iron) are restricted to the deposits of the first type.

In all the troughs and downwarps, the middle Kimmeridgian ore layer is underlain by limestones of the Pontian Stage or lower Kimmeridgian clays and is overlain by sandy or silty material of the upper Kimmeridgian stage.   The depth of occurrence of the ore layer from the outcrops on the surface along the margins of the troughs and downwarps is 70 - 100 m, and in the central parts up to 140 - 180 m (Arbuzov *et al.*, 1967).   The thickness of the ore layers in the central parts of the deposit reaches 25 - 40 m, and in the marginal parts it falls to 0.5 m (Fig. 28).

The principal types of ores are 'tobacco' and 'brown'.   The former predominate in the oxidation-reduction zone, and the latter have been formed at the expense of the former in the oxidizing zone.   Pressurized waters from the Pontian aquifer have penetrated along the joints, cavities, and capillaries in the ore layer and have preserved the tobacco ores.   This same aquifer is usually restricted to the central sectors of the troughs and downwarps, which also determines the presence of the tobacco ores in them;   along the periphery of the troughs and downwarps, the brown ores, formed in the upper part of the ore layer, have an annular arrangement.

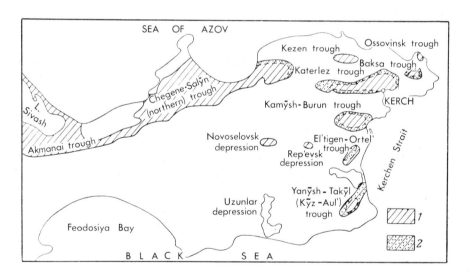

Fig. 27.   Diagram of location of ore-bearing brachysynclines and compensation downwarps of the Kerchen Peninsula (After Yu. Yurk, Yu. Lebedev, and O. Kirichenko).
1) area of distribution of tobacco ores; 2) area of distribution of brown ores.

The minor, sparsely-distributed ores are the mangansiderite-rhodochrosite concretionary ores and the manganese-iron 'roe' ores, which are distinguished from the brown ores by the increased content of manganese.

The principal minerals of the tobacco ores (Yurk *et al.*, 1960) are hydroferrichlorite, ferrimontmorillonite, and hydrogoethite (ocherous and compact varieties), and also carbonates of the manganosiderite-rhodochrosite series. Quartz and feldspar often occur in the fragments of oolites. Less widely developed are the phosphates (vivianite and kerchenite), and hydroxides of manganese (psilomelane, vernadite, and pyrolusite), and pyrite. Realgar is a rare occurence.

The principal minerals of the brown ores are hydrogoethite and ferrimontmorillonite. Psilomelane, pyrolusite, gypsum, aragonite, calcite, kerchenite, pyrite, quartz, feldspar, and glauconite occur in lesser amounts. The principal fabrics of the ores are oolitic and pisolitic.

At the present time, only the brown ores are being exploited, since the problems of economically enriching the tobacco ores have not been worked out.

The brown ores contain (in wt %); Fe, 37.7; MnO, 3; SiO_2, 18; Al_2O_3, 5; MgO, 1; CaO, 1.75; V_2O_5, 1.19; P, 1; S, 0.07; As, 0.13.

The reserves of iron ores in the principal deposits of the Kerchen Peninsula on the basis of categories A + B + C_1 are 1.7 billion tonnes, including 570 million tonnes of brown ores. The Kamysh-Burun and the Él'tigen-Ortel deposits are being worked in quarries.

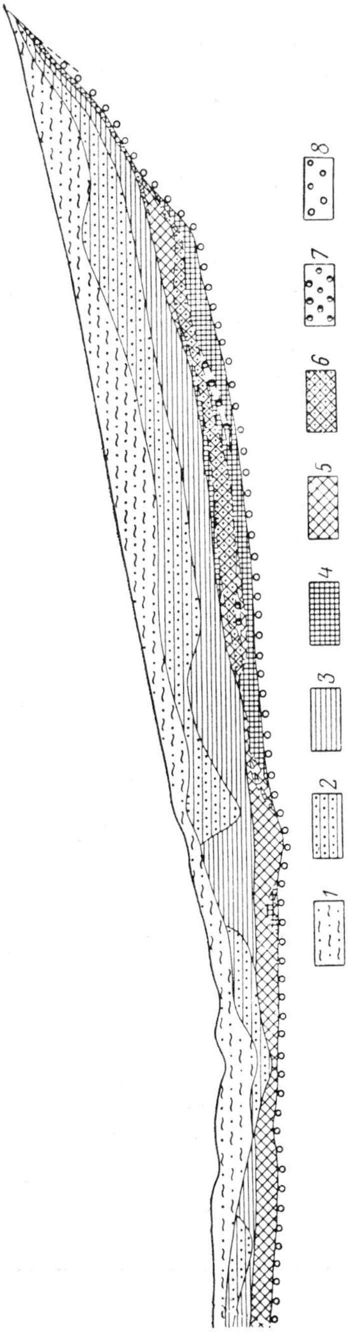

THE AYAT IRON-ORE BASIN

This basin is located 20 km north of Tobol railway station in the Kustanai district of the Kazakh SSR (Fig. 29).

At the base of the geological sequence are Palaeozoic eruptive-sedimentary rocks, consisting of basic and acid eruptives and their tuffs, gabbros, diorites, granites, serpentines, jaspers, quartzites, limestones, sandstones,

Fig. 28. Transverse geological section through the Kamysh-Burun trough (After Yu. Yurk, Yu. Lebedev, and O. Kirichenko).
1) calcareous loams; 2) sandy clays; 3) clays; 4) 'tobacco' ores; 5) 'brown' ores; 6) 'roe' ores; 7) sandy clays, and argillaceous sands with abundant bivalve shells; 8) argillaceous coquinas, sideritized at the top.

and shales.   These rocks are overlain by a thick ancient weathering crust.   In the depressions in the Palaeozoic basement there are Lower Cretaceous continental sediments of lacustrine-paludal type (clays, sands, and sandstones with commercial segregations of bauxite and refractory clays.   Higher up there are Cretaceous and Tertiary marine deposits, consisting of basal conglomerates, sideritic sandstones, quartzose and quartz-glauconite sandstones, clays and loams with phosphorites and a marine fauna (bivalves, cephalopods, and sharks' teeth).

The ore layer, consisting of oolitic, leptochlorite-siderite ores, is well maintained along the strike and in thickness, and lies flat or gently undulating with a general insignificant slope to the east and northeast.   The ore layer usually occurs in quartz-glauconite sands, and less frequently in Cenomanian clays with a clear boundary, sometimes with weakly defined erosion.   In rare cases, the country rocks of the ore layer are Palaeozoic and their old eluvium. The roof of the ore layer is usually formed of lignite clays of the lower Senonian, or overlying rocks.   The lower portion of the layer consists of weakly-cemented coarsely-oolitic ores (1 - 4 mm), the middle portion consists of seams of clays, 0.1 - 0.4 m thick, with carbonized plant remains and siderite concretions, and the upper part consists of finely-oolitic cemented ores.   The ore layer can be traced for tens of kilometres, and its thickness varies from 2 to 9 m (Fig. 30). The area of distribution of the ore layer is about 2500 km^2.

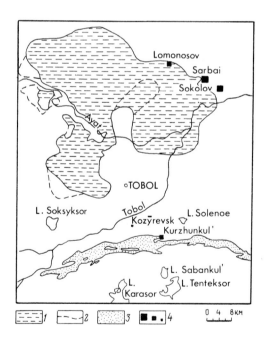

Fig. 29.
Sketch map of the region of the Ayat and Lisakovsk deposits.

1) area of Ayat deposit;   2) boundaries of redeposited ores of the Ayat deposit;   3) area of the Lisakovsk deposit;   4) deposits of magnetite ores.

The ores have been divided into oolitic and concretionary-siderite types.   The former predominate, and the latter have been observed only in the form of seams in the oolitic ore and have no independent importance.

Amongst the ores, we may distinguish primary leptochlorite-siderite types, oxidized hydrogoethite, and redeposited hydrogoethite ores.

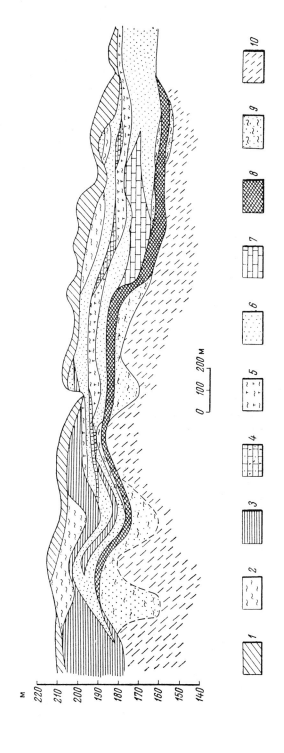

Fig. 30.   Geological section through the Ayat deposit (After A. Zhilyakov, D. Toporkov, and M. Uzbekov.

1) loess-like loams (Q);   2) quartzose clays and sands (N);   3) shaly clays (Pg$_3$);   4) quartz-glauconite sandstones (Pg$_{1\ 2}$);   5) gaize clays (Pg$_{1\ 2}$);   6) sands and marly clays (Pg$_{1\ 2}$);   7) limestones (Cr$_2$m);   8) oolitic hydrogoethite and hydrogoethite-siderite-leptochlorite ores (Cr$_2$cm);   9) sub-ore sands (Cr$_2$t – st);   10) weathering crust of Palaeozoic rocks.

The primary leptochlorite-siderite ores have a compact or granular structure, and rarely form oolitic nodules.   The principal ore-forming minerals in them are leptochlorite and siderite, cementing quartz sandstones, and segregations of glauconite.   The amount of chlorite and siderite in the matrix varies within wide limits, and they seemingly replace one another.   In accordance with the quantitative ratio of the principal minerals, we may distinguish amongst these ores, glauconitic, glauconite-sideritic, glauconite-leptochlorite- -sideritic, leptochlorite-sideritic, and sideritic varieties (*Oolitic Brown Ironstones of the Kustanai District* ... , 1956).

The oxidized hydrogoethite and redeposited hydrogoethite ores are distinguished by their oolitic structure.   The principal ore mineral in these ores is hydrogoethite, which comprises the oolites and their clasts, and also in part the cementing material, which consists mainly of extraordinarily finely- -clastic clay-silt material of friable aspect.   In lesser amounts in these ores is leptochlorite, and small amounts, glauconite, quartz, siderite, and pyrite. According to the content of leptochlorite, these ores, on the basis of mineral composition, have been subdivided into hydrogoethite, hydrogoethite-leptochlorite, and leptochlorite types.

The content of silicon and iron in the ores is in inverse relationship, and the amount of the remaining components does not change with alteration in the amount of iron.   The oxidized ores in all cases contain free silica in greater amount than the primary ores.   The silica in the primary ores is associated with chlorite and clay minerals, and its amount in the free form is insignificant.   In the oxidized ores the amount of free silica increases at the expense of the chlorites.

The maximum amounts of manganese are restricted to the peripheral parts of the deposit, and to the area of distribution of the siderite ores. The amount of phosphorus is steady within the limits of 0.3 - 0.5% and is associated with the absorption of compounds of this element by the iron hydroxides.   Sulphur enters into pyrite and gypsum;   the most commonly recorded amount of sulphur is from 0.1 up to 0.7%.

The chemical composition of the ores is characterized by the following average amounts (in wt %):   Fe, 37.1; SiO_2, 16; Al_2O_3, 7.3 - 8.6;   CaO, 1.6 - 1.8;   MgO, 0.1 - 1;   MnO, 0.5 - 5; TiO_2, 0.1 - 0.2; V_2O_5, 0.05 - 0.13; NiO, 0.004 - 0.01; Cr_2O_3, 0.052;   Co, 0.007;   Zn, 0.03;   Cu, 0.01;   As, 0.005; S, 0.35 - 0.36;   P, 0.37 - 0.4.

The reserves of the deposit on the basis of categories A + B + C_1 are 1.7 billion tonnes, and in category C_2, 5 billion tonnes.

THE WEST SIBERIAN IRON-ORE BASIN

This basin is traceable from the sources of the Rivers Bakchar, Parbig, and Anderma in the south to the upper course of the River Vakh in the north (Fig. 31).   Over an area of 66,000 km^2, drill-holes have exposed an horizon of sedimentary iron ores of Cretaceous age below thick (hundreds of metres) younger sedimentary deposits.   Within the basin, four main deposits have been recognized: the Bakchar, Narym, Kolpashevo, and South Kolpashevo.   The total reserves of ores in the basin with more than 30% of iron have been estimated at 400 billion tonnes (*West Siberian Basin*, 1964).

The Bakchar Deposit

This deposit, situated 200 km to the northwest of Tomsk, is relatively the best of the deposits of the West Siberian basin.

The ore-bearing marine deposits range in age from Coniacian to Turonian, and consist of quartz-chlorite-glauconite sandstones, sands, and siltstones with seams of gravelites (Nikolaeva, 1967).   Amongst them, four productive horizons of oolitic leptochlorite-hydrogoethite ores have been recognized (Fig. 32).   The best of them is the Bakchar horizon which is distributed over an area of about 700 km^2, with an average thickness of 26 m and an
average iron content of 37.4%, together with an average thickness of cover
rocks of 190 m (varying from 155 up to 275 m).   The amount of phosphorus in
the ores is 0.38 - 0.69%, and the average amount of vanadium is 0.13%.

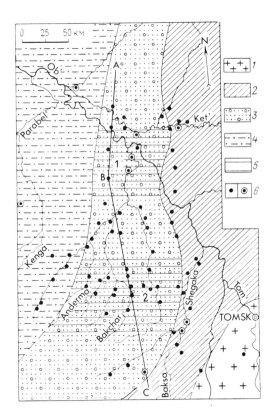

Fig. 31.
Palaeogeographical map of the Upper
Cretaceous iron-bearing sediments of
the southeastern portion of the West
Siberian iron-ore basin (After
A. Babin and I. Zal'tsman).

1) rise in the Palaeozoic basement;
2) continental sediments; 3) nearshore-marine iron-bearing deposits
with horizons of oolitic ores;
4) nearshore-marine and shallow-
-water iron-bearing deposits;
5) areas prospective for search for
new deposits; 6) drill-holes:
a) core, *b*) rotor reference.
Ore fields: *1* - South Kolpashevo,
2 - Bakchar.

The geological reserves of the deposit have been estimated at 28
billion tonnes.   The technical mining conditions of the deposit are complex,
since above the ore segregations, five water-bearing horizons have been encountered, partly under pressure, and associated with Cretaceous, Oligocene, and
Quaternary sediments.

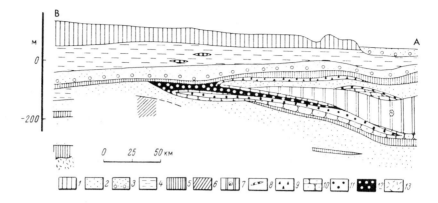

Fig. 32.   Geological section of the southeastern portion of the West Siberian
iron-ore basin (see Fig. 31).   (After A. Babin and I. Zal'tsman).

1) sands, loams, and pebble beds;   2) fine-grained sands;   3) sands
of varying grainsize, with gravel;   4) silts;   5) clays;   6) varicol-
ored clays;   7) calcareous clays with remains of molluscs;   8) brown
coals and lignites;   9) granular glauconite-siderite ore;   10) quartz-
-glauconite sandstones, and siltstones;   11) ore sandstones;   12)
oolitic ores;   13) quartz keratophyres.

DEPOSITS OF THE CONTINENTAL SEDIMENTARY GROUP

The Akkermanovsk Deposit

This deposit is involved in the Orsk-Khalilovsk group of deposits of
naturally-alloyed iron ores, located 20 km west of Orsk.   The deposit is restricted to the southwestern marginal portion of the Tanalyk-Baimak Mesozoic
depression.   The base of the depression is formed by Palaeozoic and older folded
complexes of the Urals with discordant basic and ultramafic intrusions.   The
depression itself has been filled with Mesozoic-Cainozoic continental and marine
sediments.

The Palaeozoic rocks in the vicinity of the Akkermanovsk deposit consist of Tournaisian siliceous schists and strongly karsted Viséan limestones.
At 5 - 10 km to the southwest of the deposit, they have been cut by a gabbro-
-peridotite intrusion, the peridotites of which are strongly serpentinized.   A
Triassic-Jurassic weathering crust has been developed on the Palaeozoic rocks,
including the serpentinites.

On the limestones and siliceous schists, with marked unconformity,
rest Jurassic continental deposits of the depression, consisting of rubbly deluvium, sands, clays, and pebbles, and including two ore horizons, the lower sideritic and the upper, goethite-hydrogoethitic (Fig. 33).   The ore bodies of the
lower horizon are layered; in the ores there are thinning-out argillaceous seams.
The thickness of the ore sequence is about 35 m in the central part, and towards

the periphery it decreases. The sequence consists of siderite-clay and siderite-hydrogoethite oolitic-brecciated ores, the siderite in which has been oxidized in varying degree. The sediments of the upper ore horizon fill the karst depressions in the limestones, in which respect the thickness of the ore bodies is markedly variable and varies from 1 up to 50 km, on average 14 m. The upper horizon lies partially on weakly karsted limestones and clays, having a regular layered form (Kiselev, 1963). It is evident that the upper horizon was formed later, as the region of the deposit has undergone substantial uplift, which caused increase in karst formation.

In the upper horizon, four types of ores have been distinguished: 1) predominant ochre-clay material from a fine mixture of clay matter, ferruginous chlorites, hydrogoethite, and hydrohematite; 2) blocky-rubbly ores, in which the main ochre-clay mass contains a varying number of fragments, concretions, and geodes of aphanitic hydrogoethite and hydrohematite, and also pisolite-oolitic formations of the same minerals. Such ores form thin lenses in the ores of the first type; 3) conglomerate-oolite ores of the same composition as in the ores of the second type; 4) pebbly ores, developed as a result of local erosion of ores of the preceding types.

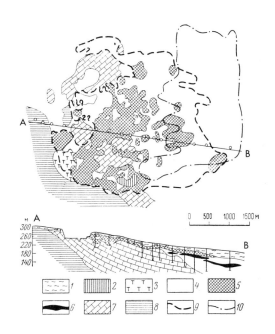

Fig. 33.
Diagrammatic geological map and section through the Akkermanovsk deposit. (After Kiselev, 1963).
1) brown and grey loams; 2) rubbly--pebbly redeposited iron ores;
3) yellowish-grey sandy clays, frequently with gypsum; 4) mottled clays, ferruginized sands, and pebble beds; 5) brown-ironstone ores (upper horizon); 6) siderite-clay ores (lower horizon); 7) Viséan limestones; 8) siliceous shales;
9) outlines of brown-ironstone ores (upper horizon); 10) outlines of siderite ores (lower horizon).

In the Akkermanovsk ores, in addition to hydroxides of iron and siderite, ferruginous chlorites have been recorded, along with a minor development of magnetite and maghematite, a large group of manganese minerals (asbolan, rancierite, vernadite, manganite, etc.), quartz, chalcedony, opal, carbonates (calcite, aragonite, dolomite, magnesite), gypsum, barite, pyrite, muscovite, hydromicas, epidote, zircon, rutile, various clay minerals, nontronite, minerals containing alloying elements (revdinskite, nickel nontronite, chrome-spinels, volkonskoite, chrome allophane, and erythrite (*Khalilovo Deposits of Complex Iron Ores*, 1942).

The average amounts in the ores of the lower horizon (in wt %) are :
Fe, 27.1; Ni, 0.29; and Cr, 1.43.   The average amount of iron in the hydrohematite-hydrogoethite ores of the upper horizon is 32.0%, and ores of type I
have been recognized with an iron content of more than 35%, nickel about 0.4%,
and chromium more than 1%.

The reserves of ores based on categories A + B + C$_1$ are 163 million
tonnes, and in category C$_2$, 120 million tonnes.   The deposit is being worked
in a quarry.

The Berezovo Deposit

This deposit is located in Eastern Transbaikalia, 12 km from the
village of Nerchinskii Zavod, near the River Argun'.   It is restricted to the
western side of the Argun' basin, the basement of which is formed of Cambrian-
-Silurian rocks (phyllitic and quartz-sericite schists, dolomites and limestones).
The basin in the vicinity of the deposit is filled with Upper Jurassic continental lacustrine sediments, in which there is, in upward succession: 1) basal
deluvial-alluvial clay breccias (up to 100 m thick); 2) an iron-ore sequence
of conglobreccias and breccias, weakly or unsorted according to clast size, and
a facies of talus and collapse material, consisting of fragments mainly of limestones and dolomites (sideritized, and in the zone of oxidation converted into
brown ironstones), and in lesser degree, shales, up to 340 m thick; 3) a siltstone sequence (up to 500 m thick) of alternating thin seams of siltstone, argillite, and sandstones; 4) sandstones and conglomerates (up to 300 m thick)
with thin layers of brown coals; 5) pyroclasts of rhyolites and rhyolite-dacites;
6) flows, dykes, and sills of andesites and basalts.

In the vicinity of the deposit, the western margin of the Argun' basin
slopes eastwards at angles of 25 - 50°; towards the axis of the basin and upwards in the sequence the dips decrease to 5°.   The western margin is complicated by faults, apparently synchronous with the sedimentation, around which
the thicknesses and facies of the sediments are changed.

The ores of the deposit consist of intensely sideritized (in the zone
of oxidation, ocherized) conglobreccias and breccias of the iron-ore sequence,
separated in the form of layer- and lens-like ore bodies.   Six ore bodies are
known (the Glavnoe (Main) and Obosoblennoe (Separate) in the southern half, and
the Pervoe (First), Vtoroe (Second), Tsentral'noe (Central), and the Severnoe
(Northern) in the northern half).   The dimensions of the ore bodies vary in
length from 3.5 km to 370 m, in width from 1000 down to 100 m, and in thickness
from 230 to 20 m.   The deposits are in part restricted by faults.

The primary siderite ores of the deposit comprise about half of all
the reserves; they occur at a depth of 50 - 140 m from the surface.   About
85% of the siderite ores have been formed after the conglobreccias, and less
than 15% at the expense of the Palaeozoic limestones of the margin of the basin.
The degree of sideritization of the conglobreccias depends on the relative amount
in them of fine- and very fine-clastic material (the greater the amount of such
material, the more intense the sideritization).   The shale clasts and sectors
of silicification in the limestone clasts have been feebly replaced by siderite.
The boundaries between the low-grade and rich ores are indistinct.   The richest siderite ores are developed mainly near the contacts between the ore-bearing
sequence and the siltstone sequence.

The oxidized ores consist of brown ironstones (compact, porous, and spongy, with inclusions of portions of clasts unaltered during sideritization). In addition to the compact brown-ironstone ores, there are in places, especially in the Central and Northern ore bodies, friable ores, consisting of rubble and small blocks of oxidized ores, associated with friable ochre-clay material. The boundary between the siderite and oxidized ores has been defined by the level of the groundwaters.   Two commercial kinds of siderite ores have been recognized: I) with an iron content of more than 30%, and II) with an iron content of from 20 to 30%.   The lower-grade ores have been assigned to the non-quality type.

The average amount of sulphur in the siderite ores is 0.55 - 0.94%, and phosphorus, 0.05 - 0.09%.   Two kinds have also been established for the oxidized ores: I) with an iron content of more than 45%, and II) with an iron content of from 30 to 45%.   The average amount of sulphur is 0.01 - 0.06%, and phosphorus, 0.09 - 0.12%.

The reserves of brown ironstones and siderites in the deposit, based on categories $A + B + C_1 + C_2$ are 50 million tonnes, including brown ironstones with an average iron content of 44.6%, 200 million tonnes, siderite ores with an average iron content of 35.7%, 235 million tonnes, and 9 million tonnes come from semi-oxidized ores.

The Lisakovsk Deposit

The Lisakovsk deposit lies 110 km southeast of Kustanai in the Kazakh SSR.

The oolitic iron ores of the Lisakovsk deposit are restricted to a Middle Oligocene river valley, extending from west to east over a distance of more than 100 km, with a width of from 2 - 3 up to 7 - 8 km (see Fig. 29).

The ore-bearing sediments rest on the eroded surface of marine clays of the Chegan Group of Late Eocene - Early Oligocene age and are overlain with a break by continental sandy and argillaceous sediments of the latest Middle and Late Oligocene (Fig. 34).

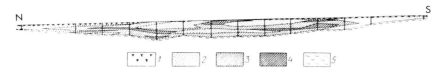

Fig. 34.   Transverse geological section through the Lisakovsk deposit of oolitic brown-ironstone ores.   (After I. Stitsenko).

1) Quaternary loams and soils; 2) quartz sands with hydrogoethite oolites; 3) oolitic hydrogoethite ore with iron content of 20 - 30%; 4) oolitic hydrogoethite ore with iron content of more than 30%; 5) clays of the Chegan Group.

The oolitic iron ores are located in Middle Oligocene gravel-pebble and sandy sediments of the Kutan-Bulak Group.   The country rocks and oolitic ores in places have been cemented by iron hydroxides, siderite, ferruginous chlorites, and calcite.   The ore-bearing sequence in the lower and middle horizons contains plant remains in the form of fragments of wood, twigs, and accumulations of plant debris, which have in places been coalified and ocherized.

The oolitic iron ores in the western part of the deposit appear on the surface, but in the eastern part are overlain with a break by sands, silts, and clays of the Chilikta Group of latest Middle Oligocene age. The ore-bearing sequence is 25 - 35 m thick.

The oolitic ores, which fill the river channels, form numerous lens--like and layer-like segregations, tens of kilometres in length with a width of several tens up to several hundreds of metres. Sometimes, several bands merge into a common band with a width of 1 - 2 km (*Oolitic Iron-Ores of the Lisakovsk Deposit ...*, 1962). The ore bodies which formed in the lacustrine-paludal basins and oxbow lakes, have been outlined in the form of individual segregations, which have an oval, lensoid, and less frequently irregular rounded shape; cross--bedding has been observed in the ores.

The oolitic ores have been subdivided into hydrogoethite and hydro-goethite-siderite-leptochlorite types. Amongst the hydrogoethite ores, several varieties have been recognized: oolitic sands, oolitic sandstones, and pebble--gravel sediments with hydrogoethite oolites. These ores occur in the sediments of the river channels and partly in the lacustrine basins.

The hydrogoethite-siderite-leptochlorite oolitic ores have been found in lacustrine-paludal sediments and the sediments of oxbow lakes, and in part in the middle horizons of river channels below a cover of clay rocks, which have preserved them from oxidation. The hydrogoethite oolites have been cemented with a leptochloritic and a siderite groundmass.

The dimensions of the oolites vary from 0.05 to 0.6 mm, the commonest having a diameter of 0.3 - 0.6 mm.

The iron content in the ores is from 34 to 42.6% (the quality ores include those with an iron content of more than 30%), phosphorus, 0.45 - 0.55%, and sulphur, 0.02 - 0.05%. The presence of phosphorus is associated with its occurrence in the iron hydroxides in the form of stilpnosiderite, and the presence of sulphur is associated with gypsum.

The reserves of ores based on categories $A + B + C_1$ are 1.7 billion tonnes with an average iron content of 35.2%.

The Taldy-Éspe Deposit

This deposit is the largest of the Near-Aral group of fluvial sedimentary iron-ore deposits. It is located near Perovsk Gulf, 45 km southwest of Saksaul'skaya railway station in the Kzyl-Ordyn district of the Kazakh SSR.

Like the Lisakovsk deposit, the ore segregations at Taldy-Éspe consist of oolitic brown-ironstone ores, occurring in the sands and sandstones of the Kutan-Bulak Group of the Middle Oligocene. The ore-bearing rocks are underlain by sandy silts, silty sands, and clays of the Chegan Group, and are overlain by clay silts and silty clays of the Chilikta Group. In the rocks of the Kutan--Bulak Group which surround the ores, there are facies transitions and thinning out of the rocks along the strike, and a freshwater fauna and flora have been found.

The ore, channel-like segregation has been traced along the strike for 22 km, with the maximum width of 3 km (Fig. 35). The thickness of the ore seg-

regation is 2.8 - 3.5 m.   The segregation has a small primary slope to the
south, and its base, as one passes southwards, gradually transects horizontal
layers of the older sediments.   In transverse section, the segregation has a
trough-shaped lower surface (Fig. 36) incised into the underlying rocks.   Numerous traces of erosion of the ore material have been found in the ores.   Amongst them are fine-pebbly, conglomeratic, pisolite-oolitic, and oolitic varieties.

On mineral composition, we may distinguish hydrogoethite and chlorite
varieties.   The chlorite ores occur in the deep-seated inundated portions of
the segregation.   In the zone of oxidation, the chlorite ores are replaced by
hydrogoethite types.   In section, seams of compact sideritic sandstone are
also found.   They have been observed within the zone of oxidation, and, near
the contact between the sandstone and the ore, its matrix is hydrogoethitic,
and within the layers, sideritic.   Sometimes the sideritic matrix is retained
only in individual lenses within the hydrogoethitic sandstone.

In the zone where the iron-ore segregation thins out in the southwestern part of the Central sector, a band of ore has been recorded, which is
enriched in oxides of manganese.   The bedding of the ores is approximately
horizontal, and cross-bedding is often observed.   The coarse-clastic composition of the ores and the cross-bedding undoubtedly demonstrate that their
deposition took place in river channels.

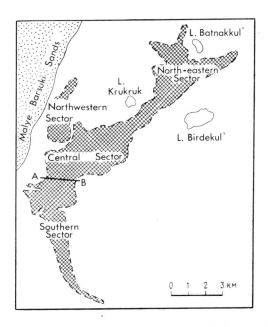

Fig. 35.
Diagram of location of sectors of
the Taldy-Éspe deposit (After
L. Formozova)

Distribution of ores is shown by
cross-hatching.

The primary minerals of the oolitic ores are chlorite and hydrogoethite.   Amongst the fragments in the centres of the oolites, we find quartz,
feldspar, magnetite, and ilmenite.   The matrix of the oolites involves diagenetic siderite, ankerite, pyrite, and marcasite.   During the process of epigen-
esis, secondary segregations of hydrogoethite after siderite and ankerite,
calcite in the matrix of the friable ores, and gypsum in the upper parts of
the exposed ore segregation, have been formed.

Peculiar diagenetic formations are the seams and lenses of massive
non-oolitic siderites and ankerites.   They have developed (Formozova, 1959)

as a result of changes in the silty and weakly-sandy clays along the contacts
between the oolitic ores and clays, owing to diffusion of bicarbonate solutions
into the latter and the subsequent formation of carbonates during diagenesis.

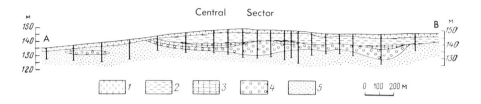

Fig. 36. Geological section through the Taldy-Éspe deposit (see Fig. 35).
(After L. Formozova).

1) loams; 2) supra-ore silty clays; 3) ferruginous sandstones and
sands with oolites of brown ironstone; 4) oolitic brown-ironstone
ores; 5) sub-ore silts.

The amounts in the oolitic ores (in wt %) are: Fe, from 26.4 to 36.4
(average for the deposit, 35.3); SiO_2, 21.2 - 32; P, 0.21 - 0.84 (average 0.5);
S, 0.37 - 0.43; MnO, 0.3 - 1.4 (average 0.8).

The reserves of ores based on categories A + B + C_1 are 100 million
tonnes.

The Alapaevo Deposits

The Alapaevo group of deposits is located near Alapaevsk in the Sverdlovsk district.

The geological structure of the region has involved rocks of Palaeozoic, Mesozoic, and Cainozoic age. The Palaeozoic sediments consist of Devonian and Carboniferous volcanogenic and sedimentary terrigenous and carbonate
rocks. The Mesozoic sediments, overlying the eroded surface of the Palaeozoic,
consist of the products of an ancient weathering crust of Palaeozoic rocks, the
ore sequence, the rocks of a mottled (turquoise) sequence. At the base of the
Mesozoic complex, located in karst limestones, is the ore-bearing sequence, consisting of variously coloured unstratified clays of the footwall, regarded as a
residual weathering crust of limestones, the ore horizon, and alluvial-proluvial
sediments of the so-called beliki sequence (Fig. 37).

The beliki sequence has been developed only on the Lower Carboniferous
limestones and consists of conglomeratic, sandy, and argillaceous rocks, containing unevenly distributed pebbles and clasts of silicified limestone, jaspers,
quartz, and siliceous shales. The beliki sequence is transgressively overlain
by a mottled (turquoise) sequence, consisting of clays, sands, pebble-beds, and
gravelites, mutually interstratified. The overall maximum thickness of the
Mesozoic sediments is 180 - 250 m. The Mesozoic rocks are transgressively overlain by Cainozoic sediments, consisting of sands, and gaize, which are overlain
by Quaternary loams and alluvial sediments.

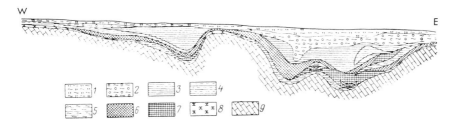

Fig. 37.   Geological section of the Alapaevo deposit, Sukhoi Log sector
(After V. Gruzdev).

1) loams;   2) turquoise deposits;   3) beliki;   4) ferruginized
brown beliki;   5) green strigovitic beliki;   6) rich ores;   7) lean
ores;   8) weathering crust of Palaeozoic rocks, and clays of footwall;   9) limestones.

Most of the deposits, including the largest, are restricted to karst
areas in the Carboniferous limestone.

The ore segregations of the Alapaevo deposits have a layer-like form.
The boundary of the zone of mineralization has been established on the basis of
the minimum content of iron (20%).   The thickness of the ore bodies, including
the sectors of non-quality ores and barren rocks, varies within wide limits
(from 0.5 up to 70 m).   The ore segregations can be traced along strike for
up to 5 - 10 km, and across the strike, for up to 0.5 - 1 km.

Two types of ores have been recognized in the Alapaevo deposits:
hydrogoethite, forming the upper horizons of the ore segregations above groundwater level, and strigovite-hydrogoethite, located below groundwater level,
where siderites also occur sporadically (Krotov *et al.*, 1936).

The ores are accumulations of predominantly hydrogoethitic nodules,
crusts, concretions, and irregular segregations, associated with an argillaceous, ocherous-clay, or ocherous-leptochlorite-clay mass, with a variable relative amount of the last, down to insignificance.

On the basis of the amount of iron in each of these types, three kinds
of ores have been recognized: I) with more than 36% of iron; II) from 30 to
36%; and III) from 20 to 36% (non-quality ores).

The content of iron in the ores varies from 20 to 58%.   In the quality ores, there is on average (in wt %): Fe, 38.5; Al_2O_3, 6.4 - 6.9; MgO,
0.3 - 0.4; CaO, 0.8; Cr_2O_3, 0.07 - 0.22; MnO, 0.15 - 0.36; Ni, about 0.1;
Co, about 0.01; calcination loss, 8.1 - 10.8; As, from traces to 0.27.   The
average amount of sulphur is 0.02 - 0.05, and phosphorus, 0.06 - 0.09%.

The reserves of brown-ironstone ores in the Alapaevo deposits (Alapaevsk, Zyryanovsk, and Sinyachikha) based on categories A + B + C_1, are 42
million tonnes.

DEPOSITS OF THE WEATHERING CRUST

The Yakovlevsk Deposit

This deposit is one of the largest of the rich iron-ore accumulations in the basin of the Kursk Magnetic Anomaly (KMA), situated 35 km north of Byelgorod, in the Byelgorod iron-ore region of the KMA, on its southwestern margin. Like all the iron-ore deposits of the KMA, the Yakovlevsk deposit is located in the Precambrian complexes of the base of the Russian Platform. The surface of the basement lies at a depth of 490 - 550 m. Such a thickness here appertains to the Carboniferous and younger sediments of the platform cover. The crystalline basement in the vicinity of the deposit consists of Archaean formations of the lower structural stage, and Proterozoic formations of the upper structural stage (Leonenko *et al.*, 1969). The Archaean formations involve ortho- and paragneisses, quartz-mica schists, and amphibolites (metabasites); the Archaean age of the latter is assigned provisionally.

The Proterozoic complex includes quartzites, phyllites, and ferruginous quartzites of the Kursk Series, which is the equivalent of the Krivoi Rog Series. Three groups have been recognized in the Kursk Series: 1) a lower (K_1), primarily sand-shale group, metamorphosed to arkosic metasandstones and black phyllitic sericite-biotite schists; 2) a middle (K_2) iron-ore group; and 3) an upper (K_3) slate sequence (Fig. 38).

The sediments of the middle group have been crumpled into a deep, compressed synclinal fold, overturned to the northeast, so that these sediments are represented in plan by two parallel belts, the Western (Yakovlevsk proper) and Eastern (the Pokrovsk).

The seven horizons recognized in the middle group are distinguished by thinly- or thickly-bedded aspect, overall colouring of the rocks (red-banded, blue-banded, and grey-banded), by the combination of the principal ore minerals (martite, specularite, and hydrohematite) and gangue minerals.

The investigated portion of the Yakovlevsk deposit lies almost completely within the deep zone of oxidation, so that oxidized iron-ore minerals and hypergene gangue minerals (ferruginous chlorites, clay minerals, etc.) predominate here in the ferruginous quartzites.

The axis of the Yakovlevsk syncline has a northwesterly strike of 320°, and the limbs dip northeastwards at 60 - 70°. The syncline closes at a depth of not less than 2 km. Its length exceeds 70 km, and its width on the surface of the basement in the north is from 0.8 to 2.0 km, and in the south, it increases to 4.5 km; within the sector of the deposit investigated, it is 1.2 - 1.6 km.

The syncline is complicated on the limbs by additional folding. There are widespread dislocations, expressed in zones of jointing, crushing, and brecciation. Interstratal movements are clearly defined. Micro-faults are widely developed with amplitudes from a few millimetres up to several centimetres.

The Yakovlevsk deposit includes the Yakovlevsk and Pokrovsk segregations of rich ores, located in belts of ferruginous quartzites of the same name. The Yakovlevsk segregation can be traced for a distance of 50 km, without thinning out, and its width in plan is from 200 to 400 m. The shape of the segregation in transverse section represents the junction of a mantle-like portion,

resting on the top of steeply-set ferruginous quartzites, with wedge- or layer-
-like apophyses, in general conformable with the bedding of the ferruginous
quartzites, extending to depth.   The thickness of the mantle-like portion of
the segregation is from 20 to 160 m, on average 30 - 60 m.   The greatest thicknesses of the ore bodies extending to depth vary from 40 to 160 m, and the depth
of distribution (from the Precambrian surface) is 300 - 600 m and more, since
not all the bodies can be traced until they wedge out.

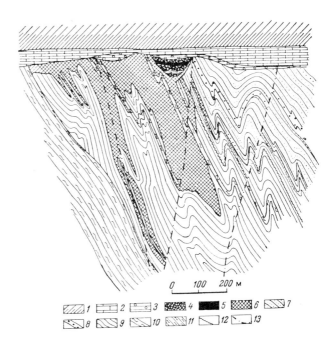

Fig. 38.
Geological section through the
Yakovlevsk deposit. (After
Chaikin & Klekl', 1965).

1) Mesozoic-Cainozoic deposits
of the sedimentary cover;   2)
Lower Carboniferous sediments;
3) allites and ferriallites;
4) ore conglomerate-breccia;
5) iron-alumina ores;   6) rich
specularite-martite, martite-
-specularite, and partly hydro-
hematite-martite, and martite-
-hydrohematite iron ores;   7)
upper slate group (K_3), predominantly phyllitic slates;
8) upper slate horizon (K_3),
conglomerates;   9) ferruginous quartzites of upper and
middle groups;   10) middle
(iron-ore) group (K_2), inter-
-ore slates;   11) lower slate
group (K_1);   12) erosion surface between middle and upper
groups;   13) faults.

The mantle-like portion of the ore segregation has been assigned to
the type of weathering crust of the same name, and the bodies, which pass to
depth, to the linear type of weathering crust.

In the Pokrovsk segregation, mainly the mantle-like type of ore bodies
is represented.   The ore bodies of the linear type are weakly manifested.   The
rich ores are friable or weakly lithified, and finely porous.   In the near-
-surface portion of the ore segregations, the ores have been cemented by late
carbonates and chlorites and have undergone compaction.

On the basis of mineral composition, five types of rich ore have been
recognized: 1) specularite and specularite-martite ('blues');   2) martite-
-hydrogoethite ('blue-reds');   3) hydrohematite-hydrogoethite ('reds');   4)
carbonatized (siderite-specularite and siderite-specularite-martite);   5) chloritized.

Ores of types 1, 2 and 3 have been formed from a definite type of
ferruginous quartzite, which has been defined on transitional varieties.   Types

4 and 5 have been formed from ores of types 1, 2, and 3 as a result of the effect of late hypergene solutions, which percolated from the cover rocks of the platform and controlled the cementation of the friable ores with calcite and chlorite.

The minerals of the rich ores are: martite, specularite (relict), dispersed and earthy hematite, hydrohematite, goethite, hydrogoethite, magnetite (relict), clay minerals (kaolinite, montmorillonite, etc.), chlorite, calcite, and quartz (relict). In the compact cemented ores, in addition, there are siderite, and ferruginous chlorite, and in subordinate amounts, hypergene pyrite, marcasite, hypergene magnetite, and galena.

In addition to the ores of the weathering crust noted, redeposited deluvial ores, and brecciated and conglomeratic ores, sometimes also having the aspect of an ore sandstone, are developed in the Yakovlevsk deposit. They have been distributed on the surface of the stony ores or at some distance from them, filling depressions in the ancient relief. The redeposited ores consist of fragments (1 - 3 cm across) of stony rich ores of various types, and less frequently slates and quartz; the matrix of the clasts is clay-ferruginous, sand-clay, and unevenly carbonatized and chloritized.

The best examined central portion of the Yakovlevsk ore segregation is characterized by the following average amounts, depending on the types of ores (in wt %): Fe, 56 - 63 (average 60.5); P, 0.14 - 0.31; and S, 0.05 - 0.11.

The proven reserves of the Yakovlevsk deposit based on categories B + + C_1 are 1.8 billion tonnes, and in category C_2, 8 billion tonnes.

The Mikhailovsk Deposit

The Mikhailovsk deposit of the KMA occurs near Zheleznogorsk in the Kursk district, 100 km northwest of Kursk.

Rocks of Archaean and Proterozoic age crop out in the area of the deposit (Fig. 39). The Archaean rocks (outside the limits of Fig. 39) consist of gneisses, plagioclase granites and their migmatites, and the Proterozoic formations are the rocks of the Mikhailovsk (provisionally) and Kursk Series.

The Mikhailovsk Series (up to 3 km thick) consist mainly of amphibolites with subordinate quartzites and metasandstones, talc-carbonate rocks, metadiabases, and serpentinites. The Kursk Series comprises a lower sand-shale group, 500 - 1000 m thick, a middle iron-ore group, consisting of specularite--magnetite, magnetite, and lean-ore quartzites, 500 - 600 m in overall thickness; the upper group, consisting of quart-sericite phyllitic and carbonaceous slates with seams of dolomites, has a total thickness of about 700 m; the Kurbakinsk Group, consisting of metamorphosed quartz porphyries, their tuffs, tuffites, sandstones, and sedimentary breccias, 1000 m in overall thickness.

Amongst the magmatic rocks of Proterozoic age, plagiogranites and migmatites (beyond the limits of Fig. 39) have been identified, occurring in the form of a layer-like segregation on the Archaean-Proterozoic boundary, and a small body of diabase porphyrites amongst the slates of the upper group of the Kursk Series, on the southeastern boundary of the deposit.

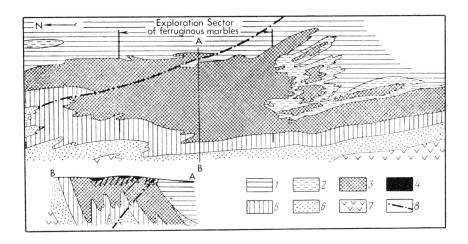

Fig. 39. Diagrammatic geological map and section of the Precambrian sequence
of the Mikhailovsk deposit. (After V. Polishchuk).

1) metamorphosed quartz porphyries, their tuffs, tuffites, sandstones,
and breccias of the Kurbaka Group; 2) slates and siltstones of the
upper division of the Kursk Series (K_3); 3) ferruginous quartzites
of the middle division of the Kursk Series (K_2); 4) rich iron ores;
5) slates of the lower division of the Kursk Series (K_1^2); 6) metasandstones and quartzites with seams of shales (K_1^1); 7) amphibolites,
metabasites, silicate shales, and seams of quartzites of the Mikhailovsk Series; 8) tectonic faults.

In structural respects, the Mikhailovsk deposit is restricted to a
large massif of ferruginous quartzites on the western limb of the Mikhailovsk
synclinal structure. Here the quartzite layers have been crumpled into a series of compressed folds with steep (60 - 80°) easterly dips of the axial planes.
From southeast to northwest in the northern portion of the massif, there is
apparently a disturbance of the fault type.

The covering sedimentary sequence of the platform consists of sediments of Devonian, Jurassic, Cretaceous, Palaeogene, and Quaternary age, consisting of clays, limestones, sands, and loams. The least thickness (35 - 40
m) of the sedimentary rocks has been observed in the central part of the deposit, over the uplifted portion of the crystalline basement, and the greatest
thickness (100 - 114 m) occurs on its margins.

Two mantle-like segregations of rich iron ores have been identified
in the area of the deposit, the Vereteninsk and Ostapovsk, with areas respectively of 8.6 and 1.7 km^2, with average thicknesses of 13 and 9.5 m, and an
average thickness for the covering sediments of 90 and 109 m. Both segregations are distinguished by sinuous outlines and a large number of non-ore
windows and pinches. Their floor in places is pocket-like.

The compact rich ores contain (in wt %):   Fe, 45 - 46.4;   S, 0.7 - 0.9; P, 0.06 - 0.08;   in the friable rich ores, Fe, 52 - 58.5;   S, 0.24 - 0.32;   and P, 0.03 - 0.05.   In calculating the reserves, account was taken of ores with minimal amounts of iron in the friable specularite-martite and martite ores of 45%, and in the compact carbonatized varieties of these same types, 35%, with a silica content respectively of not more than 25 and 20%.

The ferruginous quartzites underlie segregations of rich ores in the form of a massif, 14.8 km^2 in area.

The quartzites contain (in wt %):   Fe, 37.5 - 39;   SiO$_2$, 40 - 42;   S, 0.01 - 0.07;   and P, 0.01 - 0.06.

Ferruginous quartzites have been investigated to a depth of 280 - 330 m.   On the basis of categories A + B + C$_1$, 3.7 billion tonnes of ferruginous quartzites and 340 million tonnes of rich iron ores have been proved in the deposit.

DEPOSITS OF THE METAMORPHOSED GROUP

THE KRIVOI ROG IRON-ORE BASIN

The Krivoi Rog Iron-Ore basin (Krivbass), the second in the Soviet Union after the KMA on the basis of reserves and the first with respect to output, is located on the right bank of the River Ingulets and its tributaries, the Saksagan' and the Zhëltaya (Yellow), and extends in a northnortheasterly direction for about 100 km (Fig. 40).   The basin is involved in the Kurainian crystalline massif and is in part a Precambrian geosyncline.   Its Archaean stage consists of gneisses, granites, migmatites, amphibolites, and schists, distributed west and east of the Krivbass.   The overlying stage is formed of the Proterozoic Krivoi Rog geosynclinal series, which is divided into three sections: 1) a lower arkose-quartzite and phyllitic sequence, with an horizon of talc schists,   2) a middle, ore-bearing proper ferruginous quartzite and slate sequence;   and 3) an upper, quartzite--sandstone-slate sequence with lenses of marmorized limestones.   The complete sequence of the middle section includes nine horizons of ferruginous quartzites, interstratified with quartz-sericite, chlorite-sericite, and other slates and microquartzites (Belevtsev, 1962).

The Krivoi Rog Series forms a complex synclinorium, consisting of synclinal and anticlinal folds with dips on the limbs of 45 - 80°, for the most part with a carinate closure of the syncline (Fig. 41).   The hinges of the syncline plunge at up to 40° northwards.   Usually the limbs of the folds have been cut to some degree by longitudinal faults or overthrusts, the largest of which is the Saksagan' overthrust (Akimenko *et al.*, 1957).   As a result of these deformations, the ore-bearing structures have been segregated on the general strike of the basin (from south to north):   the Ingulets (Southern ore field), the Saksagan' (Main, or Saksagan', ore field), the Pervomaisk, Annovsk, Zheltorechensk (Northern ore field), and the Popel'nostovsk.   In the Northern ore field, the sedimentary rocks have been cut by Proterozoic granite intrusions.   The largest reserves of rich ores have been concentrated in the Saksagan' part of the basin. The ore-bearing middle section here has a thickness of up to 2000 m and contains up to eight seams of ferruginous quartzites.

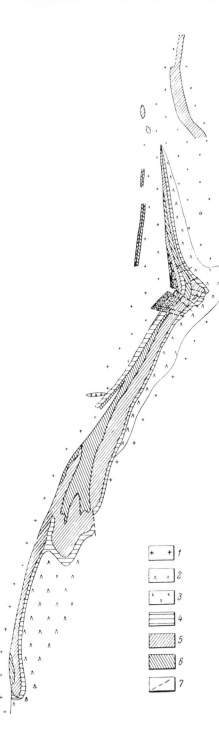

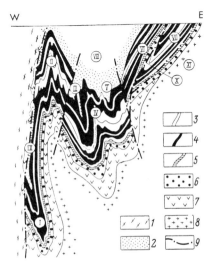

Fig. 40.
Survey geological-lithological map of
the Krivoi Rog iron-ore basin.

1) granites, gneisses, and migmatites;
2) greenstone rocks; 3) diabase dykes;
4-6) rocks of the Saksagan' Series:
4) arkoses and phyllites of the lower
group, 5) ferruginous rocks (jaspilites,
hornfelses, and slates), 6) slates of
the upper group, 7) tectonic faults.

Fig. 41.
Tectonic map of the Krivoi Rog basin
near Krivoi Rog. (After Ya. Belevtsev).

1) microcline-plagioclase granites;
2) sediments of the upper group of the
Krivoi Rog Series; 3) slate horizons
of middle group; 4) ferruginous horizons of middle group; 5) talc-carbonate horizon; 6) deposits of lower
group; 7) amphibolites; 8) plagiogranites; 9) tectonic faults.
Folded and disrupted structures:
I) Likhmanovsk syncline; II) Tarapak-
-Likhmanovsk anticline; III) West
Ingulets syncline; IV) Sovetsk anticline; V) East Ingulets syncline,
VI) Saksagan' anticline; VII) Saksagan'
syncline; VIII) Main syncline; IX)
Main Krivoi Rog fault; X) Saksagan'
thrust; XI) Eastern thrust.

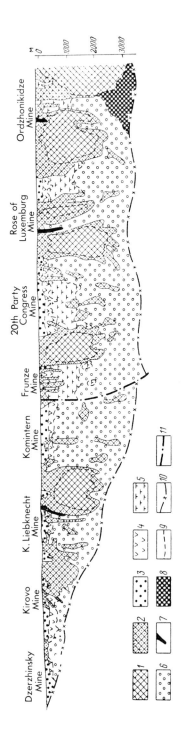

Fig. 42.
Mineralogical zonation of mineralization of deep weathering crust of Saksagan' syncline (vertical longitudinal projection).   (After V. Kravchenko).

1) unoxidized metamorphic rocks; 2) oxidized metamorphic rocks; 3) porous, friable ores of eluvial pockets of areal weathering crust; 4) porous and compact ores (in varying degree quartzified with poikilitic quartz); 5) porous, friable ores (in varying degree quartzified with poikilitic quartz); 6) porous, semi-friable ores of deep weathering crust; 7) compact ores, cemented with goethite; 8) quartz-magnetite primary ores; 9) contacts between varieties of rich ores; 10) hinge of syncline; 11) tectonic faults.

The ferruginous quartzites, the reserves of which are vast in the Krivbass, consist of magnetite, magnetite-hematite, and hematite varieties. The main commercial importance attaches to the unoxidized magnetite and magnetite-hematite varieties. The mining and concentration combines of the Krivbass (GOK) work the ferruginous quartzites in large quantities and obtain a high-quality magnetite concentrate from the quartzites.

The rich ores, consisting mainly of minerals of iron oxides and hydroxides, form layer-, column-, stock-, and lens- -like segregations and bodies among the ferruginous quartzites. The attitudes of the rich ores are somewhat varied in the different parts of the basin.   In the main Saksagan' ore field, the ore segregations are located totally within the layers of ferruginous quartzites, on the limbs and in the hinge portions of the synclines.   Especially favourable in localizing the ore bodies on the limbs of the folds are the zones of tranverse crumpling of the gentle fold and flexure type, accompanied by the development of cleavage and interstratal movements, and also zones of faults along the strike of the bedding and nodes of fractures with crushing and jointing.   As a result, the ore bodies are arranged in the layers of ferruginous quartzites as groups and chains with larger or smaller discontinuities (Fig. 42). The individual ore bodies seemingly merge at depth in the hinge portions of the synclines, forming large ore columns, often with a carinate base, coinciding in strike and dip with the fold hinges. The rich ores have been formed from ferruginous quartzites as a result of leaching of quartz and oxidation of the iron silicates (Belevtsev *et al.*, 1959).

The greatest concentration of ore bodies in the Saksagan' region has been observed in the fifth ($K_2{}^{5zh}$) and sixth ($K_2{}^{6zh}$) ferruginous horizons, which is associated with the larger amount of iron in them and with their thinner bedding, which contributed to the development of the deformations.

In the Southern ore field, the rich ores, with layer-like aspect, occur in the zone of stratigraphical unconformity between the sediments of the middle and upper sections of the Krivoi Rog Series. Gershoig (1971) and others consider that the rich ores were formed here in a Proterozoic weathering crust, and they were then metamorphosed to magnetite and specularite ores and further enriched and oxidized under conditions of a post-Proterozoic weathering crust.

Rich ores of such stratigraphical position have also been recorded at a number of points in the Saksagan' and Zheltorechensk regions.

In the Northern ore field, the ores in contrast to those in other parts of the basin, have been hydrothermally-metasomatically altered at the expense of both the ferruginous quartzites and the slates.

The shapes of the ore bodies are predominantly column-like, less so stock- and lens-like, and rarely layer-like. They rest both conformably with the replaced rocks, and frequently also cut their bedding and schistosity. The ore bodies have been controlled by zones of crushing, crumpling, tectonic seams, and nodes of fractures. The size and shape of the sections across the ore bodies is extremely variable. According to Ya. Belevtsev, the segregations of rich iron ores of the Krivoi Rog developed as the result of metamorphism.

Several principal types of rich ores have been recognised in the Krivbass group, based on the nature of the principal ore minerals.

1) martite and hematite-martite (they bear the local name 'sin'ka' ('blue');

2) martite-hematite-hydrohematite-hydrogoethite ('krasko-sin'ka' = 'red-blue');

3) hematite-hydrohematite-hydrogoethite ('kraska' = 'red');

4) magnetite and magnetite-specularite (only in the Northern ore field).

In the ores of the first type, besides martite and relict hematite, relict magnetite, dispersed hematite, chlorite, sericite, pyrite, carbonates, clay minerals, quartz, and apatite have been observed in small amounts. In the ores of the second type, in addition to those listed above in the name of the type, there are dispersed hematite, martite, sericite, quartz, clay minerals, alunite, and sphene in small and accessory amounts.

In the ores of the third type, in addition to the main minerals, there are kaolinite, clay minerals, chlorite, and carbonates.

In the fourth type (Zheltorechensk), in addition to the predominant magnetite and specularite, amphiboles (cummingtonite, grunerite, and riebeckite), aegirine, biotite, albite, quartz, carbonates, chlorite, grains of pyrite, pyrrhotite, and chalcopyrite, have been noted.

The rich iron ores of the Krivbass are characterized by the large amount of iron and the low content of gangue material (Table 3).

The overall reserves of high-grade iron ores of the Krivbass comprise, on the basis of categories $A + B + C_1$, 15.6 billion tonnes, in category C_2, 4 billion tonnes, including rich ores with an average iron content of 57.6% based on categories $A + B + C_1 + C_2$ in the amount of 1.7 billion tonnes, and ferruginous quartzites with an average iron content of 35.9% in the amount of 18 billion tonnes.

Table 3.   *Amounts of Iron, Phosphorus, and Sulphur in Rich Ores of the Krivoi Rog Basin (in wt %).*

Type of ore	Iron	Phosphorus	Sulphur
Martite and hematite-martite	63.7	0.26	0.043
Martite-hematite-hydrohematite	62.3	0.08	0.03
Hematite-hydrohematite-hydrogoethite	57.5	0.088	Thousandths and hundredths percent
Magnetite and magnetite-specularite	54.0	0.04	0.15

The Olenegorsk Deposit

The Olenegorsk deposit is situated 7.5 km northwest of Olen'ya railway station in the Murmansk district.

The ferruginous quartzites of the Olenegorsk deposit, up to 150 m thick and 2.8 km in extent, occur in the Archaean amphibole-bearing gneisses and amphibolites of the Kola Series (Fig. 43a).   Gradual transitions have been recorded between the ferruginous quartzites and the gneisses.   In plan, the segregation of ferruginous quartzites has the shape of a constricted large lens with transverse flexures.   In depth, it can be traced down to 300 m without any sign of wedging-out.

The ferruginous-quartzite sequence has been crumpled into a synclinal fold, overturned to the southwest, and complicated by anticlinal and synclinal folds of lower orders.   The strike of the synclinal structure is northwesterly with the dip of the limbs to the southwest at $50 - 80°$ (See Fig. 43b).

Overthrusts with throws of up to $50 - 55$ m also complicate the syncline, and these are accompanied by gently-dipping faults and displacements.   It has been shown that the upper (relatively to the displacement) blocks have been overthrust over the lower, that is, southwestwards.   In the rocks of the deposit, there is clear evidence of flow cleavage, linearly parallel fabrics, and jointing (Tochilin & Goryainov, 1964).

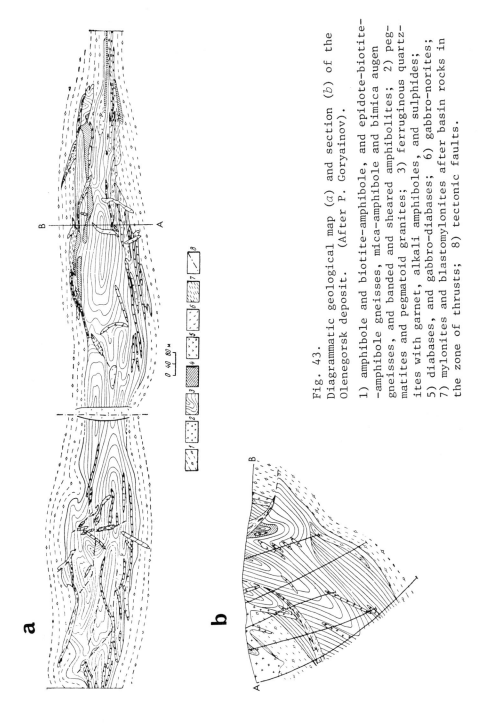

0 40 80 м

Fig. 43.
Diagrammatic geological map (a) and section (b) of the Olenegorsk deposit.   (After P. Goryainov).

1) amphibole and biotite-amphibole, and epidote-biotite-amphibole gneisses, mica-amphibole and bimica augen gneisses, and banded and sheared amphibolites;   2) pegmatites and pegmatoid granites;   3) ferruginous quartzites with garnet, alkali amphiboles, and sulphides;   5) diabases, and gabbro-diabases;   6) gabbro-norites;   7) mylonites and blastomylonites after basin rocks in the zone of thrusts;   8) tectonic faults.

The intrusive rocks in the vicinity of the deposit consist of microcline granite-gneisses, pegmatoid granites, pegmatites, and aplites.

All the intrusive rocks are younger than the ferruginous quartzites, and their apophyses cut the sequence of ferruginous quartzites and country rocks.

The ore segregation consists of magnetite quartzites with a trace of hematite.   The ratio of magnetite to hematite on average in the deposit is 3 : : 1.   The minor minerals of the ferruginous quartzites are tremolite, actinolite, cummingtonite, pyroxenes, alkali amphiboles, calcite, and siderite.   There are rare occurrences of sulphides (pyrite, pyrrhotite, and chalcopyrite).   The predominant fabrics of the ores are banded and plicated.

The ores are distinguished by the low content of gangue materials (sulphur and phosphorus).   The reserves of magnetite quartzites, on the basis of categories $A + B + C_1$ are 477 million tonnes with an average iron content of 32.3%.   The deposit is being quarried.

The Balbraun Deposit

The Balbraun deposit of the Karsakpai group lies 5 km to the south of Karsakpai in the Dzhezkazgan region of the Karaganda district.   It is the most significant in the association of ferruginous quartzites of the Proterozoic (pre--Riphean or Riphean) massif, being an uprise in the metamorphic rocks of the basement of the geosynclinal complex of the Ulutau anticlinorium (Kalganov & Finkel'stein, 1960).

The Proterozoic sequence consists of metamorphosed epidotized eruptive diabase porphyrites and their tuffs, and to a lesser degree quartzites, marmorized limestones, and metamorphic slates.   The upper group of the sequence, termed the Karsakpai iron-ore group, consists of quartzites, greenstone tuffogenic, graphitic, and quartz-sericite schists, and hematite quartzites.   The thickness of the Karsakpai Group is about 1 km.   The sediments of the group have been crumpled into a series of isoclinal folds, including a series of disrupted iron-ore segregations (Fig. 44).

Fig. 44.
Diagrammatic geological section
through the Karsakpai synclinorium
at the latitude of the Balbraun
deposit.   (After M. Uzbekov).

1) ferruginous quartzites with frequent seams of quartz-sericite schists and less frequently, non-ore quartzites;   2) quartz-sericite and quartz--chlorite schists with rare seams of green and tuffogenic shales, quartzites, and graphitic slates;   3) quartzites, often brecciated, with seams of green tuffogenic shales and quartz-sericite schists;   4) green eruptives and their tuffs.

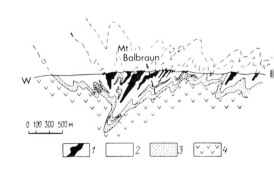

In the Karsakpai Group, about 30 ore bodies have been recognized, with varying dimensions from lenses a few metres thick and 50 m in extent up to thick segregations up to 5 km long and up to 120 m wide. These segregations have been grouped into seven ore bands. The dip of all the ore bands is westerly, with the exception of the southern portion of the first band, which stands vertically. Large faults, disrupting the continuity of the structures, occur rarely, but small local displacements are frequent, breaking the continuity of the individual layers within the structures. The structures can be traced at depth up to 200 - 250 m vertically.

The ferruginous quartzites of the Karsakpai Group belong to the hematitic variety. In the Balbraun deposit, the primarily layer-like and lens-like forms of hematitic quartzite segregations have been complicated by second- and third-order folding, and also by tectonic distortions.

The hematitic quartzites of the deposit consist in practice of two minerals, hematite and quartz. Their total amount varies within 90 - 98%. Minor minerals are sericite, chlorite, calcite, apatite, tourmaline, magnetite, pyrite, siderite, and in the zone of oxidation, martite and gypsum. The ores have a thin-banded fabric.

The reserves of hematitic quartzites in the deposit, on the basis of categories $A + B + C_1$ have been reckoned at 77 million tonnes, with an average iron content of 40.4%. A direct extension southwards is the Kerege-Tas deposit of hematitic quartzites with reserves, based on the categories $A + B + C_1$, of 49 million tonnes with an average iron content of 40.4%. The prospective reserves of all the deposits of the Karsakpai Group have been estimated at 500 million tonnes.

REFERENCES

AKIMENKO N.M., BELEVTSEV Ya.N. & GOROSHNIKOV B.I. *et al.*, (АКИМЕНКО Н.М., БЕЛЕВЦЕВ Я.Н., ГОРОШНИКОВ Б.И. и др.) 1957: Геологическое строение и железные руды Криворожского бассейна *(The Geological Structure and the Iron Ores of the Krivoi Rog Basin)*. Gosgeoltekhizdat, Moscow.

ANTIPOV G.I. *et al.*, (АНТИПОВ Г.И. и др.) 1960: Ангаро-Илимские железорудные месторождения трапповой формации южной части сибирской платформы *(The Angara-Ilim Iron-Ore Deposits of the Trap Association of the Southern Part of the Siberian Platform)*. Gosgeoltekhizdat, Moscow.

ARBUZOV V.A. *et al.*, (АРБУЗОВ В.А. и др.) 1967: Керченский железорудный бассейн *(The Kerchen Iron-Ore Basin)*. Nedra Press, Moscow.

(Assessment of Iron-Ore Deposits During Exploration and Research). Оценка железорудных месторождений при поисках и разведках. Nedra Press, Moscow. (1970).

BARDIN I.P. (Ed.) (БАРДИН И.П. (отв. ред)) 1962: Оолитовые железные руды Лисаковского месторождения Кустанайской области и пути их использования (Oolitic Iron Ores of the Lisakovsk deposit of the Kustanai district and means of utilizing them), *in* "Железорудные месторождения СССР" *('The Iron-Ore Deposits of the USSR')*, Vol.5, 235 pp. Akad. Nauk SSSR Press, Moscow.

BELEVTSEV Ya.N. (Ed.) (БЕЛЕВЦЕВ Я.Н. (отв. ред)), 1962: Геология Криворожских железорудных месторождений, Т. 1, 2 *(The Geology of the Krivoi Rog Iron-Ore Deposits)*, Vols 1 and 2.   Akad. Nauk Ukr. SSR Press, Kiev.

BELEVTSEV Ya.N. *et al.* (БЕЛЕВЦЕВ Я.Н. и др.), 1959: Генизис железных руд Криворожского бассейна *(The Origin of the Iron Ores of the Krivoi Rog Basin)*.   Akad. Nauk Ukr. SSR Press, Kiev.

BELYASHOV M.M. & PLEKHOVA K.R. (БЕЛЯШОВ М.М., ПЛЕХОВА К.Р.), 1965: Влияни осадочных ангидритов на метасоматические процессы при образовании Качарского магнетитового месторождения (Тургайский прогиб) (The effect of sedimentary anhydrites on the metasomatic processes during the formation of the Kachar magnetite deposit (Turgai downwarp)). *Geologiya rudn. Mestorozh.*, 7, No.2 pp.38-49.

BOGATSKY V.V. & KURTSERAITE Sh.D. (БОГАЦКИЙ В.В., КУРЦЕРАЙТЕ Ш.Д.), 1966: Закономерности размещения метасоматических магнетитовых месторождени Северной части Западного Саяна *(Distribution Patterns of the Metasomatic Magnetite Deposits of the Northern Part of West Sayan)*.   Nedra Press, Moscow.

CHUGUEVSKAYA O.M. (ЧУГУЕВСКАЯ О.М.), 1969: Генетические особенности Сарбайского и Елтайских магнетитовых месторождений в Тургае (The genetic features of the Sarbai and Yeltai magnetite deposits in Turgai). *Avtoref Diss. Kand. Geol.-Miner. Nauk, Alma Ata.*

DERBIKOV I.V. & RUTKEVICH I.S. (ДЕРБИКОВ И.В., РУТКЕВИЧ И.С.), 1971: Железорудные месторождения Горной Шории в свете вулканогенно-осадочн теории их происхождения (The Iron-Ore Deposits of Gornaya Shoriya in the Ligh of the Volcanogenic-Sedimentary theory of their Origin). *Trudy sib. nauchno-issl Inst. Geol. Geofiz. miner. Syr'ya, 125.*   Novosibirsk.

(Distribution Patterns and the Formation of Magnetite and Chromite Deposits of the Mugodzhary and the Turgai Downwarp) Закономерности размещения и образован магнетитовых и хромитовых месторождений Мугоджар и Тургайского прогиб Kaz. Inst. Miner. Syr'ya Press, Alma-Ata (1968).

DYMKIN A.M. (ДЫМКИН А.М.), 1966: Петрология и генезис магнетитовых месторождений Тургая *(The Petrology and Origin of the Magnetite Deposits of Turgai)*.   Nauka Press, Novosibirsk.

FORMOZOVA L.N. (ФОРМОЗОВА Л.Н.), 1959: Железные руды Северного Приаралья (Iron ores of the Northern Aral region). *Trudy geol. Inst. Mosk., 20.*

FORMOZOVA L.N. (ФОРМОЗОВА Л.Н.), 1973: Формационные типы железных руд докембрия и их эволюция (Associational types of Precambrian iron ores and their evolution). *Trudy geol. Inst., Mosk., 250.*

GAR'KOVETS V.G. (ГАРЬКОВЕЦ В.Г.), 1946: О генезисе Абаильского желе-зорудного месторождения (The origin of the Abail iron-ore deposit). *Dokl. Akad. Nauk SSSR, 54, pp.337-338.*

(Geology and Genetic Features of the Magnetite Deposits of Turgai) Геология и генетические особенности магнетитовых месторождений Тургая. *Trudy Inst. geol. Nauk Akad. Nauk Kaz. SSR, 28.*   Nauka Press, Alma-Ata (1969).

GERSHOIG Yu.G. (ГЕРШОЙГ Ю.Г.), 1971: Генетическая классификация железных руд Кривбасса (A genetic classification of the iron ores of the Krivbass). *Geologiya rudn. Mestorozh.*, *13*, No.4, pp.3-17.

GRIGOR'EV V.M. (ГРИГОРЬЕВ В.М.), 1971: Закономерности распределения и условия накопления германия в железорудных месторождениях *(Distribution Patterns and Conditions of Deposition of Germanium in Iron-Ore Deposits)*. Nedra Press, Moscow.

(Iron-Ore Basis for Heavy Metallurgy in the USSR) Железорудная база черной металлургии СССР. Akad. Nauk SSSR Press, Moscow (1957).

KALGANOV M.I. & FINKEL'SHTEIN A.S. (Eds) (КАЛГАНОВ М.И., ФИНКЕЛЬШТЕЙН А.С. (Ред.)), 1960: Железорудные месторождения Центрального Казахстана и пути их использования *(The Iron-Ores of Kazakhstan and the Means of Utilizing Them)*. Akad. Nauk SSSR Press, Moscow.

KASHKAI M.A. (КАШКАЙ М.А.), 1965: Петрология и металлогения Дашкесана и других железорудных месторождений Азербайджана *(The Petrology and Metallogenesis of the Dashkesan and Other Iron-Ore Deposits of Azerbaidzhan)*. Nedra Press, Moscow.

(Khalilovo deposits of complex iron ores) Халиловские месторождения комплексовых железных руд. *(Trudy Inst. geol. Nauk Mos.-Leningr.*, *67*. (1942).

KISELEV G.N. (КИСЕЛЕВ Г.Н.), 1963: Новые данные по Аккермановскому месторождению природно-легированных железных руд (New data on the Akkermanovsk deposit of naturally-alloyed iron ores) *in* "Кора выветривания" *('The Weathering Crust')*, Vol. 5, pp.245-256. Akad. Nauk SSSR Press, Moscow.

KROTOV B.P. *et al.* (КРОТОВ Б.П. и др.), 1936: Железорудные месторождения алапаевского типа на восточном склоне Среднего Урала и их генезис *(The Iron-Ore Deposits of the Alapaevo Type on the Eastern Slopes of the Urals and Their Origin)*, Vols. 3 and 4. Akad. Nauk SSSR Press, Moscow.

LEONENKO I.N., RUSINOVICH I.A. & CHAIKIN S.I. (ЛЕОНЕНКО И.Н., РУСИНОВИЧ И.А., ЧАЙКИН С.И.), 1969: Геология, гидрогеология и железные руды бассейна Курской магнитной аномалии, Т.3, Железные руды *(The Geology, Hydrogeology, and Iron Ores of the Basin of the Kursk Magnetic Anomaly*, Vol. 3, *The Iron Ores)*. Nedra Press, Moscow.

MYASNIKOV V.S. (МЯСНИКОВ В.С.), 1959: Некоторые особенности месторождений титаномагнетитовых руд Южного Урала и проявление в них метаморфизма (Some features of the deposits of titanomagnetite ores of the Southern Urals and the manifestation of metamorphism in them). *Geologiya rudn. Mestorozh.*, *1*, No.2, pp.49-62.

NIKOLAEVA I.V. (НИКОЛАЕВА И.В.), 1967: Бакчарское месторождение оолитовых железных руд *(The Bakchar deposit of oolitic iron ores)*. Nauka Press, Novosibirsk.

NIKOL'SKY A.P. & KAUKIN B.V. (НИКОЛЬСКИЙ А.П., КАУКИН Б.В.), 1968: Железо (Iron), *in* "Геологическое строение СССР", Т.4 *('The Geological Structure of the USSR')*, Vol.4, pp.309-327. Nedra Press, Moscow.

(Oolitic Brown Ironstones of the Kustanai District and Means of Utilizing Them). Оолитовые бурые железняки Кустанайской области и пути их использовани Akad. Nauk SSSR Press, Moscow (1956).

OVCHINNIKOV L.N. (ОВЧИННИКОВ Л.Н.), 1960: Контактово-метасоматические месторождения Среднего и Северного Урала (The contact-metasomatic deposits of the Middle and Northern Urals). *Trudy gorno-geol. Inst. ural'. Fil., 39.* Sverdlovsk.

PAVLOV D.I. (ПАВЛОВ Д.И.), 1964: Анзасское магнетитовое месторождение и участие хлора в его формировании *(The Anzas Magnetite Deposit and the Participation of Chlorine in Its Formation).* Nauka Press, Moscow.

PERVAGO V.A. (ПЕРВАГО В.А.), 1966: Алданская железорудная провинция *(The Aldan Iron-Ore Province).* Nedra Press, Moscow.

(Principal Iron-Ore Deposits of Siberia) Главнейшие железорудные месторождения Сибири *Trudy sib. nauchno-issled. Inst. Geol. Geofiz. miner. Syr'ya, 96.* Novosibirsk (1970).

SAPOZHNIKOV D.G. (САПОЖНИКОВ Д.Г.), 1963: Караджальские железо--марганцевые месторождения *(The Karadzhal Iron-Manganese Deposits).* Akad. Nauk SSSR Press, Moscow.

SHNYUKOV E.F. (ШНЮКОВ Е.Ф.), 1965: Генезис киммерийских железных руд Азово-Черноморской рудной провинции *(The Origin of the Cimmerian Iron Ores of the Azov - Black Sea Ore Province).* Naukova Dumka Press, Kiev.

SLEDZYUK P.E. & SHIRYAEV P.A. (СЛЕДЗЮК П.Е., ШИРЯЕВ П.А.), 1958: Магнетитовые руды Кустанайской области и пути их использования *(The Magnetite Ores of the Kustanai District and the Means of Utilizing Them).* Akad. Nauk SSSR Press, Moscow.

SLEDZYUK P.E. & SOKOLOV G.A. (Eds) (СЛЕДЗЮК П.Е., СОКОЛОВ Г.А. (Ред.)), 1959: Железорудные месторождения Алтае-Саянской горной области *(The Iron-Ore Deposits of the Altai-Sayan Mountain Region),* Vol.1, book 2. Akad. Nauk SSSR Press, Moscow.

SOKOLOV G.A. (СОКОЛОВ Г.А.), 1967: Закономерности размещения железорудных месторождений на территории СССР (Distribution patterns of iron-ore deposits in the USSR), *in* "Закономерности размещения полезных ископаемых" *('Distribution Patterns of Mineral Deposits'),* Vol.8, pp.79-94. Nauka Press, Moscow.

SOKOLOV G.A. & PAVLOV D.I. (СОКОЛОВ Г.А., ПАВЛОВ Д.И.), 1970: Об условиях и локализации Тургайских магнетитовых месторождений (The conditions and the localization of the Turgai magnetite deposits), *in* "Магматизм и эндогенная металлогения Зауралья" *('Magmatism and Endogenic Metallogenesis of the Trans-Urals'),* pp.111-113. Kustanai.

TOCHILIN M.S. & GORYAINOV P.M. (ТОЧИЛИН М.С., ГОРЯЙНОВ П.М.), 1964: Геохимия и генезис железных руд Приимандровского района Кольского полуострова *(The Geochemistry and Origin of the Iron Ores of the Near-Imandrovsk Region of the Kola Peninsula).* Nauka Press, Moscow.

(West Siberian Basin) Западно-Сибирский бассейн. Akad. Nauk SSSR (Siberian Division) Press, Novosibirsk (1964).

YANITSKY A.L. & SERGEEV O.P.   (ЯНИЦКИЙ А.Л., СЕРГЕЕВ О.П.), 1962:   Бакальские железорудные месторождения и их генезис *(The Bakal Iron-Ore Deposits and Their Origin)*.   *Trudy Inst. Geol. rudn. Mestorozh., 73.*

YUDIN N.I.   (ЮДИН Н.И.), 1968:   Литология железорудных месторождений Ангаро-Питского бассейна *(The Lithology of the Iron-Ore Deposits of the Angara-Pit Basin)*.   Nauka Press, Moscow.

YURK Yu. Yu. *et al.*   (ЮРК Ю.Ю. и др.), 1960:   Минералогия железорудной формации Керченского бассейна *(The Mineralogy of the Iron-Ore Association of the Kerchen Basin)*.   Krymizdat, Simferopol'.

ZAVARITSKY A.N. (ЗАВАРИЦКИЙ А.Н.), 1939:   К вопросу о происхождении железных руд Бакала   (The problem of the origin of the iron ores of Bakal).   *Trudy Inst. geol. Nauk Mosk., ser. rudn. Mestorozh., 13,* No.2.

DEPOSITS OF MANGANESE

PRINCIPLES OF CLASSIFICATION OF MANGANESE DEPOSITS

The problems of classifying manganese deposits have been considered in the works of V.A. Obruchev (1934), Betekhtin (1946), Park (1956), Roy (1969), and others. In spite of significant progress in the study of various aspects of the geology, geochemistry, and mineralogy of manganese deposits, the principles, on the basis of which a subdivision of these deposits has been carried out, have changed little in the last 50 years. Most investigators (Obruchev, 1934; Betekhtin, 1946; Park, 1956, Suslov, 1964; and Rakhmanov, 1967) classify the manganese deposits on a genetic basis.

I. Sedimentary (manganese ores precipitated in a sedimentation basin);

1) strictly sedimentary (exogenic source of ore components: weathering crust, erosion of rocks of source landmass, and submarine leaching);

2) volcanogenic-sedimentary (endogenic source of ore components: hydrotherms, exhalation, etc.).

II. Magmatogenic:

1) hydrothermal;

2) contact-metasomatic.

III. Metamorphosed (regional and contact metamorphism of sedimentary magmatogenic ore accumulation).

IV. Weathering crusts:

1) residual accumulations (laterite type, deeply leached);

2) infiltration formations[1].

A distinguishing geochemical feature of manganese is its concentration in the surface portion of the lithosphere, in the zone of free oxygen (Vernadsky, 1954*a, b*). Thus, in the Soviet Union, more than 95% of the high-grade ores of manganese have been concentrated in sedimentary deposits (Fig. 45).

THE PROBLEM OF THE SOURCE OF ORE COMPONENTS AND THE MANGANESE- -BEARING ASSOCIATIONS

The nature of the ore- and rock-forming components of the sedimentary sequences has seldom been clearly discerned. Even in present-day basins with active volcanic activity (the Pacific and Indian Oceans), where widespread devel-

---

[1] In the scheme presented, the specific features of manganese ores are weakly expressed, and the diagram bears quite a general character, and therefore it may also be used for a number of other mineral resources, in particular, for iron ores.

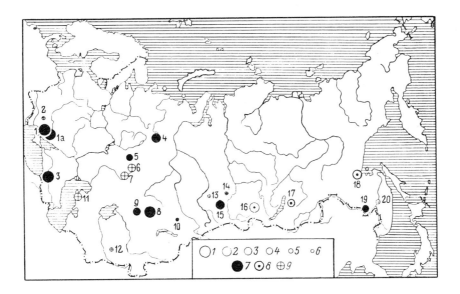

Fig. 45.   Diagram of distribution of principal sedimentary and volcanogenic-
-sedimentary manganese deposits in the USSR.

1-6) Deposits with reserves of ores in millions of tonnes: 1) more
than 100,   2) from 50 to 100,   3) from 10 to 50,   4) from 5 to 10,
5) from 1 to 5,   6) less than 1;   7-9) categories of reserves: 7)
of commercial importance,   8) promising and generally geological,
9) non-commercial.

Deposits of manganese (numbers on map):   *1*) South Ukrainian manganese
ore basin (*1a* - Bol'she-Tokmak deposit),   *2*) Khoshchevat,   *3*) Georgian manganese basin,   *4*) Northern Urals manganese basin (Polunochnoe,
Berezovsk, Novo-Berezovsk, Yekaterininsk, Yurka, Loz'va, Ivdel,
Vishersk, Marsyat, etc.),   *5*) Ulutelyak,   *6*) Near-Magnitogorsk group
(Niazgulovsk, Yalimbetovsk, Mamilya, Faizula, etc.),   *7*) Akkermanovsk,
8) Atasu group (Karazhal, Bol'shoe Ktaiskoe, Keretat, Dzhumart, Kamys,
etc.),   *9*) Karsakpai group (Dzhezda, Promezhutochnoe, Naizatas, etc.),
10) Murdzhik,   *11*) Mangyshlak,   *12*) Zeravshan group (Dautash, Takhta-
-Karacha, etc.),   *13*) Durnovsk,   *14*) Mazul,   *15*) Usa,   *16*) Near-Sayan
group (Nikolaevo, Kettsk, Kamensk, etc.),   *17*) Barguza-Vitim zone of
manganese-ore shows (Ikat, Taloisk),   *18*) Uda,   *19*) Malo-Khingan group
(South Khingan, Bidzhansk),   *20*) Vandansk.

opment of iron-manganese ores has been observed, it has not been possible un-
equivocally to distinguish components of endogenic and exogenic nature.   This
has been even more complicated when there is a significant distance between the
sectors of ore accumulation and those of the sources of ore components.   Establishing the nature of the source of the ore material for basins of the geological past is even more difficult, because the original information about the
sedimentary-ore palaeo-basin has been erased during the course of time by other
later phenomena;   the restoration of the primary genetic characteristics becomes
an exceedingly complex task, or is possible with extreme limitations.   Thus,

the absence of unequivocal proven and sufficiently complete information about the phenomena of ore-formation in ancient basins and, in particular, the nature of the ore components, leads to disputes about the origin of deposits.   This applies to the Chiatur, Usa, Karazhal, and other manganese deposits in the USSR.

In the present monograph, classification of manganese deposits has been carried out on an associational basis.   The manganese associations, as definite parageneses of ores and rocks, have been considered in the works of Kheraskov (1951) and Shatsky (1954, 1960, 1965).   Varentsov (1962*b*), Roy (1969), and Rakhmanov & Chaikovsky (1972) have attempted to systematize a large amount of factual information on the manganese deposits of the world and the Soviet Union, and have also used the associational method for this purpose.

In the USSR, the following manganese-bearing associations have been recognized.

1.   The quartz-glauconite, sand-clay association, which includes the Lower Oligocene deposits of the Ukraine (Nikopol, Bol'she-Tokmak, etc.), Georgia (Chiatur, etc.), and Mangyshlak;   the Upper Pliocene Laba deposit in the Northern Caucasus;   the group of Palaeocene deposits on the eastern slopes of the Northern Urals, the Lower Cambrian Oldakit deposit in the Baikal region, the Upper Proterozoic Nizhne-Uda deposit in the Sayan area, etc.

This association has been developed mainly in the tectonically stable areas: on the slopes of the crystalline shields, on the platforms and blocks of the epi-Hercynian Platform, and the median massifs of the geosynclines.

The scale of the ore occurrences is massive: not less than 75 - 80% of the reserves of manganese ores of the continents.

2.   The carbonate associations of the geosynclines and platforms, which include the following deposits: Usa (Lower Cambrian, Kuznets Alatau), Ulutelyak (Lower Permian, Aral region), Sagan-Zaba (Archaean, Baikal region), etc.

The manganiferous associations of this type are characterized by the moderate scale of the ore occurrences.

3.   The carbonate-chert association with the following typical deposits: Karazhal (Upper Devonian, Central Kazakhstan, miogeosynclinal region), Takhta--Karacha, etc. (Silurian, Uzbekistan, Zeravshan Range, eugeosynclinal zone).   The scale of ore occurrences is moderate.

4.   The group of volcanogenic-sedimentary associations has been subdivided into two types: the spilite-keratophyre-chert type (the deposits of the Magnitogorsk region of the Urals, Lower-Upper Devonian) and the porphyry-chert type (the Durnovsk deposit, Salair, Lower-Middle Cambrian).   These have developed in eugeosynclinal and miogeosynclinal zones.   Ore occurrences are small.

5.   The manganiferous ferruginous-siliceous association (jaspilite); in the region of the Malo-Khingan (miogeosynclinal region), Upper Proterozoic (Sinian) ore segregations of manganese have been developed, the reserves of which are moderate.

Manganese ore weathering-crusts proper, developed on rocks in which the manganese content does not exceed the clarke level, are distributed quite restrictively in the USSR.   However, we must apparently assign to segregations of this type, the zones of hypergene oxidation, manganese gossans, formed after manganite, carbonate, and silicate ores of a number of deposits: a substantial portion of the oxide ores of the Nikopol and Chiatur deposits, and the zones of oxidation of the Palaeozoic deposits of the Urals, the Usa deposit, etc.

Three geochronological intervals are clearly recognized in the stratigraphical distribution of the manganese ores of the USSR (Fig. 46).

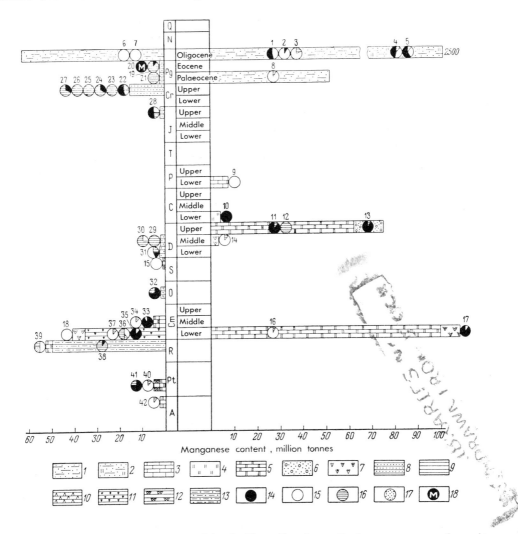

Fig. 46. Diagram of the stratigraphical distribution of the manganese deposits and ore-shows of various genetic types in the USSR.

1) sand-silt-clay rocks; 2) the same rocks with seams of spongolites and gaize; 3) manganiferous limestones; 4) cherty shales and gaize; 5) chert-carbonate rocks (limestones, cherty and clay-chert shales); 6) conglomerates and coarse-grained sandstones; 7) jaspers, tuffs, tuffites, cherty shales, and jasperoid quartzites; 8) porphyrites and their tuffs, tuffites, tuff-sandstones, and limestones; 9) porphyrites (quartz porphyries), quartzites, cherty shales, and lime-

(Continued on page 118)

1) Palaeogene, especially Oligocene. In the area of the South European portion of the USSR, colossal accumulation of manganese ores took place, reckoned in billions of tonnes;

2) Devonian; during this period, comparatively moderate amounts of manganese ores (hundreds of millions of tonnes) were accumulated in individual regions of the USSR (the Urals and Kazakhstan);

3) Cambrian-Sinian, during which time relatively large deposits formed in the Asiatic portion of the USSR (hundreds of millions of tonnes).

In spite of the fact that during these intervals, the formation of manganese ores did not have a global or at least broad regional character, and took place not strictly contemporaneously in spatially separated basins, there is a substantial difference in the specific nature of the Palaeogene and Palaeozoic-Sinian associations. The Palaeozoic-Sinian manganiferous associations are geosynclinal in nature; they include either volcanogenic-sedimentary (the Near-Magnitogorsk deposits, Durnovsk, etc.), or sedimentary associations, laterally transitional into non-ore volcanogenic-sedimentary associations (Usa and Karazhal deposits). During the Palaeogene, colossal exogenic, strictly sedimentary accumulations of manganese ores were formed in the environment of the platform or similar tectonic regime (Nikopol, Bol'she-Tokmak, and Chiatur deposits).

SEDIMENTARY DEPOSITS

The information available at the present time on manganese ores of the world convincingly indicates that all the significant deposits of manganese ores occur in sedimentary formations and are associated with sedimentary processes.

Fig. 46. (Continued from previous page)

stones; 10) jaspers, cherty tuffites, tuff-siltstones, and argillites; 11) lavas and tuffs of basic composition, jaspers, cherty and chert--clay shales, and limestones; 12) manganese-iron quartzites; 13) sand-silt-clay rocks and carbonate sediments; 14-18) ores: 14) oxide, 15) carbonate, 16) oxidized, 17) silicate, 18) mangano-magnetite.

Deposits and ore-shows (figures on diagram): 1) Nikopol, 2) Bol'-she-Tokmak, 3) Shkmer, 4) Chiatur, 5) Chkhari-Adzhamet, 6) Laba, 7) Mangyshlak, 8) Northern Urals group (Yurka, etc.), 9) Ulu-Telyak, 10) Akkermanovsk, 11) Atasu group (Karazhal, etc.), 12) Murdzha, 13) Dzhezda-Ulutauk group (Dzhezda, etc.), 14) Near-Magnitogorsk group (Niazgulovsk I, etc.), 15) Dautash, 16) Usa, 17) South Khingan, 18) Bidzhansk, 19) Aiotsdzor group (Martiros, Karmrashen, etc.), 20) Near-Sevan group (Chaikenda, etc.), 21) Tetritskaroi group (Sameb, etc.), 22) Svarants, 23) Idzhevan-Noemberyan group (Sevkar, etc.), 24) Tetritskaroi group (Tetritskaroi, etc.), 25) Gegechkhor group (Naukhunao, etc.), 26) Molla-Dzhalda, 27) Kodmana group (Kodmana, etc.), 28) Tsedis group (Tsedis, etc.), 29) West Altai (Zyryanovsk, etc.), 30) Klevaka, 31) Sapal, 32) Mugodzhar group (Kos-Istek, etc.), 33) Durnovsk, 34) Kiya-Shaltyr', 35) Udsk-Selemdzha group (Ir-Nimiisk, etc.), 36) Arga group (Mazul, Gar'sk, etc.), 37) Gornoshorsk (Kamzas, etc.), 38) Nizhne-Uda (Nikolaevo, Kettsk, Arshan, etc.), 39) Seiba, 40) Ikat-Garga, 41) Sosnovyi Baits, 42) Sagan-Zaba.

Subsequent phenomena of post-sedimentational changes and metamorphism have only led to a predominance of certain mineral forms, concentration of formerly lean ore accumulations, and under conditions of deep-seated metamorphic zones, to the dissemination of earlier formed concentrations.   The manganiferous weathering crusts occupy a special place , being developed mainly in regions of tropical climate (India, Africa, and Brazil), on rocks in which the amounts of manganese often do not exceed the clarke values.

The problem of the source of the ore components is considerably more complicated.   Geological environments exist in which the ratios between the sources of supply and the areas of accumulation of the ores have been inadequately revealed.   In such cases, disputes arise between investigators about the origin of a particular deposit, although essentially discussions are centred around problems of the nature of the source of the ore-forming components.   Such a situation is typical of a number of manganese deposits in the USSR: Chiatur, Karazhal, Usa, etc.

THE QUARTZ-GLAUCONITE SAND-CLAY ASSOCIATION

Based on the scale of occurrences, this association is huge, and not comparable with the other manganiferous associations.   It includes no less than 85% of the commercial reserves of manganese ores of the Soviet Union. Varentsov (1962*b*) and Roy (1969) have considered this association as the 'Nikopol' type (orthoquartzite-glauconite-clay).   Chaikovsky, Rakhmanov & Khodak (1972) have subdivided the quartz-sand-clay associations into the quartz-sand-clay subassociation proper and the quartz-sand-clay gondite subassociation.   This association is distinguished by extremely clear features and, as a rule, it has a threefold structure;   three portions have been recognized in it: sub--ore, ore-bearing, and supra-ore.   The association has been developed on relatively rigid, consolidated basements of the platforms, or on the stable sectors of the geosynclinal regions of the median massif type.   The association has been formed mainly of terrigenous material of quartzose and clay composition, and glauconite is quite a characteristic authigenic mineral.

Laterally, this association (more usually wedging out) is normally replaced by a non-ore, relatively coarsely-clastic association, sometimes (*e.g.* to the north of the Nikopol deposit) with seams of coal-bearing rocks.   It is clear that the non-ore association separates the ore-bearing association from the weathering crust of the supplying land area.   In the direction of distant wedging-out, the manganiferous association passes into an essentially argillaceous sequence with a subordinate amount of siltstones, in which it is extremely difficult to discern subdivisions, equivalent to the above-noted three members of the ore-bearing association.

This association is most typically developed in the vicinity of the Nikopol and Bol'she-Tokmak manganese deposits (the South Ukrainian manganese--ore basin);   it also includes the Chiatur, Mangyshlak, and Laba deposits, and the group of Palaeocene deposits on the eastern slopes of the Northern Urals, etc.

DEPOSITS OF THE SOUTH UKRAINIAN OLIGOCENE BASIN

This manganese-ore basin is the largest of those known;   its deposits include not less than 70% of the world's reserves of manganese ores.   This

region is portion of the vast South European Oligocene basin, to which also
belong such significant manganese deposits as the Chiatur and the Mangyshlak
in the USSR, and the Varentsi in Bulgaria. The boundaries of the South
Ukrainian manganese-ore basin are plotted quite arbitrarily. They include a
comparatively wide band (approximately 25 km) of development of manganese ores,
occurring in Lower Oligocene (Khar'kov) sand-clay sediments. The latter strike
along the southern slopes of the Ukrainian crystalline shield and into its extension, the western slope of the Azov massif (Figs 47, 48).

Fig. 47.   Map of distribution of
           thicknesses of Oligocene
           sediments in the South
           Ukrainian manganese-ore
           basin and the Dneprovsk-
           -Donets basin.

           1) sediments absent;
           2-5) isopachs: 2) up to
           20 m, 3) from 20 to 50 m,
           4) from 50 to 100 m, 5)
           more than 100 m; 6) seg-
           regations of manganese ores;
           7) outlines of Ukrainian
           crystalline shield.

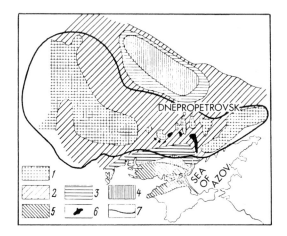

The manganese-ore layer is characterized by interstratification of
the ores with sand-silt-clay sediments; its thickness varies from 0 up to 4.5
m (on average 2.0 - 3.5 m). The outlines of the zone of deposits has been
plotted along the natural wedging-out of the ore layer. Its western boundary
is on the meridian of Kirvoi Rog, and in the east, it is approximately the meridian of Orekhov. The overall extent of the manganese-ore band is 250 km. Post-
-Oligocene erosion breaks substantially disrupt the primary nature of the distribution of the manganese-ore sediments. At the present time, only the relatively
large sectors of development of the manganese ores have been preserved. These
formations fill the erosional depressions (basins in the crystalline basement),
which have been associated with the principal deposits, ore areas, and sectors
(from west to east): the Ingulets, Vysokopol'sk, Novo-Vorontsovsk, Zapadnyi
(Western), Sulitsk, Komintern-Mar'evsk, Brushev-Basan, and Bol'she-Tokmak (Betekhtin, 1964b).

The band of manganese-ore sediments or zone of deposits occupies the
slope of the Ukrainian crystalline shield, which gradually descends southwards,
in the direction of the Black Sea basin. In the north, the manganese-ore sediments are manifested in absolute heights of their floor of +8 m (Grushev-Basan
sector) and +35 m (Ingulets deposit). The northern manganese-ore sediments
thin out, being replaced along the strike by sand-clay deposits, frequently coal-
-bearing, with a significant amount of glauconite. However, in many regions,
such a transition may be traced only with difficulty, since the Oligocene sediments have been eroded.

In the south, the manganese ores also thin out, being gradually exhausted of the clay component. In this way, there is a significant impoverishment
of the ore material in the stratigraphical interval, corresponding to the ore
layer. In other words, the thickness of this stratigraphical interval increases,

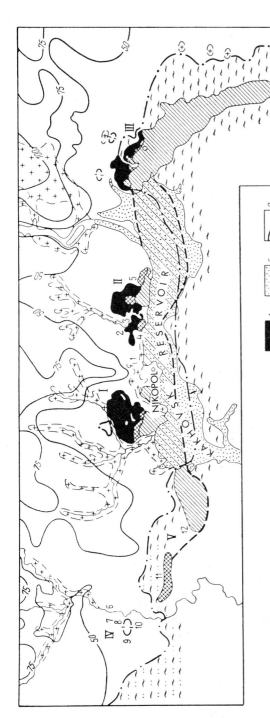

Fig. 48. Map of distribution of manganese-ore areas
in the South Ukrainian basin (After Gryaznov,
in Betekhtin, 1964*b*).

1-3) manganese ores; 1) oxide, 2) oxide and
carbonate, 3) carbonate; 4) area, in which
carbonate ores have been partly or completely
eliminated by erosion during post-Oligocene
time; 5) Dnepr terrace (Kakhovsk dam); 6)
Oligocene clays; 7) Oligocene sandy clays; 8) 'islands' of Oligocene sediments; 9) northern
boundary of continuous field of Oligocene sediments; 10) assumed southern boundary of accumulation
of Oligocene sediments; 10) assumed southern boundary of accumulation of carbonate ores; 11) outcrops of crystalline rocks on surface and below Quaternary sediments; 12) structural contours on
surface of Precambrian rocks.
  Ore-bearing areas: I) West Nikopol, II) East Nikopol, III) Bol'she-Tokmak, IV) Krivoi Rog
(Ingulets) group, V) Dnepr-Ingulets watershed. Ore sectors (figures on diagram): *1*) Maksimo-
-Timoshevsk, *2*) Znamensk and Novoselovka, *3*) Nikolaevo, *4*) Komintern-Mar'evsk, *5*) Grushev-Basan,
6) Novoselovka village, *7*) Zelënaya gully, *8*) Vizirka station, *9*) Ingulets sector, *10*) Nikolo-Kozel,
11) Vysokopol, *12*) Novo-Vorontsov.

and the amount of manganese decreases.   Decrease in the manganese content takes place relatively more slowly than in the area of the northern ore contour (near wedging-out).   The absolute heights, at which wedging-out of the ore layer occurs (far wedging-out), reaches -60.0 m in the Maksimo-Timoshevsk sector.   Consequently, the manganese-ore sediments gently descend southwards as the relief of the crystalline basement changes (Fig. 49).

The manganese-ore sediments are quite readily traced in the area of the basin as a marker horizon of definite stratigraphical position.   Frequently, a glauconitic sand, from 20 - 50 cm up to 1.5 m thick, is located at the base of the ore layer.   In many cases, the Khar'kov sediments rest with a break on the underlying rocks from the sediments of the Kiev Group (Upper Eocene) to the crystalline rocks of the basement.   The ore layer represents a relatively uneven interstratification of manganese concretions, lenses, and nodules with clay--silt material.   Frequently, the manganese concretions and lenses form vague accumulations, traceable along the strike for several tens, and sometimes a few hundreds of metres.   The amount of manganese-ore material, included in the clay--silt non-ore mass, reaches 50% by weight, and the average amount of manganese in the mass of the layer comprises 15 - 25%.

The supra-ore sediments within the zone of the deposits have been clearly separated from the underlying ore-bearing rocks.   Iron hydroxides (sub-zone of oxide ores) have usually been developed along this boundary, or there is a marked enrichment in glauconite (subzone of carbonate ores).   These rocks consist of greenish-grey, massive, laminated clays, consisting mainly of mont-morillonite, and finely-dispersed glauconite with a trace of very fine quartz grains.   Their thickness varies from a few centimetres up to 25 m.   In some sectors, they were either deeply eroded, or completely eliminated during the pre-Late Oligocene - early Sarmatian or pre-Karagan continental intervals.

Within the belt of deposits, depending on the degree of lowering of the crystalline basement, three sequential ore subzones have been recognized: oxide, mixed (oxide-carbonate), and manganese carbonate ores.

The subzone of manganese oxide ores extends from the northern contour of thinning-out of the ore segregation (approximate absolute heights of + 15 m) to the provisional boundaries with the subzone of mixed ores (absolute heights of approximately +4 - 6 m).   The manganese ores within this subzone consist of pyrolusite, polypermanganite, and manganite lumpy nodules, of irregular shape, and friable earthy segregations of manganese hydroxides, frequently impregnating and pigmenting individual parts of the layer.   It is difficult to establish a regular trend in the distribution of individual types of ores.   Sometimes, amongst the sectors of wide development of manganese oxide compounds, there are relicts of more or less deeply oxidized carbonate ores.   Gryaznov (*in* Betekhtin, 1964*b*) has pointed out the existence of a belt of manganese carbonate ores up to 1 km wide and 3.5 km long, developed near the River Solenaya (northern contour of the western part of the Nikopol deposit) and extending southwestwards. It is typical that these carbonate ores occur in the lower portion of the layer; higher up they are covered by oxide types.   In individual cases (the Aleksand-rovsk and Shevchenko quarries) the entire sequential series of alteration prod-ucts of carbonate manganese ores may be observed up to the nodule-concretionary crystalline-granular pyrolusite varieties, in which it is extremely difficult to identify traces of the primary material.   The processes of deep hypergene oxidation of the carbonate manganese ores have evidently been accompanied by a substantial redistribution of the ore material.   This is indicated by the irr-egular, streaky pigmentation of the inter-ore clays, and by sooty segregations of manganese hydroxides, the most compact sectors of which contain decomposed, markedly oxidized relicts of manganocalcite segregations. Definite criteria, leading to the clear distinction between the deeply oxidized varieties and the so-called primary oxide ores, have not so far been established, since the latter have not yet been found in clear exposures (Fig. 50).

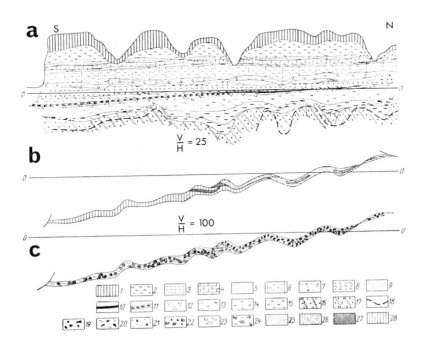

Fig. 49. Geological section (*a*), mineral composition (*b*), and fabrics (*c*) of ore layer of the Grushev-Basan sector of the Nikopol deposit.

1) loam; 2) red-brown clays; 3) dark-grey marly clays; 4) limestone-coquinae; 5) quartzose sands; 6) marly clays; 7) dark-grey detrital clays; 8) clays with marly seams; 9) greenish-grey clays; 10) manganese oxide ores; 11) manganese carbonate ores; 12) quartz--glauconite sands; 13) silts; 14) carbonaceous clays; 15) carbonaceous sands; 16) polymigmatites; 17) plagiogranites and their migmatites; 18) boundary of weathering crust of crystalline rocks; 19) manganese lump-concretion ore in sooty-clay matrix; 20) oxide lump-concretion ore in clay and sand; 21) concretion-oolite ore in sooty-clay matrix; 22) mixed oxide-carbonate ore; 23) concretionary carbonate manganese ore with manganite pisolites; 24) concretions of carbonate manganese ore in clay; 25) polypermanganite ore; 26) manganite-psilomelane ore; 27) manganocalcite-manganite ore; 28) manganocalcite ore.

The question arises: is it possible, in the absence of clear features for separating primary- and secondary-oxide manganese ores, to recognize 'subzones of oxide ores' as such? In this respect, it should be noted that

in many works (Betekhtin, 1946; Gryaznov, 1960; Gryaznov & Selin, 1959; Danilov, 1971a, b), the existence of three sequentially arranged ore subzones, based
on the degree of lowering of the crystalline basement, has been interpreted as
evidence for facies control of the original sediments of the manganese ores.

Fig. 50. Oxidized carbonate ore, completely retaining initial structural feat-
          ures. Nikopol deposit, western portion, Sample No. 7011. Natural
          size.

However, in spite of the great complexity of this problem, the recognition of a subzone of oxide ores is completely acceptable, since it represents
an actual band, within which manganese oxide ores substantially predominate.
There are two points of view concerning the origin of the manganese ores, developed within this subzone. In the opinion of some geologists, the ores of
this subzone were formed in a sedimentary environment with predominant oxidizing conditions. Only in the relatively depressed portions of this zone could
carbonate varieties be formed in a reducing medium. Other geologists consider
that these ores may be regarded as secondary-oxidation products of the carbonate
varieties. As factual information has accumulated and methods of study have
been perfected, the importance of the processes of oxidation in the formation
of ores of this subzone has increased substantially. In the present work, the
subzone of oxide ores is regarded as portion of the zone of deposits within
which mainly secondary-oxide varieties of ores have been developed.

The subzone of mixed ores is located between the sectors of development of manganese oxide and carbonate ores. The boundaries of this subzone
are quite provisional, and its width is extremely variable: from 0.3 to 4.5 km.
A characteristic feature of the manganese ores distributed in this area is the
twofold structure of the ore layer: the upper portion consists of oxide varieties: pyrolusite, polypermanganite, and manganite ores, and the lower portion.
or carbonate ores, which are similar to those distributed to the south of the
carbonate subzone.

In many cases, it has been possible to trace the relationships between the lower unaltered portion of the ore layer and the upper part, transformed by oxidation into an irregularly stratified segregation of concretions, and lens- -like, and lumpy segregations, consisting of manganese oxide compounds.   The carbonate ores of the lower part of the layer frequently contain oolites, piso- lites, or small nodules of manganite composition, which have been impregnated with manganese carbonate and are replaced by it along the joints from the peri- phery.

Thus, in the subzone of mixed ores, there is intense oxidation of the original carbonate varieties and a relatively wide development (inherent mainly in the carbonate ores of this subzone) of manganite inclusions in them.   The subzone of mixed ores is recognized quite arbitrarily, since in the other neigh- bouring subzones there are sectors with similar structuring of the ore layer. But in the subzone of oxide ores the processes of oxidation have a relatively wide and frequently complete development, and in the subzone of carbonate ores their effect is usually insignificant.   If we mentally remove the effects of hypergene alteration of the manganese ores, then a characteristic feature of the carbonate ores of this subzone may be considered the wide development of manganite inclusions in them.   Somewhat less frequently, there are inclusions of clay material, intensely impregnated with iron hydroxides, pisolites of limonitic and hydrohematitic composition, and also nest-like segregations of glauconite and clay-carbonate matter in the manganocalcite, and calcium-rhodo- chrosite groundmass of the ore.

The subzone of manganese carbonate ores is located to the south of the previous subzone and extends southwards until it wedges out completely in the Black Sea basin depending on the degree of downslope of the basement.   The sediments of this subzone are characterized by a relatively feeble hypergene alteration.   Only in the northern, relatively uplifted sectors reached by groundwaters, are traces of oxidation of the carbonate ores observed.   In individual places only the lower and upper parts of the ore layer, where they are in contact with water-bearing sands, have been subjected to oxidation. This is manifested in the presence of a brown oxidized layer of altered glau- conitic sands at the base and top and in the presence amongst them of black sooty segregations of manganese hydroxides.

In the northern part of the subzone there are mainly coarse lumpy, lens-like, carbonate ores, grey in colour, compact, tough, with a characteristic spongy and cellular nature.   They contain a marked amount (up to 25%) of sand- -silt clastic particles, fragments of bivalves, gastropods, and forams, and phosphatized fish remains.   In the overall carbonate matrix of the ores there are inclusions of manganite nodules, segregations of glauconite, and carbonate- -clay pisolites.   In the upper and lower parts of the layer there are normally irregularly angular, cavernous nodules, containing a marked quantity of a clay- -chert substance.   Towards the south compact carbonate concretions of irregul- arly nodular shape (up to 10 - 15 cm) acquire much greater importance in the ore layer, and the coarse lumpy, lens-like segregations (up to 50 - 60 cm) gradually disappear.   In the carbonate nodules themselves there is an increase in the quantity of chert-clay material, although the thickness of the ore layer does not markedly alter.   It is at a maximum in the most depressed central parts of the subzone (up to 3.5 - 4.0 m) and substantially decreases towards points of local uplift.   Towards the Black Sea basin the ore layer wedges out, but not because of a decrease in its overall thickness.   The latter, within this stratigraphical interval, gradually increases southwards.   Wedging-out is manifested in the fact that impoverishment in manganese-ore material occurs in this interval (the dimensions and total number of manganocalcite deposits decrease, and the quantity of inter-ore clays increases, which substantially diminishes the value of the ore layer).

The structure of the ore layer of this subzone is well displayed in the section through the manganese-ore layer in the southern face of the Mar'evsk quarry (Komintern-Mar'evsk sector), and the eastern part of the Nikopol deposit (in downward succession):

1. Dark-grey, thin clays, frequently with black nest-like segregations of pyrite and carbonaceous matter (lower Sarmatian).

2. Greenish-grey, sandy clays, with a glauconitic, medium-grained, quartzose sand at the top .. .. 25 cm

3. Sand-clay, greenish-grey matrix enclosing relatively large lumps of carbonate ore with nodules of manganite, and frequently nests of glauconite and clay also; abundant segregations of fine powdery pyrite are present .. .. 40 cm

4. Coarse-lumpy (up to 50 cm) manganese carbonate ore, forming a relatively well-defined layer along the strike; the carbonate matrix of the ore is steel-grey in colour, cherty, compact, with marked conchoidal fracture, and contains brown, friable, porous manganite nodules (up to 25 - 30%), measuring 3 - 70 mm (on average 25 mm) .. .. 70 cm

5. Grey clays with numerous manganite nodules measuring up to 3 - 7 cm, the number of which varies along the strike .. .. 15 cm

6. Greenish-grey clays, poorly stratified, with scarce remains of bivalve shells; there are rare manganese carbonate nodules 15 cm

7. A layer of carbonate-manganite ore; abundant reddish-brown nodules and friable, earthy segregations of manganite are enclosed in a greyish-white carbonate-clay matrix; the amount of oxide segregations is from 30 to 80% of the overall mass; the dimensions of the rounded manganite nodules is from 2 to 20 cm (average 4 - 7 cm) ..120 cm

8. Greenish-grey clays with vague micro-bedding (2 - 10 mm), emphasized by segregations in the form of seams of manganese carbonate; in the upper part there are lumpy segregations of manganese carbonate of irregular shape .. .. 40 cm

9. Greenish-grey clays, poorly bedded, and laminar in upper part, with rare nodules of manganese carbonate .. .. 30 cm

10. Manganese carbonate ores, massive, coarse-lumpy formations, compactly adjacent to one another, and traceable in the form of a layer; in the freshly broken state, the carbonate matrix is steel-grey in colour with a creamish tint; inclusions of manganite oolites are quite rare .. .. 50 cm

11. Greenish-grey clays, vaguely bedded, with streaky partings of black manganese hydroxides .. .. 20 cm

12. Carbonate ore, oxidized in places, covered with black oxide segregations, cellular, and concretionary-lumpy; irregularly shaped lumps (25 - 30 cm) form a tough chert-carbonate mass with abundant inclusions of small (2 - 25 mm) oxide conchoidal pisolites; on the surface the pisolites are covered by envelopes of black sooty pyrolusite; their internal parts consist of dense brown manganite;

these small segregations comprise 20 - 30% of the total bulk
of the ore .. .. 30 cm

13. Sand, in places compact sandstone, greenish-grey, coarse- to
medium-grained, glauconitic, gradually passing into carbonate,
in places oxidized ore .. .. 30 cm

14. The base of the layer consists of grey clays with some glauconitic
sand.

Thus, amongst the manganese ores of the South Ukrainian basin, the
carbonate varieties have been least altered by hypergene processes. These
ores have retained the features of their composition, texture, and the nature
of their relationships with the country rocks, that is, relatively complete
information, weakly obscured by subsequent post-sedimentation processes, which,
having been deciphered, may have provided the sedimentary basis for this part
of the basin, the environment, and the processes of their formation. Later
on, deposition of the material has led from comparatively simple and clear
objects in the genetic sense (the carbonate ores) to the oxide varieties, which
are in most cases the products of their alteration.

The manganese carbonate ores

The manganese carbonate ores may be regarded as concretionary in
spite of the fact that in individual cases, the relatively large (up to 60 cm),
irregularly-shaped lumps of manganese carbonate, compactly arranged one with
another, can be traced for several tens of metres along strike.

The carbonate ores, on the basis of size and morphology of the concretionary formations may be subdivided into two groups:

1. Concretionary-nodular ores, with typical irregular rounded shape,
frequently with nodular-tuberculate surface, cavernous and spongy; 1 - 25 cm
in size. These ores are usually enclosed in a predominantly clay-silt matrix.

2. Coarse-lumpy ores, usually irregular in shape, angular, sometimes with lensoid outlines; 15 - 60 cm in size. They are characterized by
coarse cavernous (up to 8 mm), and cellular texture. The cavities are frequently filled with glauconite, manganite segregations, argillaceous, and carbonate-clay material. Ores of this type almost never occur in isolated form, but
have the nature of layer-like formations, in which clay-silt material is an
insignificant contaminant.

The main bulk of these ores consists of grey and dark-grey cryptocrystalline carbonate matter, in which minute powdery particles of organic
matter are disseminated. The dimensions of the individual carbonate grains
are about 0.005 mm. In most of the concretionary-nodular and, to a lesser
degree, the coarse-lumpy ores, the dark-grey cryptocrystalline material has the
nature of streaky, clotted segregations with regularly rounded, and oval-elong-
ated outlines, 0.2 - 0.4 mm in size. Such spheroidal formations are usually
cemented by dark-grey, relatively coarse-grained (0.01 mm) manganocalcite.

In the carbonate matrix there are sponge spicules replaced by carbonate matter, remains of diatom tests, and phosphatized fish bones. In many
cases, there are brown segregations, consisting of manganite, with dimensions
from fractions of a millimetre up to several centimetres. These segregations
are characterized by a collomorphic concentric structure. The enclosing
carbonate matrix clearly corrodes, and replaces such manganite segregations.

In spite of the known similarity in the composition of both types of manganese carbonate ores, the coarse-lumpy varieties are distinguished by the following features:

1) a marked predominance (85 - 90%) of uniform dark-grey cryptocrystalline matrix;

2) the groundmass has been relatively weakly recrystallized and altered; and

3) the main carbonate matrix of the ore contains an insignificant number of inclusions of various types (glauconite, manganite, etc.).

The carbonate ores are not formed of any single definite carbonate material, but several phases of such compounds. They include: 1) a dark-grey cryptocrystalline substance with a groundmass of manganese carbonate, the amount of which is not less than 80%; 2) segregations of a light crystalline-granular manganese carbonate, enclosed in the above-mentioned cryptocrystalline groundmass; 3) a light-grey, fine-grained manganese carbonate, in many cases cementing clasts of the groundmass; and 4) a light, transparent carbonate, filling the cracks that cut the groundmass. The quantity of carbonate segregations usually does not exceed 10 - 20% (see points 2 - 4).

The polyphase nature of the manganese carbonates of these ores is indicated by thin-section studies, DTA curves, and results of X-ray investigations.

The chemical composition of the concretionary-nodular and coarse-lumpy ores is extremely similar; they are distinguished only on the basis of phosphorus content. In the coarse-lumpy varieties, from 0.024 to 0.3% phosphorus is usually present, whereas it rarely exceeds 0.07% in the concretionary-nodular ores. A substantial enrichment in phosphorus has been recorded in the dark--grey pelitomorphic ores with relatively increased amounts of clay and organic matter (up to 2.16% phosphorus). The amount of insoluble mineral residue in both varieties varies from a few up to 25 - 30%. However, as a rule, the amount of terrigenous contamination in the ore is poorly reflected in the composition of the manganese carbonate.

The carbonate ores usually contain not more than 20.0 - 41.41% MnO, which corresponds on recalculation to 32.42 - 67.10% $MnCO_3$. If, for example, the dark-grey cryptocrystalline carbonate groundmass contains 30.7% MnO (recalculated as 49.74% $MnCO_3$), then the light clay-carbonate material, playing the role of matrix, from this same sample of ore contains 19.50% MnO or 31.60% $MnCO_3$. This carbonate matter of later generation is markedly impoverished in the rhodochrosite molecule (molecular ratio $MnCO_3$: $CaCO_3$ = 0.59:1).

The amount of ferrous iron, combined in the carbonate form, in the ores described, usually does not exceed 1.56%, or on recalculation, 2.52% $FeCO_3$. There is no clear correlation between the bulk content of manganese and iron and the amounts of normative molecules of rhodochrosite and siderite. Such indefinite ratios of these two components may be explained by the twofold nature of iron. On the one hand, the solution and redistribution of iron, combined in the hydrosilicate form, during deep-seated corrosion and replacement of the glauconitic segregations of the carbonate groundmass, indicates a local enrichment of the carbonate groundmass in the siderite molecule. On the other hand, portion of the Fe^{2+} has been combined along with a predominant amount of the Mn^{2+} in the carbonates of the groundmass, which is a later formation than the glauconite segregations. Thus, the differences in the mineral forms of these elements (manganese is present almost solely in the carbonate form, and iron, partly in the carbonate form, but mainly in the form of glauconite) have determined the fundamental chemistry of the manganese carbonate ores.

The carbonate ores of the Nikopol and Bol'she-Tokmak deposits are characterized by the following amounts[1] of principal ore components (in wt %) and minor elements[2] (p.p.m.): Mn, 15.81 - 31.32 (23.73); Fe, 0.48 - 5.10 (1.31); P, 0.008 - 0.630 (0.090); CO_2, 22.56 - 35.08 (32.40); C_{org}, nil - - 0.88 (0.28); Cu, 2 - 37 (14); Ni, 17 - 147 (64); Co, nil - 29 (10); V, 11 - 123 (28); Cr, nil - 15 (10).

Carbonate-manganite ores

The fabrics of the carbonate-manganite and carbonate ores proper are extremely similar; the composition and structure of the groundmass of these ores are even more alike. The groundmass, usually predominantly over the manganite inclusions, consists of light and pale-grey manganese carbonate, consisting of unevenly distributed streaky segregations, frequently spherulitic, concentrically layered formations, enclosed in a light chert-zeolite matrix. There are less frequent varieties, in which the groundmass is formed of a multiphase mixture of microgranular carbonate minerals, among which there is a predominance of calcic rhodochrosite, manganocalcite, manganous calcite, a trace of terrigenous quartz grains, segregations of glauconite, and biomorphous phosphatized remains. The overall composition of the groundmass has been altered from manganocalcite to a calcic rhodochrosite (molecular ratios $MnCO_3:CaCO_3$ change from 0.16 : 1.00 to 3.78 : 1.00).

The manganite segregations of these ores of irregularly-rounded shape, with an uneven tuberculate surface, are from fractions of a millimetre up to 80 mm in size. The content of manganite segregations is 1 - 70%, on average about 25%. These formations are characterized by a collomorphic, zoned-concentric structure; and the concentrates possess irregularly-wavy outlines. The study of these segregations under the microscope, and also with the aid of DTA and X-rays, demonstrates that the principal phase in them is manganite with a trace of pyrolusite and manganese carbonates (Fig. 51).

The textural relationships between the manganite segregations and the surrounding carbonate matrix are of substantial importance in understanding the formation of these ores. In the samples from the ore layer, which has undergone hypergene transformations, it is frequently clearly observed that the carbonate groundmass corrodes and replaces the manganite segregations. The manganese carbonate veinlets cut a concentrically layered manganite nodule, dividing it into irregularly angular fragments, and sometimes one only. It is clear that the manganite segregations were formed earlier than the surrounding microgranular carbonate groundmass of the ore. The manganite segregations and glauconite globules are quite alike in their relationships with the carbonate groundmass, which indicates their relatively early formation in an environment with comparatively higher values of oxidation-reduction potential. The carbonate groundmass of these ores and the manganite segregations are characterized by the

---

[1] Here and later in this chapter the average amounts are given in parentheses.

[2] The minor elements include Cu, Ni, Co, V, and Cr.

following amounts of ore-forming components (in wt %) and minor elements (in p. p.m.): Mn, 6.20 - 24.49 (18.26); 33.12 - 54.47 (44.42)*; Fe, 0.72 - 3.00 (1.53); 0.60 - 2.34 (1.07)*; Mn:Fe = 11.93; 41.51*; P, 0.026 - 1.480 (0.306); 0.016 - - 1.140 (0.253)*; CO_2, 9.80 - 40.38 (25.60); 0.16 - 10.34 (3.73)*; C_{org}, 0.04 - - 0.35 (0.13); nil - 0.11 (0.06)*; Ni, 25 - 165 (67); 117 - 714 (334)*; Co, nil - 20 (9), 2 - 30 (16)*; Cu, 6 - 25 (16); 17 - 65 (37)*; V, 11 - 78 (37); 56 - 305 (118)*; Cr, nil - 21 (8); 1 - 37 (16)*.

Fig. 51.
Carbonate-manganite ore. The manganite pisolites have been corroded by a carbonate ground-mass. Nikopol deposit, Mar'evsk Quarry, south face.
Sample No. 7017. Natural size.

Products of oxidation of carbonate and carbonate-manganite ores.

Hypergene oxidation has been clearly manifested within the subzones of oxide and mixed ores, that is, in the northern part of the belt of development of the manganese-ore deposits. The process of hypergene oxidation of the carbonate and manganite ores is evidently polyphase, and it began at the outset of a break in deposition of the sediments, which preceded the accumulation of the supra-ore apple-green clays, and the last phases of hypergene oxidation were associated with the contemporary activity of the groundwaters, relatively rich in dissolved oxygen. In the sequence of the ore layer of the two northern subzones of the South Ukrain-ian manganese-ore basin it is possible to observe textural and structural relation-ships, reflecting all the stages of transition from the primary carbonate, mangan-ite ores to pryolusite concretionary formations, the final products of alteration, in which almost none of the original material of the carbonate ores has been pres-erved.

As a result of a study of the mineral composition of the ores, which consist of oxides and hydroxides of manganese, the following ore types have been recognized: 1) manganite, 2) pyrolusite-manganite, 3) pyrolusite, and 4) poly-permanganite. Each of these types has been subdivided into structural varieties (Figs 52, 53).

---

* Here and elsewhere, the asterisk denotes manganite segregations.

Fig. 52.
Manganite ore, formed as a result
of intense oxidation of carbonate
varieties;   structural features
(cavernous nature and nodular outlines) of carbonate nodule are
visible.   In the central inner
portion of the sample, relicts of
calcic rhodochrosite have been
observed.   Ingulets deposit,
Skilevatyi Quarry.   Sample No. 3141.
Natural size.

Fig. 53.
Pyrolusite oolites, segregations
in groundmass consisting of manganese oxides material and residual products of oxidation of
carbonate ore (quartz-clay, and
relict carbonate material).
Nikopol deposit, eastern part,
Grushevsky Quarry.   Sample No.
7013.   Magnification X 3.

Manganite ores

The manganite ores may be observed as oxidation products in places
which have retained the typical structural and textural features of the original
carbonate ores.   In these cases, it has been possible to trace sequential phases
of transformation from small, streaky sectors of oxidation of carbonates to relatively large segregations, almost entirely consisting of black lustrous manganite.   Such formations are frequently characterized by oolitic, rounded shapes
of the manganite segregations of concentric metacolloidal structure, which occur
in the black groundmass with brownish tint.   In the latter, there are relicts
of original manganese carbonate, pigmented with finely-disseminated powdery particles, consisting of manganite, and grains of quartz and glauconite, corroded
and replaced by manganite material.

The manganite concretions, 3 - 250 mm in size, formed as a result of
hypergene oxidation, are usually characterized by irregularly rounded shape and

with a tuberculate surface, and compressed, elongated varieties are frequently observed with fantastic excrescences and cavities (see Fig. 52).   They consist of a black, quite compact substance with clear anisotropy and birefringence in polished section.   These objects are typified by a concentric zonation, frequently complicated by deformities and dislocations of a metacolloidal nature. Large concretions are quite common, consisting of aggregates of smaller concretions.   In the central parts of such concretions there are relicts of a fine--grained manganese carbonate and residual quartz-clay material.

It is likely that in the environment of hypergene redeposition of the manganese-ore material such oxidized, intensely altered relicts of original material played the role of concretion centres, on the activated surfaces of which accumulation of manganese compounds took place, leading to the formation of the manganite concretions.   Along with the formation of the so-called residual ores, in which hypergene oxidation did not eliminate the structural and textural features inherent in the original manganese carbonates, the overwhelming majority of concretionary manganite ores were formed through the well-known transfer of manganese from the surrounding parts of the ore layer.

In places where the manganite concretions accumulate, enclosed in a light greenish-grey clay-silt matrix, in the densest sectors of the latter, there are relicts of carbonate ores and lens-like intercalations, consisting of whitish-grey, mealy carbonate matter.   A study of sections of these rocks has shown that this material is a residual carbonate-clay substance, preserved after intense decomposition of the carbonate manganese ore.   In this material, newly--formed zeolite crystals (clinoptilolite) are also observed.

Sometimes, in the manganite concretions, as in the surrounding clay--silt matrix, there are relatively large (up to 50 mm) barytes nodules, consisting of stellate and spherulitic aggregates of tabular crystals.

These Nikopol manganite ores, separated from the country rocks and earthy varieties, are characterized by the following amounts of principal ore components (in wt %) and minor elements (in ppm):   Mn, 33.13 - 59.39 (51.16); Fe, 0.16 - 2.04 (0.57);   P, nil - 0.120 (0.022); CO_2, nil - 5.54 (0.60); C_{org}, nil - 0.48 (0.12);   Cu, 10 - 70 (41);   Ni, 80 - 875 (328);   Co, 6 - 100 (34); V, 33 - 134 (68);   Cr, 1 - 28 (10).

Pyrolusite-manganite ores

The pyrolusite-manganite ores, on the basis of textural and structural features, are extremely close to the above-mentioned manganite varieties.   They are distinguished from the latter only by the markedly large amount of pyrolusite, which is present in the form of intercalations, reaction rims after manganite segregations, and in the form of separate segregations (products of a more intense oxidation of the original material.

Pyrolusite ores

The pyrolusite ores, under hypergene conditions, are terminal and stable products of oxidation of carbonate or manganite varieties.   In texture, they are similar to the oxidized manganite ores.   However, oolitic, pisolitic, and fine-concretionary varieties are quite widely distributed among the pyrolusite ores (see Fig. 53).   They consist of nodules of irregularly rounded shape, from fractions of a millimetre up to 100 mm in size, occurring in a residual earthy black matrix, consisting of manganese hydroxides, quartz-clay material, and relicts of carbonate matter.   Relatively less frequent are pyrolusite concretions (up to 25 cm) in the clay-silt matrix of the ore layer.   Frequently, in the central parts of the pyrolusite segregations there are clear textures of metacolloidal crystallization, in the core of which there are relict sectors of manganite and carbonate composition.

Within a relatively restricted portion of the ore layer it is possible to trace all stages of transition from weakly altered carbonate ores through manganite varieties to almost pure pyrolusite types (*e.g.* the Mar'evsk Quarry in the western part of the Nikopol deposit).   The pyrolusite ores, as a rule, are a mixture of pyrolusite proper, as the predominant phase, and markedly subordinate amounts of compounds of polypermanganite and manganite composition.

It is difficult to evaluate the true location of the pyrolusite ores within the Nikopol deposit.   There are usually less altered, or not so intensely oxidized manganite and polypermanganite varieties, and relicts of carbonate material in the ore layer along with the pyrolusite ores.   However, in the northern part of the subzone of oxide ores, where the absolute heights of the floor of the manganese-ore layer are characterized by relatively high values (usually +6 to +10 m above sea level), pyrolusite ores are more widely developed than in the more depressed sectors of this floor.   It is evident that the sectors of development of the pyrolusite ores were more affected by the agents of hypergene weathering.   The main water-bearing horizon of the deposit coincides with the Oligocene sub-ore sands and silts, but in some sectors (the West Nikopol ore-bearing area, and the Grushev-Basan sector), the ground waters of this horizon intensely saturate the ore layer, consisting of oxide ores.   The supply of water to this horizon comes from the supra-ore, old-alluvial sands.

These ores, as distinct from the surrounding sequence, are characterized by the following amounts of the principal ore components (in wt %) and minor elements (in p.p.m.):   Mn, 31.59 - 59.21 (47.53);   Fe, 0.36 - 1.95 (0.92); P, nil - 0.040 (0.016); CO_2, nil - 1.96 (0.31); C_{org}, nil - 0.27 (0.09);   Cu, 10 - 125 (38);   Ni, 156 - 500 (306);   Co, nil - 60 (20);   V, 22 - 190 (91); Cr, nil - 18 (7).

Polypermanganite ores

According to Rode (1952), the polypermanganite ores include the manganates, the anhydrite of which is manganese dioxide.   They include minerals, ascribed by a number of authors to psilomelane.   However, at the present time, under the name psilomelane we understand compounds of essentially different composition: Ramsdell (1932) regarded as 'real psilomelane' a potassium variety, termed by Richmond & Fleischer (1942) cryptomelane.   Stunz (1957) considers psilomelane as a barium-bearing polypermanganite, along with other varieties of the present modification MnO_2.

Amongst the manganese ores of the South Ukrainian basin, the polypermanganite formations are relatively widely distributed.   However, ores consisting exclusively of polypermanganite minerals are relatively rare.   There is a significantly greater distribution of polypermanganite segregations, which occur in a mixture with manganite, pyrolusite, and residual carbonate-clay material.

The polypermanganite ores most commonly occur in the form of black earthy aggregates, or lustrous, tough oxidation crusts after manganite and carbonate varieties.   Reniform and sintery formations, consisting of tough dark--grey material are sometimes observed with clear reflective capacity and typical red and brown internal reflexions.   X-ray structural, D.T.A., and chemical data suggest that cryptomelane (potassium) and psilomelane proper (barium) varieties occur most often amongst the minerals of this group.   Ores have also been recorded that consist of todorokite and birnessite.

The polypermanganite formations are an oxidation product of more reduced manganese compounds.   However, pyrolusite is frequently developed after polypermanganite segregations during their dehydration.   In addition, various ratios have been observed: in sectors of abundant water circulation, friable collomorphic incrustations are formed after pyrolusite, which consist of hydrated varieties of MnO_2 with absorbed cations.

The observed relationships of the oxidation products of carbonate and manganite varieties, and experimental data from the synthesis of polypermanganite compounds (Rode, 1952) provide grounds for believing that the polypermanganite ores were formed mainly as a result of absorption of ions of hydrated MnO_2 from solutions of alkaline, alkaline-earth, and other metals.   The hydrated MnO_2 was formed during oxidation of more reduced compounds.

The polypermanganite ores are characterized by the following amounts of principal ore components (in wt %) and minor elements (in p.p.m.): Mn, 25.17 - - 48.14 (35.32);   Fe, 0.45 - 2.40 (1.43);   P, 0.011 - 0.090 (0.031); CO_2, nil - - 2.86 (1.02); C_{org}, nil - 0.30 (0.12);   Cu, 15 - 125 (48.1);   Ni, 70 - 352 (190.7);   Co, 6 - 146 (71.4);   V, 6 - 200 (81.4);   Cr, nil - 20 (11.0).

Conditions of formation of deposits

The present manganese ores were formed in an environment of an extremely stable tectonic regime.   They have been located in the uppermost portion of the southern slope of the Ukrainian crystalline shield (the northern margin of the Black Sea basin) in the immediate vicinity of areas of erosion.   Sedimentation in the region of accumulation of the manganese-ore sediments was characterized by relatively low rates (the thickness of the Oligocene does not exceed 20m) as compared with the neighbouring regions of the Black Sea and Dneprovsk-Donets basins where the thickness of the synchronous sediments is 10 - 20 times greater. An analysis of the palaeogeographical situation of Early Oligocene time has shown that the manganese-ore sediments were deposited in the shallow-water part of a vast basin amongst an archipelago of islands, extending between large continental sectors for approximately 250 km.   To the north of the deposits, there were regions of depressed, littoral swampy plains, in which coal deposition occurred. Sectors of the shallow sea were incised in them in the form of narrow gulfs and estuaries.   Farther south the environment of littoral swampy plains was replaced by a shallow-water marine type.

Within the zone of the deposits, the Oligocene sediments are characterized by a threefold construction.   The sub-ore sediments consist mainly of sands and silts of orthoquartzite composition with a significant amount of glauconite. Normally, these sediments rest with clearly-defined erosion on various horizons of the underlying rocks.   The ore layer consists of quartzose sands, silts, and clays in which there are lensoid layers and concretions of manganese ores.   The

supra-ore deposits consist mainly of clays with a trace of glauconite, and finely disseminated plant detritus.   The time of accumulation of the manganese ores
corresponds to the beginning of the Early Oligocene transgression.

As the floor on the crystalline shield of the Black Sea basin sinks,
the thickness of the deposits, which consist mainly of clays and silts, increases to 200 - 500 m.   In this direction a transition to the open-sea Oligocene
basin of Southern Europe is observed.

An analysis of the data, reflecting the relationships between the
forms of relief of the crystalline floor and the thicknesses of the ore layer,
definitely demonstrates that the manganese-ore sediments were accumulated, as a
rule, in basins and erosional depressions in the basement, thinning out towards
its local rises.   This process went on against a background of markedly enfeeb-
led terrigenous sedimentation, as a result of which the ore components in the
zone of the deposits were not deprived of clastic, clay material, as has been
observed, for example, in the zone of distant thinning-out.

It is clear that only the carbonate and carbonate-manganite ores may
be regarded as primary formations, having remained unchanged under the effect
of hypergene transformation.   Corrosion and replacement of the manganite pisolites by a carbonate groundmass indicate the earlier phase of generation of this
hydroxide stage.   The question of the primary origin of the pyrolusite and polypermanganite ores remains in dispute and has been considered above.

Data from a study of the processes of ore-formation in shallow-water
internal seas, experimental investigations of these phenomena, and also an analysis of actual material from the Oligocene basin suggest a common method of formation of the ores.   From the weathering crust, developing mainly on Precambrian
metabasites, manganese was removed in the form of metal-organic compounds along
with organic suspensions.   Comparatively weak breakdown of the associated clastic minerals and the hydromica-montmorillonite composition of the clays mostly
indicate acid-organic leaching of the parent rocks (mainly, Lower Proterozoic
metabasites) rather than deep-seated development of a weathering crust.   In
sectors of erosional depressions of shallow-water marine type, manganese compounds accumulated, apparently primarily in the form of hydroxide-manganites and
possibly polypermanganites, not now preserved.

Amongst the present-day reservoirs, where processes of formation of
manganese ores occur, there are no known equivalents similar to those of the
Oligocene basin.   However, in the Gulfs of Riga and Finland in the Baltic Sea,
significant accumulations of iron-manganese hydroxide ores have been formed,
with partial development of carbonates in them.   Results of a study of the
processes of ore formation in the present-day reservoirs and observations in
the zone of development of ores of the Nikopol deposit suggest that the primary
ores in the Oligocene basin were manganese hydroxide formations.   These latter,
as a result of diagenetic processes, in the presence of considerable amounts of
organic matter, were transformed into carbonates of these metals.   Small
amounts of iron were combined in the hydrosilicate form (glauconite), and the
principal masses of this component were disseminated in the form of clay and
terrigenous minerals.

Thus, formation of the primary ores in the South Ukrainian basin
took place in two phases: 1) accumulation in the form of hydroxide compounds,
in an environment of oxidizing E_h values and with a relatively intense ingress

of ore-forming components from the near-bottom water; 2) transformation into
carbonate forms, during the phase of diagenesis, and with transformation of the
primary precipitate of manganese hydroxides in a reducing medium.   As a result
of post-Oligocene continental breaks, under the influence of groundwaters in
the relatively uplifted sectors of the deposits, hypergene processes took place,
and formation of oxidized ores occurred: pyrolusite, polypermanganite, and manganite types.

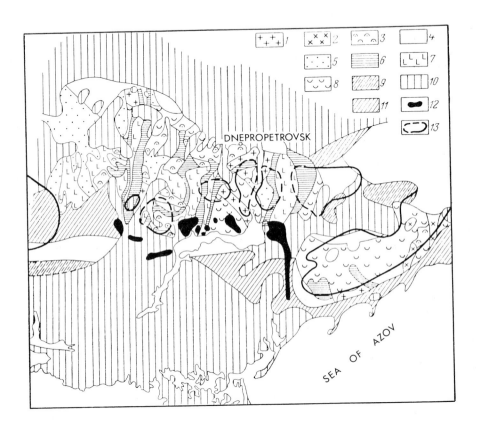

Fig. 54.   Diagram of areas of development of the metabasite and Saksagan' series on the pre-Late Eocene surface and Lower Oligocene manganese-ore
deposits

1) Novo-Ukrainsk granites of the Korosten' Complex; 2) rapakivi and
granites of the Korosten' Complex; 3) granitoids of the Bukova-Osna
Complex; 4) Archaean and Proterozoic formations; 5) grey medium-
and coarse-grained porphyroid biotite granites, less frequently aplite-
-pegmatite granites; 6) Saksagan' and metabasite series; 7) gneisses
and migmatites of the Ingul'-Ingulets gneiss series; 8) gneiss-mig-
matite sequence; 9) Upper Cretaceous sediments; 10) sediments of
the Buchak Group; 11) assumed sediments of the Buchak Group; 12)
Lower Oligocene manganese-ore segregations; 13) assumed Early Oligocene landmass.

In the formation of the vast manganese ores of the South Ukrainian Oligocene basin specific parent rocks played an extremely significant role.   It has been established that the boundaries of the belt of manganese-ore sediments and Lower Proterozoic basic metavolcanics approximately coincide (Fig. 54). The manganese deposits correspond approximately to sectors of metabasites located farther north (palaeotopographically higher up):   the Ingulets deposit and a sector of the Apostolov-Sholokhov deposit, is located to the south of the Krivoi Rog – Kremenchug downwarp;   the Nikopol deposit, to the south of the Verkhovtsev- – Nikopol downwarp;   the Bol'she – Tokmak deposit, to the south and southeast of the Orekhov – Pavlograd, Belozero – Zaporozh'e, and Korsak – Gulyaipol synclinoria (Bondarchuk, 1960).

Whereas, according to a number of authors, the amount of manganese and iron in the granitoids of the Ukrainian crystalline shield does not respectively exceed 0.03 and 2.79%, the amount of manganese in the amphibolites, and chlorite and talc schists increases to 0.09, and iron up to 26.13%.   Thus, the parent rocks, supplying the ore-forming components to the South Ukrainian Oligocene basin, were predominantly metabasites and rocks of the iron-ore – chert – shale series.   This conclusion is also supported by the relatively increased concentrations of nickel, cobalt, and copper in the Oligocene sediments.   These characteristics are substantially different from those of the synchronous sediments of similar facies in the Dneprovsk – Donets basin, which are the products of decomposition of granite-gneisses of the northeastern margin of the Ukrainian Shield (Varentsov, 1964;   Varentsov *et al.*, 1967).

THE GEORGIAN LOWER OLIGOCENE BASIN (THE CHIATUR DEPOSIT)

The Chiatur deposit of manganese ores is one of the largest in the Soviet Union.   In reserves of high-grade ores it is surpassed only by the Nikopol and Bol'she – Tokmak deposits of the South Ukrainian Oligocene iron-ore basin.

The Chiatur and other Lower Oligocene deposits of Georgia, extremely similar in age and conditions of formation (Chkhari – Adzhamet, Shkmersk, etc.), are located mainly near the Dzirula crystalline massif.   The areas of accumulation of the manganese-ore sediments may be regarded as tectonically relatively stable, belonging to the stable middle sector of the geosyncline (the Georgian block).

The geological structure of the area of the Chiatur deposit involves a stratigraphically extensive assemblage of rocks: granitoids, gabbroids, and schists of the Precambrian and lower Palaeozoic;   Liassic sandstones and limestones;   Bajocian porphyritic formations;   carbonate-terrigenous sediments of Cretaceous age, and Tertiary predominantly clastic sediments (Palaeocene – – Sarmatian, Fig. 55).

The manganese-ore horizon, stratigraphically corresponding to the lower strata of the Oligocene, rests transgressively on the underlying rocks. Within the deposit, the manganese-ore sediments rest gently on Upper Cretaceous limestones with an overall vaguely-defined easterly dip, not exceeding a few degrees (Fig. 56).   In the marginal northern parts of the deposit (the Rgani, Mgvimevi, Darkveti, and Sareki sectors) steep dips have been observed: in some sectors (Perevesi, Zeda-Rgani, etc.) the ore horizon has been dislocated by faults with amplitudes of a few metres.   A study of the relief of the surface of the ore horizon reveals a feeble wavy aspect with gentle brachyanticlines; the formation of these structures has been dated as Pliocene (Fig. 57).

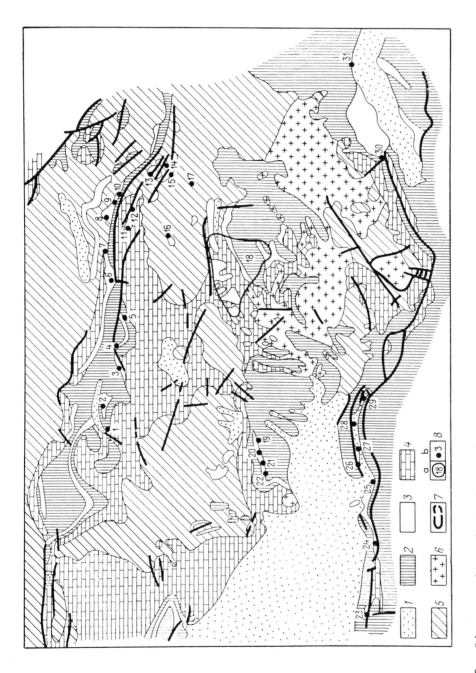

Fig. 55. Diagrammatic geological map of the Dzirula manganese-ore region (After Avaliani, 1967).

1) Quaternary sediments; 2) Palaeogene and Neogene clays, sandstones, marls, and volcanogenic formations; 3) clays and sandstones of the Maikop Series; 4) Cretaceous limestones, marls, clays, sandstones, and volcanogenic formations; 5) Jurassic sandstones, calcareous sandstones, limestones, clays,
and volcanogenic formations; 6) Precambrian and Palaeozoic intrusive and eruptive rocks of the Dzirula
massif; 7) tectonic disturbances; 8) manganese deposits (a) and ore-shows (b): 1–17) Shkmersk group,
18) Chiatur, 19–22) Chkhari-Adzhamet, 23–29) Myakov belt, 30–31) Kvena-Tkotsa.

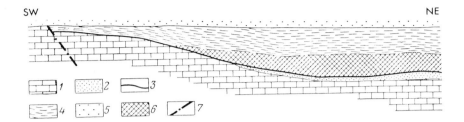

Fig. 56.   Diagram of structure of Oligocene sequence.

1) Upper Cretaceous limestones;   2) sub-ore quartzose sands and sand-stones;   3) manganiferous horizon;   4) schistose spongolite rocks;
5) Chokrak sands;   6) major fault.

The Oligocene sediments of the deposit are clearly subdivided into sub-ore, ore horizon, and supra-ore (Fig. 58).

The sub-ore sediments rest in most cases unconformably on the eroded surface of Upper Cretaceous limestones.   There is a basal conglomerate 0.15 -
- 0.50 m thick;   above come quartz-arkosic and micaceous sandstones.   To the east and northeast the thickness of the sandstones increases to 30 m.

The manganiferous horizon consists of an interlayering of ore strata and seams of gaize-like sands and clays.   In the productive part of the ore horizon the number of ore strata varies from 3 to 18, reaching 25 in places (Itkhvisi, Darkveti, and other plateaux).   The thickness of the ore strata is 1 - 50 cm, and the non-ore strata, up to 1 m.   The strata are like compressed lenses, traceable along strike for 200 m and more.   The total thickness of the ore horizon varies from 0 to 14 m, on average 4.2 m.   The number and thickness of the non-ore layers vary widely in different sectors of the deposit (in a northeasterly direction they increase markedly;   the thickness of the ore strata, on the other hand, increases in a westerly direction).

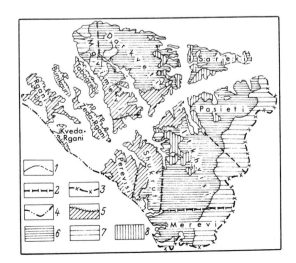

Fig. 57.
Diagram of thicknesses of manganese-bearing horizon of the Chiatur manganese deposit.

1) visible and assumed outcrop of manganese horizon;   2) boundaries of plateaux;   3) lines of zero thickness of manganese horizon;
4) faults;   5) areas under exploitation;   6-8) thicknesses of ore layer: 6) from 0 to 2 m,   7) from 2 to 6 m,   8) from 6 to 14 m.

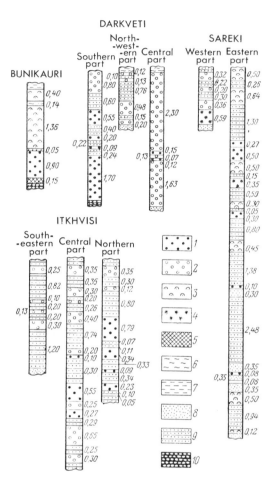

Fig. 58.
Typical sections in the manganiferous horizon based on individual
sectors in the Chiatur deposit
(Betekhtin, 1964a).

1) washing oxide ores;   2) carbonate ores;   3) oxidized ores;   4)
manganiferous sandstones;   5)
'mtsvari' ores (oolitic oxide ores,
cemented with crystalline calcite
of hydrothermal origin);   6) clays;
7) spongolite sandstones;   8) sands;
9) sandstones;   10) conglomerates.

Two parts are clearly recognizable in the ore horizon: a lower part,
including the bulk of the commercial ores, and an upper part, relatively poor
in such ores.   Between these parts of the ore horizon is a ferruginous seam,
comprising in the open sectors hydroxides of iron, and in the concealed sectors
glauconite and chlorite minerals.

In addition to the quartzofeldspathic components, a significant amount
of petroclastic material (fragments of granitoids, eruptive rocks, and slates)
has been recorded in the sandstones surrounding the ores.   Makharadze & Chkheidze
(1971) have drawn attention to the presence of glauconite tuffs, tuffites, seams
of gaize, spongolites, and chert-zeolite tuffites in the ore horizon.

A mineral zonation of the ores is clearly defined in the deposit, the
features of which, as in the surrounding sediments, reflect the facies environment of their formation.   Thus, in the relatively nearshore manganese-ore sediments of the western sectors of the deposit (the Rgani and in part, the Western
Perevisi plateaux) oxide ores are developed, consisting of pyrolusite and polypermanganite minerals.   On moving to relatively deeper-water sectors, the quantity of manganite ores (brown belt) increases in the ore layers.   The latter

are widely developed in the Zeda-Rgani, Northern Perevisi Shukruti, and Mgvimevi plateaux.   Depending on the relative lowering of the floor of the Oligocene ore--bearing basin (northeastern and northern margins of the deposit), carbonate manganese ores appear in the form of nodules, oolites, and sandstone matrix, and in the relatively deep-water sectors they are more widely developed.

Commercially the most valuable oxide ores are developed mainly in the western and especially in the central parts of the Chiatur deposit (Perevisi, Shukruti, Zeda-Rgani, and other plateaux).   Towards the east, substantial impoverishment of these ores takes place, and in the eastern and northern parts of the deposit they completely thin out.

Such an association between composition and structure of the ore horizon and the facies environment of a definite sector of the deposit relates mainly to its lower, most productive portion.   For the upper, leaner portion of the layer, a general tendency to facies change is maintained, although the transition from some types of sediments and ores to others occurs over relatively shorter distances.   The primary-oxide ores, consisting of manganite, are developed only in individual sectors (Rgani and Perevisi), and along with the oxide ores there are also nodular carbonate and oxidized varieties.   Features of facies discrepancies between the lower and upper portions of the ore horizon have been noted, for example, within a number of sectors (Mgvimevi, Darkveti, Itkhvisi, etc.) the ore strata of the upper portion of the productive horizon consist almost entirely of carbonate varieties, below which there are strata consisting of primary-oxide manganese ores.   In those cases when carbonate manganese ores crop out on the surface, or are susceptible to the effect of groundwater, a zone of oxidation has been developed in them, the width of which reaches several tens of metres.

Above the ore-bearing horizon in the western part of the deposit (in facies relatively shallow-water, close to the sources of the ore and sedimentary material) there are clay and spongolite sandstones up to 30 m thick.   In the eastern and northeastern parts of the deposit between the manganese-ore horizon and the spongolite sandstones, which are gradually replaced by clays (up to 10 m), clays of the Maikop type (Middle Oligocene; Laliev, 1964) appear, up to 10 m thick.   The principal clay mineral in these rocks is montmorillonite with a trace of hydromica.   Makharadze & Chkheidze (1971) have pointed out the wide development of chert-zeolite tuffites with spongolites in the supra-ore rocks, which are replaced laterally eastwards by Maikop clays.

The total thickness of the Oligocene sediments is 110 m.

In the Chiatur deposit, the principal commercial reserves consist of primary-oxide and carbonate ores.   A significant part (18%) in the reserves quota is assigned to oxidized ores, and the products of hypergene alteration of manganite and carbonate varieties.   Metamorphosed ores are insignificantly developed (Perevisi plateaux), being located in the contact zones of basaltoid dykes.   Near the deposit, so-called infiltration ores have been recognized, consisting of manganese hydroxide.   They include segregations of manganese hydroxide, impregnating seams of sandstones and cherty rocks in the oxidation zone of the carbonate and manganite ores.   Segregations of pyrolusite, and psilomelane, playing the role of matrix, occur quite often, or are clearly defined, distinct seams in the sub-ore sandstones which underlie the productive horizon.

Primary-oxide ores

The primary-oxide ores are developed mainly in the western and central parts of the deposit.   Several subtypes have been recognized amongst them.

Oolitic (segregated) ores markedly predominate and comprise 47% of the reserves.   Two varieties may be distinguished in them: 1) ore, consisting of hard oolites of pyrolusite and psilomelane composition; and 2) ore, consisting of relatively soft oolites (H = 1-3), consisting of brown manganite material ('brown belt') and black pyrolusite material ('black belt').   The size of the oolites varies from a fraction of a millimetre up to 20 mm, on average 2 - 5 mm. The matrix of the oolitic (segregated) ores is a cherty material with a trace of quartz and feldspar particles, fragments of sponge spicules, and plant detritus.

The raw ore contains from a few up to 35% Mn, and 25 - 55% SiO_2.

The brown and black varieties of the ore 'belt' consist respectively of manganite and pyrolusite matter, little distinctive in composition; they contain (in %): Mn, 45 - 52; Fe, 0.7 - 1.2; P, 0.1 - 0.22; SiO_2, 7 - 12. The 'plasti' variety, which has a relatively limited distribution in the deposit, is a massive black ore of oolitic texture.   It usually forms clearly defined seams.   The principal minerals are pyrolusite and psilomelane.   The amounts of basic components in this ore vary within quite wide limits (in %): Mn, 45 - - 58; Fe, 0.5 - 1.0; P, 0.10 - 0.18; SiO_2, 3 - 18.

The 'satskhrili' ore differs from the widely developed oolitic varieties in the presence of lumps of the massive variety ('plasti layer'), the dimensions of which are up to several centimetres.   Such ores form independent seams, frequently alternating with other varieties.

The 'zhgali' ore differs little from the normal oolitic ores, and is characterized by a tougher siliceous matrix.

'Mtsvari' is one of the local varieties of oolitic ores with a matrix consisting of crystallo-granular calcite of hydrothermal origin.

These varieties of ores occur mainly in the sectors of intense development of hydrothermal calcite veins (Rgani and Perevisi plateaux).

Carbonate ores

These comprise 39% of the reserves of all ores in the deposit.   They are mainly developed in the northeastern sectors, where they occur principally in the upper portion of the ore-bearing horizon.   The carbonate ores form seams (up to 0.5 m), alternating with sand-chert, opal, and carbonate-opal rocks.   The thickness of the upper part of the ore-bearing horizon, which contains the carbonate ores, is 0.5 - 2.0 m.   Two varieties have been recognized: 1) oolitic ores, in which the oolites consist of manganese carbonates; and 2) complex ores of oolitic or massive fine-grained construction.   The dimensions of the oolites are from one to several millimetres.   The shape and structure of the oolites of carbonate and oxide ores are extremely close.   The principal minerals of these ores are manganocalcite, calcic rhodochrosite, calcite, opal, and barytes; gypsum, and iron sulphides have also been recorded.   The main components are present in the following amounts (in %): Mn, 10 - 30; Fe, 2 - 4; P, 0.1 - 0.3; SiO_2, 5 - 40; CaO, 10 - 35; CO_2, 20 - 32.

Oxidized ores

The oxidized ores are the products of hypergene oxidation, and they occur in sectors where carbonate ores are exposed on the surface.   They consist of hydrated manganese oxides, sometimes with iron hydroxides, opal, and relict remains of the original ores.   The amounts of the principal components in these ores are as follows (in %):   Mn, 30 - 35;   Fe, 2 - 5; SiO_2, 8 - 35. We must also assign to the oxidized ores the pyrolusite formations ('black belt'), developed after manganite ores ('brown belt').

Conditions of formation of the deposit

The Chiatur deposit has been assigned to those relatively uncomplicated types in respect of geological construction and possibilities of reconstructing the conditions of formation.   Investigators who have studied the various aspects of the geology of Chiatur, in essence do not disagree in principle that the deposit consists of sedimentary formations accumulated in Early Oligocene time in the littoral zone of a shallow-water marine basin.   An analysis of the palaeogeographical environment demonstrates that ore-deposition occurred in the marginal portion of the reservoir, in a wide embayment, passing eastwards and northeastwards into a relatively open basin.

According to a number of investigators (Betekhtin, 1946, 1964*a*; Avaliani, 1964, 1967; Strakhov *et al*., 1968; Ikoshvili, 1971;   etc.), the source of the manganese-ore and sedimentary material could mainly be the rocks forming the cover of the Georgian block: Liassic arkosic sequences, Bajocian volcanic suites, and the crystalline rocks proper of the Dzirula massif. Dzotsenidze (1965), and Makharadze & Chkheidze (1971) have suggested that the sources of the manganese were hydrothermal solutions, entering the floor of the basin from the Adzhar - Trialet geosyncline.   The volcanic foci and centres of hydrothermal activity were located to the south and westsouthwest of the ore--bearing basin.

A clearly-defined geochemical zonation has been observed in the ore--bearing horizon, which conforms well with changes in the sedimentary environment.   As one passes from relatively littoral shallow-water conditions into deeper water a sequential replacement of the oxide ores (pyrolusite and psilomelane) by manganite and then carbonate types takes place.   In addition, an overall increase in thickness of the ore horizon has been recorded and an increase in the number of terrigenous - cherty non-ore seams.   On the basis of mechanism of formation, the Chiatur ores are similar to those of the Nikopol deposit.   The carbonate varieties were formed during a phase of diagenesis, when the original accumulations of oxides of manganese in a relatively reducing environment, with sufficient amounts of CO_2, produced as a result of decomposition of organic matter, were combined in carbonate compounds.   The oxide ores were preserved in the relatively well aerated parts of the basin, and the leading role in the formation of ore concentrations of oxide compounds of manganese has been ascribed to processes of autocatalytic oxidation of dissolved forms of this metal.

DEPOSITS OF THE NORTH URALS PALAEOCENE BASIN

The glauconite-quartz, sand-clay association is associated with relatively small, but numerous deposits of Lower Palaeocene carbonate manganese ores of the Northern Trans-Urals (the Burmantovo, Tyn'insk, Loz'va, Yurka,

Polunochnoe, Berezovo, Ivdel, Vishersk, Marsyat, Kolinsk, etc.), located in the
marginal part of the epi-Hercynian plate.   Involved in the single North Urals
manganese-ore basin, they can be traced in the form of a narrow discontinuous
belt of ore-bearing sediments along the eastern slope of the Northern Urals for
more than 200 km.   Two structural stages are readily discerned in the development of the basin: a lower stage, consisting of sedimentary, eruptive, and intrusive Palaeozoic (Middle Devonian and Upper Silurian) rocks of the basement
with weathering crusts developed on them, and an upper stage, in which Mesozoic-
- Cainozoic sediments of the cover can be recognized.

The rocks of the basement in the northern part of the ore-bearing belt
(the Burmantovo deposit) consist mainly of pyroxene-plagioclase porphyrites and
their tuffs, interlayered with siliceous-clay slates and limestones; in its
central part (the Tyn'insk and South Ivdel deposits), there are porphyrites,
pyroclastic formations with seams of shales, limestones, albitophyres, and diabases, in the southern part of the belt (near the Marsyat and Kolinsk deposits),
there are serpentinites, gabbros, diorites, granites, and granodiorites.   They
have all been intensely deformed, crumpled into tight folds, and broken by a
system of fractures.

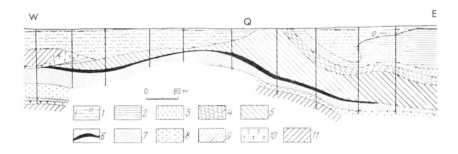

Fig. 59. Geological section of the Ivdel deposits   (After Rabinovich, 1971).

1) Quaternary sediments; 2) Middle and Upper Eocene argillaceous
rocks; 3) Lower Eocene gaize, gaize sandstones, and diatomites; 4)
upper Palaeocene gaize-like clays and argillites; 5) lower Palaeocene
diatomites and argillites; 6) lower Palaeocene ore-bearing (Polunochnoe
member; 7) Santonian glauconite-quartz, clay, montmorillonite argillites and clay diatomites; 8) Aptian - Albian sands, silts, and clays;
9) Jurassic pebble-beds, sands, silts, and clays; 10) weathering
crust of Palaeozoic rocks; 11) Palaeozoic rocks.

The Mesozoic - Cainozoic sediments during the process of their formation often developed inherited structures, dependent on the palaeo-relief of
the Palaeozoic rocks.   A geological section through the Ivdel manganese deposit,
in which the structure of the covering of a rise in the relief of the Palaeozoic
basement is seen, has been shown in Figure 59.   The weathering crusts (Triassic -
- Jurassic) on the basement rocks were unevenly developed, and their thicknesses
vary from 1 - 2 up to several tens of metres.

In the southern part of the basin (Serovsk area) deposition of ocherous and clay-ocherous lateritic products, characterized by an increased amount

of iron, nickel, and cobalt, took place on the serpentinites in the upper zone of the weathering crust (Rabinovich, 1971).

Both continental and marine sediments have been recognized in the Jurassic, Cretaceous, and Tertiary rocks.   The total thicknesses of sediment increase from south to north and from west to east, reaching 700 m in the Burmantovo deposit.   The Jurassic sediments, comprising all three divisions, have been recognized on the basis of spore-pollen assemblages and are distributed throughout the entire area of the North Urals basin.

The Lower Cretaceous deposits are characterized by a wide development of terrigenous, both marine and continental sediments, from clays to sands and gravelites of varying grainsize with numerous plant remains, segregations of ferruginous pisolite-oolite ores, and layers of bauxite-like rocks.

The Upper Cretaceous (Senonian) continental sequence (Mysovsk Group) consists of a typical lacustrine facies, in which sand-silt-clay sediments with seams of siderite and goethite-chamosite-leptochlorite iron ores have been recognized.   The Santonian sediments in the northern part of the manganese-ore basin consist of montmorillonite argillites and clays, and in the southern part, sands and sandstones.   The Maastrichtian includes quartz-glauconite sandstones, siltstones, silty clays with glauconite, and argillites distributed in the uppermost parts of the sequence in the south (Serovsk area).

Palaeogene sediments are widely distributed in the Northern Trans--Urals.   At their base is the lower Palaeocene, ore-bearing, Polunochnoe member up to 20 - 30 m thick, to which all the principal manganese-ore segregations of the basin are restricted.   The ore member consists of layers, intercalations, and lenses of gravelites, sandstones, siltstones, gaize-like clays, argillites, cherty formations, and manganese carbonate ores.   In the structure of each of the three ore horizons recognized in the member (Fig. 60) (montmorillonite, beidellite, and diatomite) there is a well-defined rhythm, characterized by a sequential replacement of the lithological varieties of rocks, surrounding the manganese mineralization (in upward succession):   from gravelite-psammitic coarse-grained varieties to sandy, sand-clay, clay, and cherty types, which has found reflexion in the segregations of different types of ores: concretionary-clay, sandy, sand-clay, clay, and chert types.

At present, about 20 medium and small manganese deposits are known, arranged in a chain, extending submeridionally for distances of from 5 - 8 up to 35 km.   The largest can be traced for 10 - 12 km (the Ivdel and Marsyat deposits) with a width of the ore segregation of up to 5 - 6 km (Kolinsk deposit).

The shape of the ore segregations is compressed lensoid, strongly--elongated along one axis (Fig. 61), with the possible exception of the Kolinsk deposit, which has an equant ore seam (7.6 x 6.4 km).

The ore-bearing members in the deposits of the northern part of the basin have a complex structure;   three manganiferous horizons have been recognized here (in upward succession):   montmorillonite, beidellite, and diatomite (Yurka, Ivdel, and other deposits).   Southwards there is a gradual thinning--out of the manganese-bearing horizons, and apparently, only the middle (beidellite) horizon is maintained.   The saturation of the ore horizons in manganese carbonate is also variable.   They often include non-ore seams from 0.1 - - 0.2 up several metres thick.   Along the strike and dip, the ore bodies are frequently replaced by non-ore sand-clay sediments.

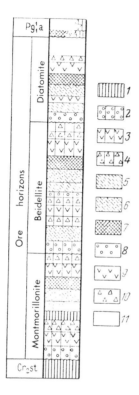

Fig. 60.
Distribution of various types of manganese carbonate ores in composite section through the manganese-
-ore (Polunochnoe) member of the North Urals basin
(After Rabinovich, 1971).

1) montmorillonite and beidellite argillites; 2) gravelites or conglomerates with argillite matrix; 3) concretionary-clay ores; 4) argillites with ore inclusions; 5) sandy ores; 6) sand-clay ores; 7) cherty ores; 8) gravelites or conglomerates with diatomite matrix; 9) diatomites with inclusions of concretionary-clay ore; 10) diatomites with ore segregation; 11) grey, non-ore diatomites.

The depth of occurrence of the ore seam varies from its outcrop on the present surface down to 100 - 300 m in those deposits, complicated by folding.

The mineralogy and geochemistry of the manganese ores of the North Urals basin have been considered in detail in the works of Andrushchenko (1954), Shterenberg (1963), and other investigators.

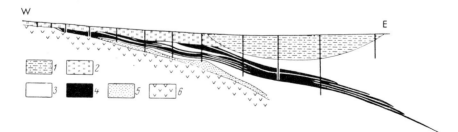

Fig. 61. Geological section through the Polunochnoe deposit (After Rabinovich, 1971).
1) alluvial sediments; 2) drift; 3) gaize clays; 4) manganese ores; 5) quartz-glauconite sandstones; 6) pyroxene-porphyrite tuffs.

It has been established that the manganese carbonate ores of most deposits in the basin consist mainly of calcic rhodochrosite with an insignificant trace of manganocalcite and oligonite (Table 4).

Table 4. *Composition of Manganese Carbonates in the North Urals Deposits (in wt %)*

(After Rabinovich, 1971)

Ore Minerals	$MnCO_3$	$CaCO_3$	$MgCO_3$	$FeCO_3$
Calcic rhodochrosite	62.9 - 82.3	10.2 - 24.6	3.3 - 17.8	1.7 - 10.9
Manganocalcite	46.7 - 64.9	26.2 - 35.3	7.2 - 13.0	1.7 - 5.7
Oligonite	67.0	6.8 - 9.0	6.7 - 9.0	15.0 - 25.5

All the structural varieties of manganese carbonate ores consist of calcic rhodochrosite: bedded, concretionary, and nodular.

The following gangue minerals must be recorded: quartz, opal, chalcedony, glauconite, and also beidellite, montmorillonite, pyrite, marcasite, and calcium phosphate. The minor minerals include: plagioclase, hornblende, epidote, chlorite, zircon, muscovite, and sometimes ilmenite and magnetite.

The most widely distributed textures of the ores are: spherulitic (with radiating and concentrically-zoned structure of individuals), and less frequently crystalline (from very fine- to medium-grained).

The oxidized manganese ores are developed predominantly in the Novo--Berezovo and Polunochnoe deposits. Manganite plays an essential role in them (Novo-Berezovo deposit), as well as pyrolusite (Polunochnoe deposit), minerals of the psilomelane group, vernadite, and oxides and hydroxides of iron.

Lumpy, concretionary-earthy, and sometimes bedded sectors of oxidized ores often consist of manganite and pyrolusite. Psilomelane is predominantly developed in the matrix of the sandy ores in the form of collomorphic-sintery, reniform, and veinlet formations. The texture of the ore aggregates, consisting of oxide and hydroxide minerals of manganese, is amorphous (cryptocrystalline and finely crystalline).

The carbonate ores of a number of deposits contain (in wt %): Mn, 17.51 - 21.92; Fe, 3.26 - 6.57; SiO_2, 17.98 - 33.90; CaO, 2.38 - 6.70; P, 0.1 - 0.3. It has been established that the most manganese-rich (more than 18%) are the sandy and sand-clay manganese carbonate ores as compared with the concretionary-clay, clay, and chert varieties.

In the deposits of the basin from north to south there is a marked increase in the content of manganese, iron, and phosphorus along with a decrease in the amount of silica; the maximum concentration of lime occurs in the central part of the ore-bearing belt. These geochemical features are associated in the basin with a zoned distribution of various structural and mineralogical types of ores: concretionary-clay types in the north and sandy and sand-clay types in the south, with manganocalcite ores in the centre and oligonite types in the south.

The manganese carbonate ores of the North Urals deposits are in nature sedimentary and sedimentational-diagenetic.

Location of numerous deposits of manganese carbonate ores occurred in a narrow littoral belt on the floor of the transgressing early Palaeocene sea. Favourable areas for manganese-ore deposition were the depressions in the floor of the basin; depressional lows, basins, and trough-like hollows, in which accumulation of manganese-ore sediments took place.

The sources of the manganiferous solutions were the rocks of the land area being weathered during the Mesozoic: pyroxenites, gabbros with increased amounts of manganese (Mn up to 0.32%), skarned porphyrites (Mn up to 0.26%), magnetite ores (Mn, 0.24 - 0.97%), and volcanogenic-sedimentary sequences of the Silurian and Devonian.

During a study of the manganese ores of the Polunochnoe deposit Betekhtin (1946) suggested that, depending on the depth of the floor of the Palaeocene basin, there was a gradual facies change in the composition of the ore--bearing sediments. The littoral facies of the primary-oxide ores, as the depth of the basin increased (through 300 - 400 m), was replaced by the carbonate-ore facies. A number of investigators, however, believe that the primary ores consist of the carbonate facies only, affected near the surface by oxidation (Rabinovich, 1971, etc.). The oxidized ores have a limited distribution in the deposits of the basin; they have to a substantial degree been worked out.

The reserves of proven carbonate manganese ores in the North Urals manganese-ore basin (Mn, 21.2%) comprise about 50 million tonnes; prospective reserves are 100 million tonnes. Attempts at enrichment of the carbonate manganese ores of the North Urals deposits have shown the possibility of obtaining quality manganese concentrates from them for the production of ferromanganese and silicomanganese.

OTHER DEPOSITS OF THE QUARTZ-GLAUCONITE SAND-CLAY ASSOCIATION

The Mangyshlak Deposit

The Mangyshlak deposit lies 2-0 km east of Fort Shevchenko. It lies on the northern limb of the Chakrygan syncline, formed on an epi-Hercynian platform base (Stolyarov, 1958; Tikhomirova, 1964; Tikhomirova & Cherkasova, 1967). The relatively older rocks in the vicinity of the deposit are the Upper Eocene marls. On them with a clear lithological contact but without visible features of a break, rest the deposits of the Golubaya (Blue) Group of the Lower Oligocene (in upward succession): silty clays (up to 8 m), and sands and clay silts (40 m). The rocks that form the Golubaya Group may be regarded as sub-ore. The deposits of this group are gradually replaced upwards by sands, silts, and clays (up to 42 m). In the upper part of this member are concentrated the manganese ores. This horizon is 3 - 8 m thick. Above the manganiferous rocks come Middle Oligocene clays (up to 35 m), above which, transgressively and with a conglomerate at the base, come Middle Miocene sediments. According to Stolyarov (1958), the Golubaya and manganese-bearing groups are the stratigraphical equivalents of the Uzunbas and Kuyulus groups, and the manganese-ore horizon occurs at the top of the Kuyulus Group.

Considerable concentrations of manganese have been located over a relatively limited area (up to 50 km^2) of the northern edge of the Chakrygan syncline, where sand-silt, littoral-shallow-water sediments have developed rel-

atively close to the source of supply, the Karatau meganticline.   The ore horizon has a lens-like construction.   Lumpy and concretionary ores are the most
common.   In the northern part, the ores consist of manganese hydroxides, farther south, mixed (oxide and carbonate) ores are distributed, and still farther
south, carbonate ores.   Tikhomirova & Cherkasova (1967) believe that such zonation has resulted from the oxidation of primary carbonate ores.

The oxide ores consist mainly of pyrolusite, manganite, and minerals
of the polypermanganite group (psilomelane, cryptomelane, etc.).   The varieties of these ores (lumpy, concretionary, cementation, and earthy) on the whole
are characterized by the following ranges of amounts of principal ore components (in wt %) and minor elements (in p.p.m.):   Mn, 3.84 - 46.90;   Fe, 1.40 -
- 6.34;   P, nil - 1.91; C_{org}, nil - 0.68; CO_2, nil - 7.74;   V, 22 - 157;   Cr,
nil - 50;   Ni, 38 - 346;   Co, nil - 180;   Cu, 26 - 116.   The largest amounts
of ore-forming elements occur in the concretionary varieties.

The carbonate ores, consisting of calcic rhodochrosite, manganocalcite, and calcite, have the following ranges (as above):   Mn, 15.96 - 22.60;
Fe, 1.28 - 3.95;   P, nil - 1.69; C_{org}, 0.14 - 0.36; CO_2, 6.40 - 31.34;   V,
22 - 68;   Cr, 12 - 29;   Ni, 14 - 92;   Co, nil - 10;   Cu, 3 - 25.

In the Mangyshlak deposit, the quantity of manganese carbonate ores
with an average manganese content of from 5 - 6 up to 10%, exceeds 30 million
tonnes.

The Laba Deposit

This deposit occurs in the Northern Caucasus, in the vicinity of the
Perepravsk, Gubsk, Khamketinsk, and Novo-Svobodnaya mesas.   The manganese-ore
sediments extend in the form of an east-west belt for 25 km between the left-
-hand tributaries of the Laba (the Rivers Khodz' and Fars).   They are involved
in the sandy group (200 - 220 m thick) of the Upper Oligocene - Lower Miocene
sequence;   in the western part of this region, the equivalents of the ore-bearing deposits belong to the Maikop Series.

The deposit is located on the tectonically rigid Adygei block, which
belongs to the larger epi-Hercynian platform structure.   The manganese-bearing
sediments of the Laba deposit were formed in the middle relatively depressed,
downwarped part of this block.   They consist of sediments of deltaic facies,
which consist mainly of clastic quartzose material.   The ore-bearing sequence
is characterized by a clear threefold construction:

1)   sub-ore clay-silt sands;

2)   the ore horizon which consists of flat ore lenses, from a few
centimetres up to 5 m thick, extending for 250 m, formed of manganese carbonate.
The ore lenses are interbedded with sandy, silty, and gravelly sediments.   The
minimum thickness of the ore horizon (5 - 12 m) has been identified in sectors
of the deposit close to the shore, and in the centre it reaches up to 20 m and
markedly increases towards the relatively deep marine sectors.   In construction
and facies characteristics, the ore horizon corresponds to the beginning of a
transgressive series of sediments;

3)   supra-ore sediments, consisting of clays and silts.

Kalinenko *et al.* (1967) have recognized three types of ores.

1. Manganese carbonate (calcic rhodochrosite) with the following amounts of principal components (in wt %) and minor elements (in p.p.m.): Mn, 24.95; Fe, 2.75; P, 0.020; C_{org}, 0.28; V, 37; Cr, 19; Ni, 151; Co, 13; Cu, 22.

2. Ferruginous-manganese carbonate (lean) ores with Mn, 8.81; Fe, 7.44; P, 0.081; C_{org}, 0.50; V, 61; Cr, 48; Ni, 51; Co, 9; Cu, 16.

3. Oxidized ores, developed after carbonate varieties: Mn, 25.16; Fe, 2.02; P, 0.035; C_{org}, 0.33; V. 72; Cr, 14; Ni, 271; Co, 12; Cu, 30.

The reserves of manganese ores (Mn, 7.2 - 12.7%) comprise about 30 million tonnes.

The Nizhneudinsk (Lower Uda) group of deposits (Sayan region).

Within the Upper Proterozoic (Riphean) complex of rocks of the Sanyan region, which is associated with gently-dipping (5 - 7° SW) carbonate-terrigenous sediments of the Karagas and Oselochnaya series, a manganese-bearing glauconite-quartz sand-clay association, including numerous small deposits and ore-shows of the Lower Uda region (Nikolaevo, Shungulezh, Kettsk, etc.), is clearly recognized. The deposition of the sediments, which comprise the association, took place under sub-platform conditions in the marginal zone of the southwestern part of the Siberian Platform, which became stabilized during the Riphean.

The manganese ores are restricted to the lower portions of the Izan Group, in the base of which horizons of clastic and coarse-grained sediments (conglomerates, gravelites, sandstones, and siltstones) are developed.

The pyrolusite-psilomelane, psilomelane (composed of cryptomelane and hollandite), and psilomelane-vernadite ores form nests, layers, and lenses within the manganiferous horizons, which extend for 1.5 - 2 km with thicknesses varying from 0.5 to 10 - 15 m, and rarely more. Down dip, the ores can be traced for 400 m. Amongst the friable and compact ores, the following fabrics are discerned: veinlet-cementation, nodular, sinter-conchoidal, collomorphic, crusty, and powdery. The manganese ores contain (in wt %): Mn, 13.6 - 17.7; Fe, 2.7 - 4.3; P, 0.05 - 0.2. In the iron-manganese ores, the quantity of iron increases to 12.7%, and the content of manganese varies from 9.5 to 12.4%.

In the Arshan ore-show, a seam of braunite-hausmannite ores is also known extending for 130 m and up to 1 m thick, occurring in coarse-grained sediments (gravelites and sandstones) of the Izan Group, the formation of which has been associated with diagenetic and metamorphic transformation of the manganese--hydroxide compounds.

It is likely that the primary sedimentation-diagenetic ore accumulations of the Sayan region consisted mainly of manganese and iron-manganese carbonates, affected by oxidation during Mesozoic and Cainozoic time. The reserves of such oxidized residual and residual-infiltration manganese and iron-manganese ores of this region have been estimated at 50 million tonnes.

The Ilikta Deposit (Near-Baikal region)

The manganese-ore concentrations of the Ilikta deposit of the Western

Near-Baikal region are also associated with the manganese-bearing quartz-sand-
-clay sediments (of miogeosynclinal type) of the Golousten Group of the Upper
Proterozoic Baikal Series formed in the platform marginal foredeep of a miogeosyncline. Ore bodies in the form of lenses and nests extending for several
metres and with a thickness of tens of centimetres, rest conformably with the
surrounding carbonate-terrigenous rocks.

Amongst the residual and residual-infiltration ores discovered here,
there is a widespread development of compact and powdery-earthy pyrolusite,
pyrolusite-psilomelane, and psilomelane varieties of ores, formed under weathering crust conditions. The amount of manganese in individual pyrolusite
and psilomelane nests reaches 54 - 60%.

The Oldakit Deposit

The Oldakit deposit of manganese carbonate ores is located on the
northeastern end of the Angara Range and is restricted to a zone of longitudinal
intermontane basins (gräben), developed in the central part of the Baikal-Vitim
uplift during Early Cambrian time. The surrounding Lower Cambrian sediments
consist of greywacke and arkosic sandstones, siltstones, and slates, to a certain degree carbonatized. Several horizons of rhodochrosite-manganocalcite ores
have been recognized in the deposit; the thickness of the largest of these is
20 - 70 m, and it can be traced along strike for 250 m. The manganiferous
horizons are separated by seams of non-ore sand-silt-clay and carbonate rocks.
The ores have a bedded structure, and their texture is microspherulitic and
allotriomorphogranular. The average amount of manganese in the ore horizons
varies from 6.6 to 14%, iron 6 - 7%, and phosphorus 0.06 - 0.11%. The ore
reserves are estimated at 20 million tonnes.

MANGANIFEROUS CARBONATE ASSOCIATIONS

The most widely developed here is the association of rocks and ores,
assigned to the group of limestone-dolomite associations (Varentsov, 1962b;
Roy, 1969; Chaikovsky et al., 1972). The manganese-ore carbonate associations
proper are most commonly relatively minor subdivisions of larger non-ore carbonate associations. As they thin out the manganese-ore associations usually
pass laterally in relatively large non-ore associations.

The features of the tectonic position, structure, composition, and
lithological-geochemical characteristics of the rocks and ores, comprising the
manganiferous associations, enable us to recognize two types in this group.

1. A dolomite-limestone manganiferous association of the geosynclines
(predominantly eugeosynclinal). A typical example of this associational type
is the Usa deposit.

2. A limestone-dolomite association of the rigid platform basement,
passing laterally into the evaporite association. In the USSR, a typical ill-
ustration of this type is the manganiferous group of deposits which include the
Ulutelyak deposit.

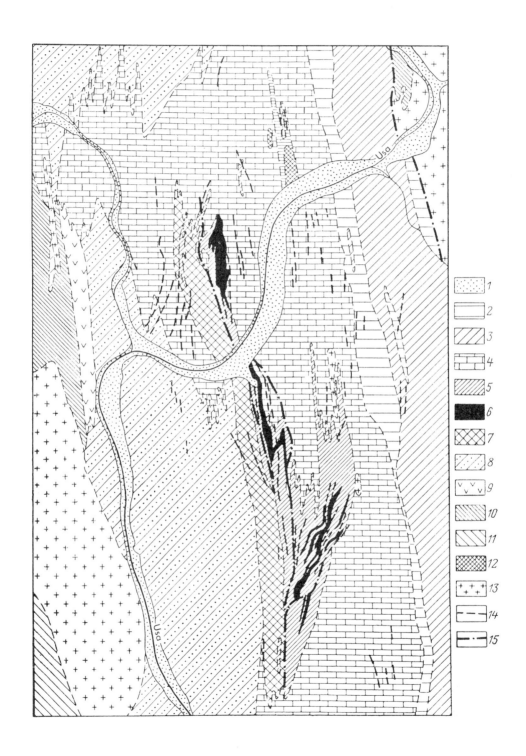

THE DOLOMITE-LIMESTONE ASSOCIATION OF THE GEOSYNCLINES

The Usa Deposit

The Usa deposit of manganese ores is the largest in Siberia.   It
lies in the Kemerovo district near the middle course of the River Usa, a right-
-hand tributary of the Tom'.   The deposit belongs to the limestone-dolomite
eugeosynclinal association, which is characterized by quite a complex and
laterally impersistent composition.   This association corresponds to the Lower
Cambrian Usa Group, which extends in the form of a submeridional belt in the
central part of the Kuznets Alatau Range (Mukhin & Ladygin, 1957; Varentsov,
1962a; Mirtov et al., 1964; Khodak et al., 1966).

The deposit lies on the steep (70 - 90°) western limb of a syncline;
the eastern limb of this fold has been destroyed by granitoid and diorite intrusions.   The rocks which make up the association have been disturbed by numerous
small fractures, complicated by local folding, and cut by diorite dykes.   In
the vicinity of the deposit, three portions can clearly be distinguished in the
association: sub-ore, ore-bearing, and supra-ore (Figs 62, 63).   Outside the
deposit these subdivisions rapidly pass into a single carbonate association,
difficult to subdivide.

The rocks of the lower sub-ore portion apparently lie conformably on
a sequence of greenstone slates which are altered volcanic rocks of basic composition.   Sometimes at the base of the association there are red tuffite sandstones up to 200 m thick, rapidly thinning out along strike.   The sub-ore sediments consist of limestones, dolorites, and their breccias, with a total thickness of 1000 m.

The ore-bearing portion of the association has been complicated by
unevenly interstratified, predominantly manganese limestones, manganocalcite
and calcic-rhodochrosite ores with dark, weakly manganiferous limestones, and
black slates.   These ore-bearing rocks rest conformably on the underlying
limestones.

The supra-ore portion of the association consists mainly of light-
-grey limestones.   Dark streaky limestones and black pyrite-bearing slates
have been recorded in considerable amounts.   These rocks are associated through

Fig. 62.
Diagrammatic geological map of the Usa manganese deposit.   (After V. Kurshs).

1) alluvial sediments; 2-3) rocks of the Mundabysh Group: 2) lilac and greenish-grey tuffogenic shales, sandstones, and conglomerates, 3) porphyrites and
mindalephyres; 4-10) rocks of Usa Group: 4) light-grey limestones with arch-
aeocyathids, 5) calc-silicate schists, silicified limestones, and manganiferous
shales and limestones, 6) manganese-ore bodies; 7) dark-grey silicified limestones; 8) grey, bedded dolomites with seams of sedimentary breccias and flints;
9) lilac and grey tuffogenic shales and sandstones; 11-13) rocks of Kondoma
Group: 11) metamorphic, greenish-grey amphibole, chlorite-sericite, and other
schists, and sheared porphyrites, 12) ferruginous dolomites, 13) diorites;
14) diabase and porphyrite dykes; 15) tectonic faults.

a gradual transition with the ore-bearing sediments. The thickness of the supra-
-ore sediments reaches 1500 - 2000 m.

The deposit has been formed of three lens-like segregations, elongated
parallel to the general strike in a northnorthwesterly direction for 4.6 km. The
Northern (Pravoberezhnaya) segregation consists of an asymmetrical lens up to
215 m thick, gradually wedging out southwards to a few metres (see Fig. 63).
The Central, or Levoberezhnaya segregation is located somewhat farther south.
It has been formed of an uneven interstratification of manganese carbonate ores
and markedly predominant manganese limestones and black slates up to 170 m in
overall thickness. Farther south, the ore-bearing member is seen to split and
the ores of the Central segregation are replaced by dark manganiferous limestones
and black pyritic slates. The latter in turn, in a distance of approximately
200 - 300 m to the southeast, pass into the Southern (Azhigol) segregation. This
segregation consists of a series of ferrorhodochrosite and manganocalcite lenses
(up to 25), interstratified with essentially dominant dark manganese limestones
and black siliceous-sericite schists. In comparison with the two more northerly
segregations, a substantially larger quantity of lean manganocalcite ores, manganiferous limestones, and black slates, have been recorded in the Southern segregation, as a result of which the overall thickness of the ore-bearing member
increases to 370 m.

A zone of oxidation is developed in the deposit, and this is associated
with quite significant (6%) ore reserves, consisting of psilomelane, pyrolusite,
and other manganese oxide compounds. The Levoberezhnyi and Azhigol sectors of
the deposit are characterized by relatively intense hypergene oxidation, reaching
125 m.

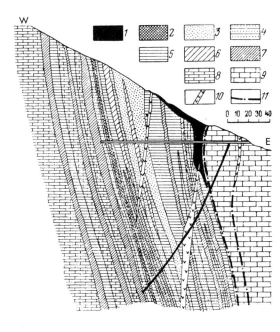

Fig. 63.
Geological section of ore segregation
(Usa deposit, Pravoberezhnyi sector)
(based on data from the Tom'-Usa Exp-
edition of West Siberian Geological
Survey).

1) psilomelane-pyrolusite oxidized
ores (Mn, 30% and more); 2-5) primary ores: 2) rhodochrosite (Mn, 30%
and more), 3) manganocalcite-rhodo-
chrosite (Mn, 24 - 30%), 4) rhodochrosite-manganocalcite (Mn, 20 - 24%),
5) manganocalcite (Mn, 10 - 20%); 6)
chert-carbonate rocks; 7) manganiferous limestones (Mn, 5 - 10%); 8-9)
non-ore rocks: 8) dark-grey and grey
limestones, 9) light-grey limestones;
10) diabase dykes; 11) tectonic
faults.

The main commercial reserves of the Usa deposit consist of carbonate
ores (Fig. 64). The oxide ores of the weathering crust are also of interest
for exploitation (Fig. 65).

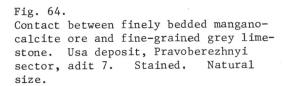

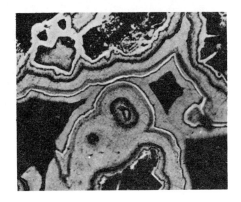

Fig. 64.
Contact between finely bedded manganocalcite ore and fine-grained grey limestone. Usa deposit, Pravoberezhnyi
sector, adit 7.   Stained.   Natural
size.

Fig. 65.
Collomorphic segregations of man-
ganese hydroxides.   Black – pores
and cavities.   Oxidation zone of
Usa deposit.   Polished section.
Magnification X 120.

The calcic-rhodochrosite and ferrorhodochrosite varieties are the
richest among the carbonate ores.   They are usually characterized by a grey
colour, quite a high density (up to 3.5), and frequently a very fine micro-
-layering, consisting of minute (0.1 – 1.5 mm) stratification of the micro-
seams of manganese carbonate and a ferro-manganese layered silicate mineral,
on the basis of X-ray data assigned to the manganostilponomelane group.   The
groundmass of these ores consists of microspherulites, and frequently two-three
concentred microlites, the refractive index of which is greater than in the
microgranular manganese carbonate cementing them, and usually containing a
trace of very finely powdery carbonaceous particles.   In a number of cases,
the carbonate ores have been intensely boudinaged and mylonitized, and stylolite seams have been observed in them (Figs 66, 67).

The manganostilpnomelane mineral occurs mainly in the rich calcic-
-rhodochrosite ores.   Manganostilpomelane is atypical of the manganocalcite
varieties of manganese limestones (see Fig. 66).

Organic matter has been metamorphosed up to the anthracite-graphite
grade; it has been finely disseminated in the carbonate groundmass, sometimes
forming segregation accumulations along the stylolite seams.   The rhodochrosite
ores are characterized by segregations of small crystals of pyrite and pyrrhotite.

The manganocalcite ores are usually dark-grey, black, finely bedded,
and often have a clastic, brecciated texture.   They consist of unevenly recrystallized fine-grained carbonate matter.   The manganic and manganous limestones
are usually similar to the non-ore dark fine-grained limestones, and may be
distinguished from the latter only on the basis of chemical-sampling data.

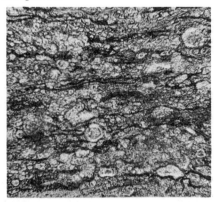

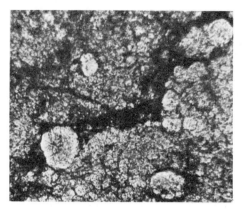

Fig. 66.
Microspherulitic fabric of finely-
-bedded rhodochrosite ore.   Dark
bands - carbonaceous matter.   Individual seams (light-grey) consist
of manganese hydrosilicate.   Usa
deposit, Pravoberezhnyi sector,
adit 7.   Thin section, ordinary
light.   Magnification X 100.

Fig. 67.
Microspherulites of rhodochrosite
in rhodochrosite-manganocalcite ore.
Usa deposit, Pravoberezhnyi sector,
adit 7.   Thin section, ordinary
light.   Magnification X 40.

The largest portion of the manganese (70 - 100%) and a substantial
portion of the iron (up to 47.6%) in the ores of the Usa deposit are present in
the form of carbonate minerals of the calcite-rhodochrosite-siderite series.
The amount of manganese in silicate form (mainly as manganostilpnomelane) reaches
30% of the bulk composition, and for iron, up to 80.6%.

From north to south (from the Northern through the Central towards the
Southern segregations), a decrease is recorded in the average-weighted components
(in %):   Mn, 11.12 → 9.98 → 3.91;   CaO, 24.24 → 22.74 → 21.22;   MgO, 3.58 → 2.34 →
→ 1.90.   However, in the same direction, an increase has been noted in the amount
of Fe, 3.13 → 5.12 → 4.04; SiO_2, 13.97 → 21.68 → 30.44; Al_2O_3, 1.61 → 1.93 →
→ 2.48; P_2O_5, 0.122 → 0.157 → 0.173.

In connexion with the fact that at 7 - 12 km to the north of the deposit, an interlayering of dark limestones and porphyroids, tuffs, and tuff-sandstones, has been observed in the stratigraphical interval equivalent to the Usa
Group, and farther to the north (Belaya Usa), the quantity of volcanogenic material increases.   Some investigators consider that the ingress of manganese is
associated with volcanic activity.   However, the absence of any features of volcanism in the limestone-dolomite association itself, as with the nature of the
distribution of the manganese, iron, silica, alumina, and phosphorus in the manganiferous association itself, suggests that the exogenic source of the ore components was not situated to the north, but to the south of the deposit and was
not associated with volcanism.

The Sagan-Zaba Deposit

The Sagan-Zaba manganese deposit is located on the western margin of
Lake Baikal and is genetically associated with an Archaean marine geosynclinal
carbonate facies.   It is restricted to the belt of development of metamorphosed

limestones of the Ozero Group, assigned to the Ol'khon Series of the Upper Archaean.   The principal member of the manganiferous association consists of white
and grey crystalline limestones with inclusions of seams of manganous limestones,
and seams and layers of quartzites, and gneisses (biotite, amphibole, pyroxene,
etc.), amphibolites, and porphyrites of varying composition.

Geological work has shown that the manganese ores in the form of
three seams of manganous limestones are localized in a 100-metre member of grey
bedded, sometimes massive crystalline limestones, traceable from the shore of
Lake Baikal eastwards for 600 - 700 m.   The ore seams lie conformably with the
surrounding limestones, dip steeply to the north (75 - 80°), and consist of an
alternation of ore and non-ore layers and seams, with a thickness of from a few
centimetres up to 1 - 2 m (Fig. 68).   The Northern ore seam can be traced along
the strike for 70 m.   To the east, it gradually thins out, and in the west is
cut by a porphyrite intrusion.   Its thickness varies from 15 to 18 m.   The
average amount of manganese is 6.67%.

The Central ore seam may be traced from Lake Baikal westwards for
380 m where it thins out.   The thickness of the seam varies from 4 to 17 m.
The average amount of manganese is 6.58%.   The Southern ore seam extends for
325 m and is from 10 to 30 m thick, with an average manganese content of 3-93%.

The deposit is in general characterized by an uneven content of manganese in the ore seams and stringers (0.72 - 22.52%), which is associated with
the uneven distribution of calcium and manganese carbonates in them.   It has
been established that manganese in the form of an isomorphous additive enters
the calcite lattice, forming in the primary ores a continuous series from manganous calcite to manganocalcite.   The minor minerals in the manganous limestones are quartz, garnet, biotite, muscovite, hornblende, and sphene.   The
fabric of the manganese ores is usually bedded, and the texture is fine- and
medium-crystalline.

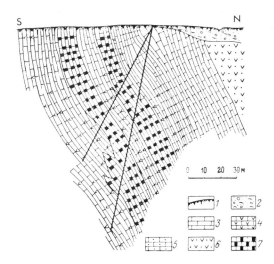

Fig. 68.
Geological section through the Sagan-Zaba
deposit (based on data of the Sagan-Zaba
party, 1957-1958).

1) soil-plant layer and manganese-ore
gossan; 2) alluvial-deluvial sediments;
3) grey fine- and medium-grained crystalline limestones; 4) crystalline limestones with porphyrite intercalations;
6) metamorphosed porphyrites; 7) manganiferous limestones.

In the primary carbonate manganese ores there are seams consisting
of manganese-bearing silicates (garnets, pyroxenes, and amphiboles).   The
amount of manganese in them reaches 10%.

The manganiferous limestones of the Sagan-Zaba deposit are characterized by the following chemical composition (in wt %): Mn, 3.93 - 6.67; Fe, 1.47 - 3.0; P, 0.09 - 0.11; S, 0.05 - 0.13; SiO_2, 7.5 - 8.1; Al_2O_3, 0.95 - - 1.57; CaO, 39.6 - 41.7; MgO, 2.74 - 3.5; calcination loss, 32.60 - 35.37. On the surface, the carbonate manganese ores have been oxidized. The thickness of the manganese-ore gossan is 2.5 - 3.0 m. In its uppermost zone, ores of pyrolusite-psilomelane, psilomelane, and vernadite composition have been developed, with a typical blackish-brown and cherry-red colour. The amount of manganese in the remaining ores of the weathering crust increases to 40 - 44%.

The primary manganese sedimentation-diagenetic carbonate ores (manganous limestones) have been recrystallized under conditions of regional metamorphism with the formation of typical crystallogranular textures and mineral ore associations: manganous calcite, manganocalcite, and less frequently a manganiferous garnet, and manganous amphiboles and pyroxenes.

The reserves of manganous limestones in the Sagan-Zaba deposit are estimated at 1 million tonnes (Mn, *ca* 6%). They may be used in part as basic fluxes for manufacturing some kinds of steel, and for preliminary mixing during the process of producing cast iron from iron ores.

THE LIMESTONE-DOLOMITE ASSOCIATION OF THE PLATFORMS

The Ulutelyak Deposit

The limestone-dolomite manganiferous association, which includes the Lower Permian Ulutelyak deposit of manganiferous limestones, marls, and carbonate manganese ores, is located on the gently dipping slope of the southeastern portion of the Russian Platform in the zone where it merges with the Pre-Urals marginal foredeep. Here, in the Karatau ore region, in the area of the Ulutelyak deposit, there are manganous limestone-marl, chert-limestone, and dolomite-anhydrite facies, assigned on the basis of age to the Filippovsk horizon of the Kungurian Stage of the Lower Permian.

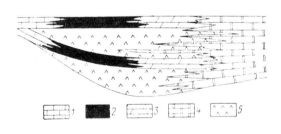

Fig. 69.
Diagram of the structure of the manganiferous limestone-dolomite association of the Ulutelyak type (After E. Gribov, 1972).

1) onkolitic and oolitic limestones; 2) manganese carbonate ores — manganous limestones; 3) manganous dolomitic marls; 4) dolomites; 5) anhydrites

The sequence of the manganiferous association in the vicinity of the deposit (Fig. 69) consists of the following rock types (in upward succession):

1. A seam of bluish-grey, massive anhydrites resting transgressively on Artinskian limestone-marl and limestone-dolomite rocks .. .. ..   30 - 40 m.

2. A manganiferous marl-limestone layer (lower) with seams of

carbonate manganese ores, up to 1 m thick.   Manganese content in
individual stringers, 12 - 18% .. ..   6 - 10 m.

   3. An anhydrite layer with intercalations of an argillaceous
dolomite (0.3 - 0.4 m) at the base   .. .. 40 - 50 m.

   4. Manganiferous marl-limestone layer (upper) . ..   6 - 10 m.

   5. Anhydrites of Irensk horizon with seams up to 3 m of
dolomites and dolomitized limestones.

   Laterally, in the direction of the Ufa Plateau, the manganiferous
marl-limestone sediments pass into very finely bedded pelitomorphic, often
oolitic dolomitized limestones, dolomites, and marls.   Towards the Karatau
Rise, there is a gradual thinning-out of the manganiferous carbonate layers
and their transgressive overlap on to limestone-clay rocks and marls of the
Artinskian Stage.   The marl-limestone manganiferous facies are associated in
the region with the transgressive nature of the development of the basin (Makushin, 1970), and the anhydrite and dolomite facies with the regressive type.

   Two types of ore have been recognized in the deposit: 1) carbonate
manganese ores, consisting of manganous marls and limestones with bedded, oolitic, lumpy, and sometimes brecciated structure.   Manganese is present in the
ores in the form of a manganous calcite and manganocalcite;

   2) vernadite and chert-vernadite (rarely psilomelane) ores, friable
and weakly compacted, formed in the weathering crust after carbonate ores.

   The amount of basic components in the layers of manganiferous marl-
-limestone rocks consists, for the upper layer (in wt %): Mn, 6 - 9; SiO_2,
7 - 9; CaO, 36 - 39; and for the lower layer: Mn, 2 - 3; SiO_2, 10 - 15;
CaO, 31 - 36.

   In the western sector of the deposit, the upper layer of manganiferous rocks, 4.3 - 7.7 m thick, can be traced over an area of 4.5 km^2, and the
amount of manganese in it varies from 3 to 9.47%, SiO_2, 8 - 12.9%, and CaO,
33.6 - 44.6%.

   The formation of the limestone-dolomite manganiferous association
of the Ulutelyak type took place in a semi-enclosed lagoonal marine basin
under conditions of an arid climate.   The sources of manganese were the weathered Palaeozoic rocks of the Ulutau.

   The reserves of manganous limestones with an average amount of manganese of 10% comprise 8 - 10 million tonnes.   These limestones may be used
for the manufacture of steel in open-hearth furnaces as a basic manganous flux,
and also during the production of spiegeleisen and specular iron.

The Burshtyn Deposit

   This Miocene manganese deposit occurs in the Ivan-Franko region of
the Ukrainian SSR.   In structural respects it is located in the zone where
the Russian Platform merges with the Pre-Carpathian marginal foredeep.   The
association rests on the eroded surface of upper Tortonian gypsum beds and the

limestones overlying the gypsum. The ore portion of the association consists of layers of manganiferous marl-clay rocks with seams of calcareous clays and volcanic tuffs.

The manganese-bearing minerals are: manganous calcite, manganocalcite, and calcic rhodochrosite, often containing cations of magnesium and iron in the form of an isomorphous additive. The amount of manganese in the primary ores varies from 0.5 to 30%, and in the oxidized vernadite ores, from 30 to 40%. The source of the manganese was probably the rocks of the islands and platform landmasses.

The manganese ore accumulation in the Carpathian region and in the Ulutelyak deposit occurred under conditions of an arid climate, as indicated by the lithological similarity between the manganiferous carbonate associations developed in both places.

OTHER DEPOSITS OF CARBONATE ASSOCIATIONS

The Khoshchevat Deposit

The Khoshchevat manganese deposit occurs on the left bank of the River Bug in the Odessa district of the Ukrainian SSR. The manganese-bearing rocks are Archaean marmorized crystalline limestones, in which two ore-bearing zones may be traced in an east-west direction: one with a thickness of 35 - 100 m, and other, 5 to 25 m. The primary ores consist of manganous carbonates and, in part, manganese silicates, hematite, and magnetite (Betekhtin, 1946). The deposit has been oxidized to a depth of up to 40 m. Friable iron-manganese ores have been developed in the weathering crust, consisting of pyrolusite-psilomelane concretions, brown manganous ironstones, and hematite. The best-quality residual ores contain about 30% manganese, 20% iron, 2 - 4% silica, less than 1% alumina, and 0.04 - 0.25% phosphorus.

The primary iron-manganese ores of the Khoshchevat deposit belong to the sedimentary metamorphosed type, genetically associated with the manganiferous carbonate association. The rich residual ores formed after it in the weathering crust may be regarded as the raw material for the manufacture of specular iron.

THE CARBONATE-CHERT ASSOCIATION

For the carbonate-chert associations, a clear paragenetic association has been established between the cherty, calcareous, manganiferous, and sometimes ferruginous sediments that form them. Their conditions of accumulation are characterized by the relatively shallow-water nature of the basin, in which, however, the supply of terrigenous material was markedly restricted. The abundance of silica, calcium carbonate, and manganese and iron components in the solutions created favourable prerequisites for the formation of ore concentrations.

The structural position of the palaeo-basin in which the manganiferous chert-carbonate associations were deposited, was distinguished by the stability of the tectonic regime, usually following a period of violent volcanic activity (the Karazhal deposit and others).

It is likely that post-volcanic exhalations, introducing silica, manganese, iron, and other components into the sedimentary basin, to a certain degree affected the geochemical features of the ore-bearing sediments.

The Karazhal Deposit

Large segregations of Upper Devonian manganese and iron ores are included in the chert-limestone manganiferous association of the Uspensk tectonic zone in Kazakhstan.   They are restricted to its western part, the Atasu ore region (the Karazhal deposit, with the Western, Eastern, Northern, Southern, and Far Eastern sectors; the Bol'shoi Ktai, Ushkatyn I, II, and III, deposits, etc.).

In structural respects, the Atasu ore region is located in a transitional sub-platform region, characterized during the period of accumulation of the upper Famennian chert-limestone ore-bearing sediments by a relatively stable tectonic regime.

Numerous deposits and ore-shows of iron and manganese ores have been restricted to the limbs of the complexly constructed Dzhail'ma synclinorium, the asymmetrical structure of which was affected by the Ustanynzhol granitoid intrusion which was injected during the Middle Devonian into Ordovician rocks. The synclinorium extended for 180 km, with a width of 30 - 40 km.   In its eastern part, on the northern margin, is the complex Karazhal polymetallic and iron-manganese deposit, the largest in Kazakhstan, consisting of marine and continental sediments of the Upper Devonian and Lower Carboniferous.   The rocks have been complicated by folding of higher orders, and have been broken by a system of fractures, dipping towards the centre of the trough at 45 - 50°,

The deposits of the Karazhal ore field have been studied by many investigators (Sapozhnikov, 1963; Kalinin, 1963; Rozhnov, 1967; Novokhatsky, 1967; etc.), who have recognized here two productive suites in the manganiferous chert-limestone association (thickness up to 1.5 - 2 km), carrying iron, manganese, and lead-zinc mineralization (Fig. 70).

The rocks of the association are underlain by pre-Famennian Devonian basic, intermediate, and acid eruptives and their tuffs, and terrigenous sediments.

Fig. 70.
Diagram of structure of the manganiferous chert-limestone association of the Karazhal type (After E. Gribov and Yu. Khodak).

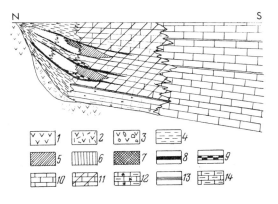

1) intermediate eruptives; 2) acid eruptives; 3) Frasnian conglomerates, sandstones, and tuffs; 4) clay-chert slates; 5-9) ores: 5) hematitic, 6) magnetitic, 7) sideritic and magnetite-sideritic, 8) oxide (braunite, hausmannite, and jacobsite), 9) carbonate (manganocalcite); 10) limestones; 11) hematitic limestones with layers of hematitic ores; 12) limestones with lead-zinc mineralization; 13) clay slates; 14) argillaceous limestones.

The lower Karazhal (Nizhnekarazhal'sk) Group consists mainly of clay-
-chert carbonate sediments.   In its sequence in West Karazhal, the following
lithological varieties of rocks have been distinguished (in upward succession);

1. Carbonaceous-chert conglomerate-like limestones with seams of
   clay-chert slates, with thin seams of magnetite and manganese
   carbonate ores, and a segregation of galena and sphalerite .. 3.5 - 6 m.

2. Carbonaceous-chert limestones, and carbonate-chert rocks with
   seams of argillite and a conglomerate-like limestone .. .. .. 25 - 30 m.

3. Mottled limestones, cherty limestones, sometimes ferruginous-
   -chert rocks, and chlorite-magnetite ores, associated with
   jaspers .. .. 15 - 18 m.

4. Dark-grey and black carbonaceous-chert limestones, carbonate-
   -chert rocks with grey chert nodules, seams of wavy-bedded limestones, and stringers of tuffogenic material   .. .. 25 - 30 m.

5. Mottled (grey, greenish, and violet) chlorite-bearing limestones
   with rare seams of tuffites, and ferruginous and manganiferous
   limestones .. .. 5 - 7 m.

The upper Karazhal (Verkhnekarazhal'sk) Group consists of mottled chert-
-carbonate rocks, and limestones, including layers and lenses of ferruginous and
manganese ores, and jaspers.   The sequence is as follows (in upward succession):

1. At the base is an horizon, consisting of ferruginous-chert and
   manganous limestones, carbonate-chert rocks, and mottled limestones.
   Individual stringers consist of braunite and hausmannite .. ..   12 m.

2. Manganese-iron horizon consisting of an alternation of ore strata,
   the number of which reaches 20, and braunite, braunite-hausmannite,
   hematite, and less frequently magnetite and jacobsite ores, ferruginous jaspers, and manganiferous limestones .. ..   50 m.

3. Horizon of thinly-bedded manganese-iron cherty limestones, limestone-
   -chert rocks, siliceous schists, and jaspers (ore minerals consist of
   hematite, braunite, and rarely hausmannite .. ..   14 m.

4. Horizon, in which there is an alternation of limestone-chert rocks,
   cherty limestones, ferruginous jaspers, and siderite-chlorite-
   -magnetite and hematite ores.

The segregations of iron and manganese ores are restricted to different
stratigraphical levels.   The largest of them are located mainly in the middle
part of the red limestones of the upper Karazhal Group and in their lowermost
horizon.   Small lens-like bodies of iron and manganese ores occur also in the
chert-carbonate rocks of the lower Karazhal Group.

The Karazhal ore field is oriented east-west and can be traced for 12
km.   It consists of five large sectors, representing independent segregations
of iron and manganese ores.   The ore member consists of a complex alternation of
layers, seams, and lenses of chert-carbonate rocks, jaspers, hematitic, hematite-

-magnetitic, braunitic, hausmannite-braunitic, and manganocalcite ores, and manganous limestones.   It extends for 5 km with a width of 600 - 700 m and more.   The thickness of the layers of iron ores is 40 - 50 m, and of the manganese ores, 10 m.

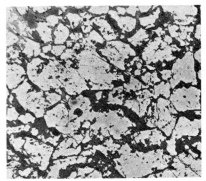

Fig. 71.
Cataclased hausmannite ore.
Cracks filled with calcite (black).
Karazhal deposit.   Polished
section.   Magnification X 70.
(After Gribov, 1966).

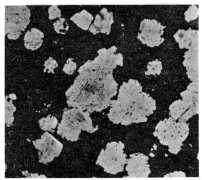

Fig. 72.
Reniform segregations of braunite
(light-grey) amongst chert-
-limestone rock (dark-grey).
Karazhal deposit.   Polished
section.   Magnification X 120.
(After Gribov, 1966).

Of commercial importance in the deposit are the ores formed under conditions of regional metamorphism (braunitic, hausmannite-braunitic, hematitic, and magnetite-hematitic) and oxidized ores, developed in the weathering crust (psilomelane and psilomelane-vernadite).

The fabric of the primary sedimentation-diagenetic ores, recrystallized under conditions of metamorphism are usually bedded: finely-bedded, lens-like, and banded.   Their textures are crystallogranular, crystalloblastic, and cataclastic (Figs 71, 72).   The fabrics of the residual ores of the weathering crust are: collomorphic, sintery-conchoidal, concentrically-zoned, etc.; their textures are cryptocrystalline, gel-like, etc.

The best quality ores are the psilomelane types.   They contain the following components (in wt %): Mn, more than 50; Fe, 0.8; P, 0.03; S, 0.28; SiO_2, 2.32; Al_2O_3, 2.14; CaO, 1.12; MgO, 0.09; BaO, 0.44; CO_2, 0.28; H_2O, 4.

In the semi-oxidized hausmannite-braunite ores there is approximately 40% of manganese; Mn + Fe, 45 - 50%; SiO_2, not more than 10%; Al_2O_3 + SiO_2, 10 - 12%; CaO + MgO, 5 - 6%; S, tenths of a percent; P, hundredths of a percent.

In regard to the origin of the Karazhal and other deposits of the Atasu region, opinions have been expressed that the ingress of the manganese and iron-manganese solutions into the sedimentary basin took place as a result of eruptive explosions in association with volcanic activity and was achieved along a system of faults in the basement of the Dzhail'ma synclinorium.   Individual investigators consider the principal source of the manganese to have been the weathering crust of Ordovocian and Lower and Middle Devonian eruptive--sedimentary rocks.

The reserves of manganese and iron-manganese ores in the deposits of the Dzhail'ma synclinorium comprise about 100 million tonnes, with an average amount of manganese in the primary braunite and hausmannite-braunite ores of 30%, in the semi-oxidized braunite-psilomelane ores, about 25%, and in the oxidized iron-manganese ores, 19%.

The Takhta-Karacha Deposit

The chert-limestone association of the Takhta-Karacha type, assigned in age to the early-middle Ludlovian, can be traced in a sublatitudinal direction in the western part of the Zeravshan Range for 250 km.   It includes small deposits and ore-shows of manganese and iron-manganese ores: the Takhta-Karacha, Dautash, Kzyl-Bairak, Ziaétda, etc.

The Takhta-Karacha deposit is restricted to an isoclinal synclinal fold with a normal northern and an overturned southern limb.   The limbs of the fold consist of sand-shale sediments of the Lower Silurian, with Upper Silurian carbonate rocks outcropping in the trough.   The dips vary from 55 to 80°.

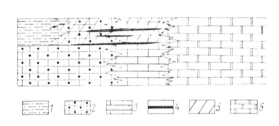

Fig. 73.
Diagram of the structure of the manganiferous chert-limestone association of the Takhta-Karacha type.   After Gribov (1972).

1) siliceous slates;   2) chert-limestone rocks;   3) limestones;   4) manganese carbonate rocks;   5) manganese-bearing rocks;   6) dolomites.

In the construction of the manganiferous association developed in the Takhta-Karacha deposit, a sequential lateral replacement of the chert and chert-limestone rocks with inclusions of layers of manganese carbonate ores and manganous limestones by limestones and dolomites has been recorded (Fig. 73).

The sequence of the association, according to Sokolova (1963) is as follows (in upward succession):

1. On the sand-shale sequence of the Lower Silurian rest grey, fine-grained, platy limestones, with calcareous shales sometimes at the base of the sequence .. .. 200 - 250 m.

2. White, fine-grained, massive limestones .. .. 30 - 40 m.

3. Greenish-grey sheared eruptives and their tuffs .. 200 - 220 m.

4. Light-coloured crystalline limestones .. .. 80 - 90 m.

5. Manganese-ore member of cherty, dark-grey and black platy, finely banded and massive limestones with layers of rhodochrosite and manganocalcite ores, and manganous limestones;   thicknesses of two exposed seams of manganese carbonate ores vary.. from 6 to 8 m

6. Marmorized, light-coloured, massive limestones .. .. up to 500 m.

In the sub-ore and ore-bearing sequences, there are seams of dolomites and dolomitic rocks, and pyroclastic material plays a markedly subordinate role (Gribov, 1972).

Carbonates of manganese and iron participate in the construction of the primary manganese and iron-manganese ores, for example, rhodochrosite, calcic and ferruginous rhodochrosite, manganocalcite, manganous calcite, and siderite. The minor minerals include pyrite, marcasite, and chalcopyrite.

A manganese gossan has been formed on the manganese carbonate ores under the conditions of surface weathering.   The residual oxidized ores consist mainly of vernadite and concentrically-zoned and collomorphic structure, and also minerals of the psilomelane group, and pyrolusite (ramsdellite).   The metamorphic minerals include spessartine.

The layers of rhodochrosite and rhodochrosite-manganocalcite ores contain the following components (in wt %): Mn, up to $14 - 16$; Al_2O_3, $0.2 - 0.7$; Fe_2O_3, traces; FeO, $2 - 4$; CaO, $16 - 20$; MgO, $1 - 3$; insoluble residue, $28 - 40$; CO_2, $22 - 28$; C_{org}, up to 1.4; P, hundredths of a percent.

Some investigators consider that the ore concentrations of the Takhta-Karacha deposit are of volcanogenic-sedimentary origin (Sokolova, 1963). Others (Gribov, 1972) give preference to the chemogenic-sedimentary theory of formation of the chert-carbonate rocks and manganese ores, associating the ingress of manganese with the weathering of the rocks of an island land area.

THE VOLCANOGENIC-SEDIMENTARY ASSOCIATIONS

The group of manganiferous volcanogenic-sedimentary associations has been subdivided on the basis of composition into an association of the greenstone series, associated with spilite-keratophyre volcanism, and an association of the porphyry series, related to volcanism of trachy-rhyolite composition (Shatsky, 1954).   The spilite-keratophyre-chert manganiferous association has been formed mainly in the eugeosynclinal zones, and the porphyric association, in the miogeosynclinal and only partly in the eugeosynclinal zones.

The spilite-keratophyre-chert manganiferous association mainly includes lean silicate, sometimes carbonate, and rarely oxide manganese ores.   The association of this type is characterized by increased silica content in the sediments.   They usually contain copper, nickel, cobalt, vanadium, and chromium; their concentrations are substantially greater than the clarke values.   Greenstone volcanism is associated with a significant group of small South Uralian manganese deposits and ore-shows, located amongst Devonian jaspers and cherty slates, which form the western edge of the Magnitogorsk Synclinorium.

The porphyry-chert association is distinguished by the significant distribution of oxide manganese ores (pyrolusite-psilomelane, braunite, and hausmannite-braunite).   There are often high clarke concentrations of lead, zinc, barium, copper, and other elements in the oxide ores and their country rocks.   An example of a volcanogenic-sedimentary deposit of the porphyry series is the Durnovo, located in Salair.

THE SPILITE-KERATOPHYRE-CHERT ASSOCIATION

The Near-Magnitogorsk Group of Deposits

The spilite-keratophyre-chert manganiferous association, developed on the eastern slope of the South Urals, forms the western edge of the Magnitogorsk Synclinorium and can be traced in a submeridional direction for almost 300 km. It includes: 1) the Irendyk Group of the Lower Devonian (diabase-porphyrite tuff-chert sequence, including the Urazovsk manganiferous horizon); 2) the Karamalytash Group of Eifelian (Middle Devonian) age (spilite-keratophyre sequence, with the Bikkulovsk manganiferous horizon, occurring amongst tuffs, and pyroxene-plagioclase porphyrites and albitophyres); 3) the Ulutau Group of Givetian (Middle Devonian) age (andesite-dacite sequence, with the most productive Bugulugyr chert-jasper manganiferous horizon). The association is 5 m thick. All the sequences have been strongly deformed, thrown into steep folds, and broken by a system of fractures. Large regional, pre-Riphean faults of deep--seated occurrence have also been recognized, which have exerted a considerable influence on the development of the Devonian eugeosynclinal marine basin and its volcanism.

The manganese bodies of all the ore-bearing horizons are genetically associated with cherty rocks, grey and red jaspers, jasperoid quartzites, and jasperoid slates, occurring amongst basic and intermediate eruptives and their tuffs. The thickness of the horizons varies from a few metres up to 80 - 130 m (the Bugulugyr horizon in the regions of the Gubaidullinsk and Mamilinsk deposits).

The shape of the ore segregations is layer- and lens-like. Being followed for 300 - 500 m and more (the Mamilinsk and Kusimovsk deposits), the ore layer often pinches out, forming a chain of interrupted lenses, sometimes flat, sometimes thickening in the central parts. The ore layers (lenses) rest conformably with the stratification of the country rocks, repeating their complex structure (Fig. 74). The thickness of the ore seams and layers varies from a few centimetres up to 4 - 5 m, comprising on average in the deposits restricted to the Bugulugyr horizon, 1 - 2 m. The bedding of the primary manganese ores is the result of interlayering of minute braunite bands (1 - 3 cm) with non-ore cherty jasperoid seams (2 - 5 cm).

The mineral composition of the ores depends on the intensity of the regional and contact metamorphism. Along with the widely distributed silicate manganese ores, consisting mainly of rhodonite and bustamite, and less frequently tephroite, pyroxmangite, piémontite, and severginite (Kozhaevo deposit), there are often braunite and hausmannite-braunite ores (Faizulinsk and other deposits), and only occasionally carbonate ores. Along the strike and dip of the ore strata, facies transitions of oxide and silicate manganese ores into carbonate types are observed (Yalimbetovsk deposit).

Manganese gossans have been developed on all the deposits. The oxidized ores, formed in the uppermost zone of the weathering crust after silicate and carbonate manganese ores, are the richest (Mn from 35 up to 50%, SiO_2, up to 5%, and P in hundredths of a percent) and consist of manganese oxide and hydroxide minerals (psilomelane, pyrolusite, and vernadite). Amongst them there are friable (sooty, and powdery) and weakly compacted (porous, cellular, and box types) varieties. These latter, distinguished by their lumpy nature, may be used for the production of standard grades of ferromanganese.

The quality of the silicate ores is high: Mn, 15 - 25% and less; SiO_2, 45 - 52%; CaO, 5 - 7%; and P, 0.02 - 0.05%.   In the monomineralic seams, consisting of rhodonite, tephroite, pyroxmanganite, and piémontite, the amount of manganese increases up to 30 - 35%.

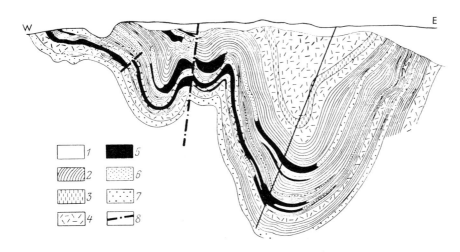

Fig. 74. Geological section through the Kusimovo deposit (After Betekhtin, 1946).

1) drift; 2) jaspers and cherty rocks; 3) tuffites; 4) tuffs;
5) high-grade ores; 6) low-grade ores; 7) mineralized tuffs;
8) faults.

The chemical composition of the braunite and psilomelane-braunite semi-oxidized ores (in wt %) is distinguished by its variability and depends on their depletion in cherty and jasperoid layers: Mn, 31.6 - 45 (in the assessed blocks, it does not exceed 37 - 33), SiO_2, 11.5 - 45; Fe_2O_3, 1.5 - 7.5; and P, hundredths of a percent.

Almost all the rich manganese-ore gossans located near Magnitogorsk were worked out during World War II.   The amounts of V, Co, Ni, Cr, Pb, Zn, Be, Ga, Mo, and W in the primary oxide and silicate ores do not exceed the clarke values.

Most investigators consider that this type of manganiferous deposit has been of volcanogenic-sedimentary metamorphosed origin (Kheraskov, 1951; Nestoyanova, 1963; Gavrilov, 1972; etc.).

The total reserves of semi-oxidized and primary sedimentary metamorphosed ores (mainly silicate) have been estimated at 3 million tonnes.   The separation of the ore segregations, their small dimensions, and the low quality of the ores are an obstacle to the commercial exploitation of the deposits.

The Klevakinsk Deposit

The Middle Devonian Klevakinsk deposit of manganese silicate and carbonate ores occurs in the Middle Urals in the Kamensk region.   It is located within the Alapaevo-Techen megasynclinorium, in its central zone, and is characterized by the development of lavas and tuffs of andesite-basalt composition, and cherty rocks.

Two ore bodies have been identified in the deposit, occurring in Givetian clay-chert slates.   The major body (lower), occurring at the base of a clay-shale horizon, can be traced along strike for 160 m and down dip for several tens of metres.   It consists of silicate-carbonate manganese ores (oligonite-rhodochrosite, sometimes with the addition of manganocalcite, rhodonite, and spessartine), oxidized in the roof of the ore segregation and in its surface outcrops.   The ore layer has the shape of a compressed lens with its margins disrupted by tectonic disturbances.   Its thickness in the thickened portion reached 9 m, and it decreases with depth.   The dips of the ore layer do not exceed 50°.   The amount of manganese in the ores varies from 8.7 up to 44.2%.

The Ir-Nimiisk Deposit

The Lower Cambrian Ir-Nimiisk manganese deposit occurs in the Uda--Shantar ore region of the Okhotsk area.   In structural respects, it is located on the northeastern margin of the Mongolo-Okhotsk fold district (Shantar Anticlinorium).

The rocks of the Lower Cambrian volcanogenic-sedimentary manganiferous association recognized here consist mainly of chert formations (jasperoid quartzites, jaspers, and chert-clay slates), interstratified with eruptives of basic composition (diabases, diabase porphyrites, and spilites) and their tuffs, tuff--breccias, and less frequently, limestones, sandstones, and siltstones.   The manganese ores have been localized in cherty sediments (jaspers, and clay-chert slates).   Frequent interlayering of seams of braunite ores with non-ore jasperoid quartzites have been observed.   On the basis of structural features, amongst the braunite ores, we have been able to recognize massive and fine-grained types with a manganese content of from 32 up to 55%, and lensoid-bedded types, in which the quantity of manganese does not normally exceed 28 - 30%.   Sampling data indicate that the manganese ores are of quite high quality, enabling us to place them in the low-phosphorus (P, 0.04 - 0.06%) and low-iron (Fe, up to 1.5 - - 2.0%) varieties.   The ore bodies can be traced along strike for hundreds of metres, with a thickness of from 1 - 2 up to 5 - 10 m and more.

Less widespread in the deposit are the carbonate (rhodochrosite and manganocalcite) ores, spatially separated at the contact between the cherty and limestone layers.   The amount of manganese in them varies from 22 up to 46%.

The reserves of ores in the Ir-Nimiisk deposit and the entire Uda--Shantar manganiferous region, allowing for the considerable extent (over 100 km) of the Lower Cambrian volcanogenic-sedimentary ore-bearing sediments, are estimated at 100 million tonnes.

THE PORPHYRY-CHERT ASSOCIATION

The Durnovsk Deposit

The porphyry-chert association, including the Durnovsk deposit of manganese ores, is located in Northeastern Salair. It crops out in the core of the Ura-Bachat anticline in the zone of marked narrowing of the folded sedimentary and volcanogenic-sedimentary rocks of the Glavrilovsk and Pecherka groups of the Lower-Middle Cambrian, which have been complicated by numerous disjunctive fractures, giving this sector a fault-block structure.

The manganese-bearing member proper of this association includes weakly ferruginous and manganous quartzites, often of jasperoid aspect, interstratified with layers and intercalations of braunite ores, and eruptive quartz porphyries (Fig. 75).

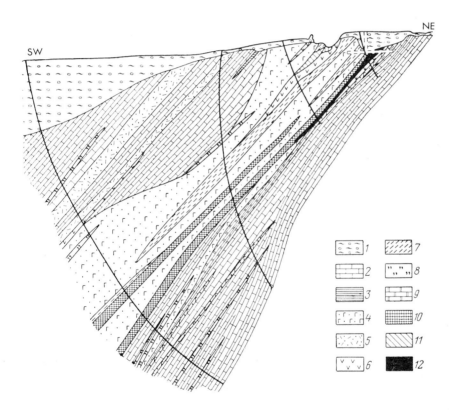

Fig. 75. Geological section through the Durnovsk deposit (After Suslov, 1964).
1) clays and pebble-beds; 2) limestones; 3) sericite schists;
4) quartz porphyries; 5) tuffs and tuffites; 6) diabases; 7) jaspers; 8) quartz veins; 9) carbonate ores; 10) braunite ores;
11) hematite ores; 12) psilomelane ores.

The rocks of the ore-bearing member are overlain without visible unconformity by a sequence of limestones, amongst which are distributed, layer by layer, horizons of pyroclastic formations, consisting of acid tuffs, tuff-sand-

stones, tuff-shales, and tuff-limestones. The fabric of the volcanics is sheared, and their texture is blastopsammitic; there is a predominance of quartz in
the clasts, along with felsites, volcanic glass, ash particles, and relicts of
altered plagioclase grains. Hematitization is almost ubiquitous.

Amongst the sub-ore rocks there is a predominance of light-coloured,
bedded, and fine-crystalline limestones. The transition of the manganese-ore
member into the underlying limestones is gradual, through thin quartzite, jasperoid, cherty, and chert-limestone layers.

Two principal types of commercial manganese ores have been recognized:
1) primary braunite ores of massive, bedded, and brecciated structure, forming
all the manganese-ore bodies; 2) secondary psilomelane ores, which are the products of alteration of braunite ores under hypergene conditions and consist of
sintery-conchoidal, reniform, and crust formations.

The ore member of the deposit is not a single manganese-ore body. It
consists of individual ore segregations of layered and lensoid shape, interstratified with weakly ferruginized and manganesed quartzites and porphyroids. Its
thickness is about 40 - 50 m. The ore bodies can be traced from southeast to
northwest for 300 m, with dips of 45 - 50°.

In this complexly constructed member, nine layers of lensoid braunnite
bodies may clearly be recognized. They are characterized by an uneven spatial
development: sometimes they thin out rapidly, and sometimes they may be traced
for up to 180 - 200 m along strike and for more than 150 m down dip with an
average thickness of about 4 m. The thickness of the ores in the central parts
of the lenses varies from 4 to 5 m.

In the braunite ores there is (in wt %): Mn, 6.5 - 26.5; Fe, 2.2 -
- 13.3; P, 0.01 - 0.11; S, 0.09 - 0.17; SiO_2, 15.5 - 48.6; and in clarke
concentrations: Cu, on average 2.7; Pb, 48; Zn, 36; As, 60; Ba, 33; Ag,
150 - 200; and Sb, up to 1000 and more.

The reserves of the primary oxide (braunite) and oxidized psilomelane
ores of the deposit are estimated at 0.6 million tonnes with an average manganese
content of about 20%.

THE FERRUGINOUS-CHERT (JASPILITE) ASSOCIATION

The manganiferous ferruginous-chert association is laterally associated
with the extraordinarily large ferruginous-chert (jaspilite) associations proper.
It is characterized by a clear paragenesis of manganese and iron oxide ores
(braunite, hausmannite, hematite, and magnetite) and cherty rocks.

This association has been quite poorly studied. A representative
example in the USSR is the Riphean manganiferous ferruginous-chert association of
the Malyi Khingan and a number of other deposits. However, in spite of the relatively modest scale of ore occurrence in this association in the USSR, in other
regions of the world it is associated with comparatively large manganese deposits
in Brazil (the provinces of Minas Gerais, Matu Grosu, and Bahia), South Africa
(Postmasburg and Kalahari regions), etc.

The Malyi Khingan Deposits

An entire series of deposits, characterized by a mixed iron-manganese mineralization (Poperechnoe, Okhrinsk, Serpukhovo, etc.), is restricted to the Riphean ferruginous-chert manganiferous association, distributed in the southern portion of the Malyi Khingan in the Soviet Far East.   The manganiferous association occurs in an Upper Proterozoic sequence, forming the southern part of the Khingan-Bureya uplift, and consists of an assemblage of strongly deformed (folded and broken by disjunctive faults), sedimentary-metamorphic rocks, traceable from the Amur River for tens of kilometres towards Kimkan railway station.

In the sequence of the association, which stratigraphically corresponds to the concept of the 'ore-bearing group', sub-ore, ore, and supra-ore members have been recognized, with an overall thickness of 300 - 400 m.

The sub-ore, cherty-clay slate member, with seams and lenses of sandstones, dolomites, and dolomitic breccias, rests without visible unconformity on the dolomitic deposits of the Murandavsk Group.

The ore member, 30 - 35 m thick, consists of cherty rocks (cherty slates, jasperoid quartzites, and manganous and ferruginous quartzites), retaining the original finely-bedded fabric.   In the lower part of the member there are predominantly cherty manganese ores, and in the upper part, iron ores.   The ore member is underlain by a seam of dolomitic breccias, from 1 - 2 up to 30 m thick.

The supra-ore, slaty, argillaceous, and carbonaceous-argillaceous member includes seams of dolomites and dolomitized limestones.   The manganiferous association is overlain by a sequence of bituminous limestones and carbonaceous-clay slates of the Londokovsk Group.

In the Poperechnoe, Serpukhovo, and Stolbukha deposits at a depth of 200 - 400 m, there are lateral transitions from the oxide manganese ores (braunite, braunite-hausmannite, and braunite-hematite) into oxide-carbonate and carbonate types.   The seam of manganese ores can be traced along strike for several hundreds of metres (Ignat'eva, 1962).   In the western ore-bearing belt, the thickness of the iron-manganese layer is 8 - 9 m.   In the central and eastern ore belts, it almost completely thins out.

Amongst the primary-oxide ores there are: braunite, hausmannite--braunite, and hematite-braunite types.   In the carbonate ores, the principal ore-forming minerals are rhodochrosite and oligonite.   Near the granitoid massifs there are silicate varieties of ores: rhodonite, containing bustamite, tephroite, spessartine, and piémontite in varying amounts.   Pyrolusite-psilomelane ores are developed in the weathering crust.

The amount of manganese in the primary oxide ores which are the most widely distributed in the deposits varies within wide limits, reaching 35% and more.   The average amount of manganese, owing to the high silica content of the ores, falls to 21%, but may be increased by means of enrichment procedures. Their reserves are estimated at 10 million tonnes.

The manganese and iron-manganese ores of the Malyi Khingan belong to the primary-sedimentary formations, recrystallized under conditions of regional and contact metamorphism.   The formation of the ferruginous-chert manganiferous association took place in Riphean-Cambrian, meridionally-oriented, narrow down-warps on a rigid (median massif type) Proterozoic substrate.   The source of the silica, manganese, and iron could have been the nearby weathering sectors of the landmass.   Portion of the ore-bearing solutions together with the silica could have been introduced along zones of deep-seated faults from a magmatic focus.

This same association includes the Bidzhan and other deposits in the north of the Malyi Khingan and the Khankai massif.

DEPOSITS OF THE WEATHERING CRUST

Manganese-ore weathering crusts are widely developed on many continents; especially large accumulations of rich manganese ores have been formed on the ancient platforms and shields, located in the tropical equatorial zone (the manganese deposits of India, Gabon, Ghana, South Africa, Brazil, etc.).   The principal amount of extractable commercial manganese ores of foreign countries is associated with this type of deposit;   the proven reserves of manganese ores of the weathering crust comprise from 85 to 90% of all the investigated reserves of foreign countries.

Small deposits of the weathering crust are also known in the Soviet Union.   The proven reserves of oxidized manganese ores, forming under weathering-crust conditions, comprise about 2% of the total of all known reserves in our country.   Oxidized manganese ores are mined at the present time in the USSR only in passing during the working-out of other types of manganese ores.

CONTACT-METASOMATIC AND HYDROTHERMAL DEPOSITS

The post-magmatic contact-metasomatic and hydrothermal deposits, known in the Soviet Union, include insignificantly small reserves of manganese ores and constitute practically no commercial value.   On the basis of physicochemical conditions of mineral formation and the spatial location of these hypogene ore occurrences, two types of deposit are recognized: contact-metasomatic, with predominant development of small stock, nest and lensoid forms of segregation, and veined, with complexly changing outlines of the ore occurrences.

The contact-metasomatic deposits have usually been formed in the zone of contact between the manganiferous carbonate rocks (limestones, dolomites, marbles, and limestone-chert-clay slates) and intrusions of basic, intermediate, and less frequently acid composition.   The contact-metasomatic ores are known in the Levoberezhnyi sector of the Ikat-Garga Upper Proterozoic deposit (Transbaikalia), where they are restricted to the contacts of Caledonian granites. The primary chert-carbonate manganese ores (mainly manganocalcite), containing up to 15% of manganese, have been converted in the contact zone to quartz-rhodonite-bustamite ores (with a manganese pyroxene), and form layer-lenses, nest--like, and veinlike ore segregations.   The thickness of such lenses in the central part is 30 m, with an extent of up to 1 - 2 km.   They lie conformably with the surrounding manganiferous chert-carbonate-slate rocks.   In the direction of the granite massif, silicate mineralization in the ores increases up to the complete disappearance of manganocalcite.   The content of manganese in the uniform ores varies from 20.2 up to 24.5%.

The formation of the vein deposits is associated with the circulation of post-magmatic hydrothermal solutions, which, passing through the sedimentary limestone, limestone-chert-clay, and cherty rocks with their increased manganese content relative to the clarke values, may partially dissolve and redeposit the manganese in the form of nests and vein-like bodies.   Numerous both small and also significantly large quartz veins are known, with coarsely crystalline segregations of pyrolusite, braunite, hausmannite, jacobsite, and rhodochrosite (the Tsedis and Kodman Upper Cretaceous manganese deposits of Georgia).

The hydrothermal (Betekhtin, 1946) manganese deposits and ore-shows in the Tagil-Kushva region (the Sapal, Lipovsk, Kazan, and others), located 100 - 120 km to the north of Sverdlovsk on the eastern slope of the Urals, are genetically associated with Silurian-Lower Devonian volcanogenic-sedimentary deposits of the Tura Group.   Hydrothermal manganese mineralization is known in the Karazhal deposit (friedelite, rhodochrosite, and manganese-bearing calcite), and in the Dautash manganse deposit, located in the Zeravshan Range (hauerite and spessartine), and in a number of others.

REFERENCES

ANDRUSHCHENKO P.F.   (АНДРУЩЕНКО П.Ф.), 1954: Минералогия марганцевых руд Полуночного месторождения (The mineralogy of the manganese ores of the Polunochnoe deposit). *Trudy geol. Inst. Mosk., 150.*

AVALIANI G.A. (АВАЛИАНИ Г.А.), 1964: Марганценосность Кавказа (The manganese occurrences of the Caucasus), *in* "Вопросы геологии Грузии" (*'Problems of the Geology of Georgia'*), pp. 447-456.   Metsniereba Press, Tbilisi.

AVALIANI G.A. (АВАЛИАНИ Г.А.), 1967: Генетические типы марганцевых месторождений Грузии (Genetic types of manganese deposits of Georgia) *in* "Марганцевые месторождения СССР" (*'The Manganese Deposits of the USSR'*), pp. 225-237.   Nauka Press, Moscow.

BETEKHTIN A.G.   (БЕТЕХТИН А.Г.), 1946: Промышленные марганцевые руды СССР *(The Commercial Manganese Ores of the USSR).*   Akad. Nauk SSSR Press, Moscow-Leningrad.

BETEKHTIN A.G. (Ed.) (БЕТЕХТИН А.Г. (ред.)), 1964a: Чиатурское месторождение марганца *(The Chiatur Manganese Deposit).*   Nedra Press, Moscow.

BETEKHTIN A.G. (Ed.) (БЕТЕХТИН А.Г. (ред.)), 1964b: Никопольский марганцерудный бассейн *(The Nikopol Manganese-Ore Basin).*   Nedra Press, Moscow.

BONDARCHUK V.G. (Ed.) (БОНДАРЧУК В.Г. (отв. ред.)), 1960: Атлас Палеогеографічних карт Українськоі і Молдавськоі РСР (з елементами літофацій) *(Atlas of Paleogeographical Maps of the Ukrainian and Moldavian SSR (with elements of lithofacies).*   Akad. Nauk Ukr. SSR Press, Kiev.

CHAIKOVSKY V.K., RAKHMANOV V.P. & KHODAK Yu.A. (ЧАЙКОВСКИЙ В.К., РАХМАНОВ В.П., ХОДАК Ю.А.), 1972: Принципы составления прогнозно-металлогенических марганценосных формаций *(The Principles of Compiling Prediction-Metallogenic Maps of the Manganiferous Associations').*   Nedra Press, Moscow.

DANILOV I.S. (ДАНИЛОВ И.С.), 1971a: Вещественный состав и геохимия пиролюзито-псиломелановых окисленных руд Никопольского месторождения (The composition and geochemistry of the pyrolusite-psilomelane oxidized ores of the Nikopol deposit). *Sb. nauch. Trud. nauchno-issled. Inst. Geol. Dnepropetrovsk* vyp. 4, pp.62–72.

DANILOV I.S. (ДАНИЛОВ И.С.), 1971b: Происхождение рудной зональности Никопольского марганцевого месторождения (The origin of the ore zonation in the Nikopol manganese deposit). *Byull. Mosk. Obshch. Ispyt. Prir., 46*, No.6, pp.131–132.

DZOTSENIDZE G.S. (ДЗОЦЕНИДЗЕ Г.С.), 1965: К вопросу о генизисе Чиатурского месторождения марганца (The Problem of the origin of the Chiatur manganese deposit). *Litologiya polezn. iskop.*, No.1, pp.3–17.

FLEGONTOVA E.I. (ФЛЕГОНТОВА Е.И.), 1969: Микроэлементы в нефтях (Microelements in petroleum). *Trudy vses. nauchno-issled. geol.-razv. Inst., 279*, geokhim. Sb. No.10).

FOMINYKH V.G., SAMOILOV P.I., MAKSIMOV G.S. & MAKAROV V.A. (ФОМИНЫХ В.Г., САМОЙЛОВ П.И., МАКСИМОВ Г.С., МАКАРОВ В.А.), 1967: Пироксениты Качканара *(The Pyroxenites of Kachkanar).* Ural'sk Fil. Akad. Nauk Press, Sverdlovsk.

GAVRILOV A.A. (ГАВРИЛОВ А.А.), 1972: Эксгаляционно-осадочное рудонакопление марганца (на примере Урала и Казахстана) *(Exhalation-Sedimentary Ore Deposition of Manganese (as Exemplified by the Urals and Kazakhstan)).* Nedra Press, Moscow.

GRIBOV E.M. (ГРИБОВ Е.М.), 1966: Петрографические типы руд каражальского месторождения (Petrographical types of ores of the Karazhal deposit), *in* "Марганцевые и железорудные концентрации Джаильминской мульды" *('Manganese and Iron-Ore Concentrations in the Dzhail'minsk Trough')*, pp.34–48. Nauka Press, Moscow.

GRIBOV E.M. (ГРИБОВ Е.М.), 1972: Улутелякское марганцевое месторождение (Башкирское Приуралье) (The Ulutelyak manganese deposit (Bashkirian Urals)). *Geologiya rudn. Mestorozh., 14*, No.6, pp.95–101.

GRYAZNOV V.I. (ГРЯЗНОВ В.И.), 1960: Закономерности размещения марганцевых руд на территории Украинской ССР (Distribution patterns of manganese ores in the Ukrainian SSR), *in* "Закономерности размещения месторождений в платформенных чехлах" *('Distribution Patterns of Deposit in Platform Covers')*, pp.160–172. Akad. Nauk Ukr. SSR Press, Kiev.

GRYAZNOV V.I. (ГРЯЗНОВ В.И.), 1971: Закономерности размещения высококачественных марганцевых руд на Никопольском месторождении (Distribution patterns of high-grade manganese ores in the Nikopol deposit). *Sb. nauch. Trud. nauchno-issled. Inst. Geol. Dnepropetrovsk*, vyp.4, pp.42–53.

GRYAZNOV V.I. & SELIN Yu.I. (ГРЯЗНОВ В.И., СЕЛИН Ю.И.), 1959: Основные черты геологии Больше-Токмакского марганцевого месторождеиия (УССР). (The principal features of the geology of the Bol'she-Tokmak manganese deposit (Ukr. SSR)). *Geologiya rudn. Mestorozh.*, No.1, pp.35–55.

IGNAT'EVA L.A.   (ИГНАТЬЕВА Л.А.), 1962:   Геологические условия формирования пород рудоносного комплекса малого хингана   (The geological conditions of formation of the rocks of the ore-bearing complex of the Malyi Khingan), *in* "Очерки по металлогении осадочных и осадочно-метаморфический пород" (*'Reviews on the Metallogenesis of Sedimentary and Sedimentary Metamorphic Rocks'*), pp.212-220.   Akad. Nauk SSSR Press, Moscow.

IKOSHVILI D.V.   (ИКОШВИЛИ Д.В.), 1971:   К литологии олигоценовых отложений Чиатурского марганцевого месторождения (The lithology of the Oligocene sediments of the Chiatur manganese deposit).   *Trudy kavkaz. Inst. miner. Syr'ya,* vyp.9, pp.189-193.

KALINENKO V.V., SHUMIKHINA I.V. & GUSAREVA A.I.   (КАЛИНЕНКО В.В., ШУМИХИНА И.В., ГУСАРЕВА А.И.), 1967:   Марганценосные отложения Лабинского месторождения и распределение в них ванадия, хрома, никеля, кобальта и меди (The manganiferous sediments of the Laba deposit and the distribution of vanadium, chromium, nickel, cobalt, and copper in them), pp.242-256.   Nauka Press, Moscow.

KALININ V.V.   (КАЛИНИН В.В.), 1963:   Железо-марганцевые руды месторождения Каражал *(The Karazhal Iron-Manganese-Ore Deposits).*   Akad. Nauk SSSR Press, Moscow.

KHERASKOV N.P. (ХЕРАСКОВ Н.П.), 1951:   Геология и генезис восточно-башкирских марганцевых месторождений (The geology and origin of the East Bashkirian manganese deposits), *in* "Вопросы литологии и стратиграфии СССР.   Памяти академика А.Д. Архангельского" (*'Problems of the Lithology and Stratigraphy of the USSR.   Memorial to Academician A.D. Arkhangel'sky').*   Akad. Nauk SSSR Press, Moscow.

KHODAK Yu.A., RAKHMANOV V.P. & YEROSHCHEV-SHAK V.A.   (ХОДАК Ю.А., РАХМАНОВ В.П., ЕРОЩЕВ-ШАК В.А.), 1966:   Месторождения марганца Кузнецкого Алатау *(The Manganese Deposits of the Kuznets Alatau).*   Nauka Press, Moscow.

LALIEV A.G.   (ЛАЛИЕВ А.Г.), 1964:   Майкопская серия Грузии *(The Maikop Series of Georgia).*   Nedra Press, Moscow.

MAKHARADZE A.I. & CHKHEIDZE R.G.   (МАХАРАДЗЕ А.И., ЧХЕИДЗЕ Р.Г.), 1971: Литология олигоценовых отложений Квирильской депрессии и о генезисе связанных с ними полезных ископаемых   (The lithology of the Oligocene deposits of the Kviril' depression and the origin of their associated mineral deposits).   *Trudy kavkaz. Inst. miner. Syr'ya,* vyp.9, pp.177-188.   Tbilisi.

MAKUSHIN A.A.   (МАКУШИН А.А.), 1970:   О марганце в нижнепермской галогенной формации Башкирского Приуралья (Manganese in the Lower Permian halide association of the Bashkirian Ural region).   *Dokl. Akad. Nauk SSSR, 191,* pp.1381-1384.

MIRTOV Yu.V. *et al.*   (МИРТОВ Ю.В. и др.), 1964:   Марганценосные и фосфоритоносные формации нижнего кембрия и верхнего докембрия (синия) Западной Сибири.   Осадочные формации Сибири   (The manganese- and phosphorite-bearing associations of the Lower Cambrian and Upper Precambrian (Sinian) of Western Siberia.   The sedimentary associations of Siberia).   *Trudy 5th Vses. Litol. Soveshch., 2,* pp.110-118.

MUKHIN A.S. & LADYGIN P.P. (МУХИН А.С., ЛАДЫГИН П.П.), 1957: Новые данные геолого-промышленной характеристике Усинского месторождения марганцевых руд (New data on the geological-industrial characteristics of the Usa deposit of manganese ores). *Vestn. zap.-sib. geol. Upr.*, No.2, pp.29-37.

NESTOYANOVA O.A. (НЕСТОЯНОВА О.А.), 1963: Вулканизм восточного склона Южного Урала. Магматизм, метаморфизм (The volcanism of the eastern slopes of the Southern Urals. Magmatism and Metamorphism), *in* "Металлогения Урала" (*'The Metallogenesis of the Urals'*), Vol.2, pp.27-47. Sverdlovsk.

NOVOKHATSKY I.P. (НОВОХАТСКИЙ И.П.), 1967: Успехи в изучении рудной базы черной металлургии Казахстана (Successes in studying the ore basis for heavy metallurgy in Kazakhstan). *Izv. Akad. Nauk Kaz. SSR, ser. geol.*, No.5, pp.36-43.

PARK, C.F., 1956: On the origin of manganese. *Int. geol. Congr.*, 20, Sympos. Manganese, 1, pp.75-98.

RABINOVICH S.D. (РАБИНОВИЧ С.Д.), 1971: Северо-Уральский марганце-рудный бассейн (*The Northern Urals Manganese-Ore Basin*). Nedra Press, Moscow.

RAKHMANOV V.P. (РАХМАНОВ В.П.), 1967: Марганцевые руды (Manganese ores), *in* "Успехи в изучении главнейших осадочных полезных ископаемых" (*'Successes in the Study of the Principal Sedimentary Mineral Deposits'*), pp.80--100. Nauka Press, Moscow.

RAKHMANOV V.P. & CHAIKOVSKY V.K. (РАХМАНОВ В.П., ЧАЙКОВСКИЙ В.К.), 1972: Генетические типы осадочных марганценосных формаций (Genetic types of sedimentary manganiferous associations). *Sov. Geol.*, No.6, pp.22-32.

RAMSDELL, L.S., 1932: An X-ray study of psilomelane and wad. *Am. Miner.*, 17, pp.143-149.

RICHMOND, W.E. & FLEISCHER, M., 1942: Cryptomelane, a new name for the commonest of the "psilomelane" minerals. *Am. Miner.*, 27, pp.607-610.

RODE E.Ya. (РОДЕ Е.Я.), 1952: Кислородные соединения марганца (искусственные соединения, минералы и руды) (*The Oxygen Compounds of Manganese (Artificial Compounds, Minerals, and Ores)*). Akad. Nauk SSSR Press, Moscow.

ROY, S., 1969: Classification of manganese deposits. *Acta miner. petrogr., Szeged, 19*, 1, pp.67-83.

ROZHNOV A.A. (РОЖНОВ А.А.), 1967: О геолого-генетических особенностях марганцевого оруденения западной части Джаильминской мульды и месте марганцевого оруденения в ряду проявлений железа и полиметаллов месторождения района. Геологическое строение западной части Джаильминской мульды (The geological-genetic features of the manganese deposit of the western part of the Dzhail'ma trough and the place of manganese mineralization in the series of shows of iron and polymetals of the deposits of the region. The geological structure of the Dzhail'ma trough), *in* "Марганцевые месторождения СССР" (*'Manganese Deposits of the USSR'*), pp.311-324. Nauka Press, Moscow.

SAPOZHNIKOV D.G. (САПОЖНИКОВ Д.Г.), 1963: Караджальские железомарганцевые месторождения *(The Karadzhal Iron-Manganese Deposits)*. Akad. Nauk SSSR Press, Moscow.

SHATSKY N.S. (ШАТСКИЙ Н.С.), 1954: О марганценосных формациях и о металлогении марганца. Статья 1 (The manganiferous associations and the metallogenesis of manganese. Report 1). *Izv. Akad. Nauk SSSR, ser. geol.*, No.4, pp.3-37.

SHATSKY N.S. (ШАТСКИЙ Н.С.), 1960: Парагенезы осадочных и вулканогенных пород и формаций (Parageneses of sedimentary and volcanogenic rocks and associations). *Izv. Akad. Nauk SSSR, ser. geol.*, No.5, pp.3-23.

SHATSKY N.S. (ШАТСКИЙ Н.С.), 1965: О геологических формациях (Geological associations), *in* "Избранные труды. Т. 3. Геологические формации и осалочные полезные ископаемые" *('Published Works. Vol.3. The Geological Associations and Sedimentary Mineral Resources')*, pp.16-51. Nauka Press, Moscow.

SHTERENBERG L.E. (ШТЕРЕНБЕРГ Л.Е.), 1963: Очерк геохимии Северо-Уральских марганцевых месторождений (Review of the geochemistry of the Northern Urals manganese deposits), *in* "Геохимия осадочных месторождений марганца" *('The Geochemistry of Sedimentary Manganese Deposits')* *Trudy geol. Inst. Mosk.*, *97*, pp.9-71.

SOKOLOVA E.A. (СОКОЛОВА Е.А.), 1963: Формационная характеристика и генезис марганцевого месторождения Тахта-Карача (Зеравшанский хребет) (The associational characteristics and origin of the Takhta-Karacha manganese deposit (Zeravshan Range)). *Litologiya polezn. iskop.*, No.3, pp.64--80.

STOLYAROV A.S. (СТОЛЯРОВ А.С.), 1958: Новые данные по стратиграфии олигоценовых отложений Мангышлака (New data on the stratigraphy of the Oligocene deposits of Mangyshlak). *Byull. nauchno-issled. Inform. Minist. Geol. SSSR*, No.3, pp.8-10.

STRAKHOV N.M., SHTERENBERG L.E., KALINENKO V.V. & TIKHOMIROVA E.S. (СТРАХОВ Н.М., ШТЕРЕНБЕРГ Л.Е., КАЛИНЕНКО В.В., ТИХОМИРОВА Е.С.), 1968: Геохимия осадочного марганцерудного процесса (The Geochemistry of the Sedimentary Manganese-Ore Process). *Trudy geol. Inst. Mosk.*, *185*.

STRUNZ, H., 1957: *Mineralogische Tabellen.* 3. Auflage. Akad. Verlagsgesellschaft, Leipzig.

SUSLOV A.T. (СУСЛОВ А.Т.), 1964: Марганец (Manganese), *in* "Металлы в осадочных толщах" *('Metals in Sedimentary Sequences')*, pp.100-170. Nauka Press, Moscow.

TIKHOMIROVA E.S. (ТИХОМИРОВА Е.С.), 1964: Палеогеография и геохимия нижнеолигоценовых марганцевоносных отложений Мангышлака (The palaeogeography and geochemistry of the Lower Oligocene manganese deposits of Mangyshlak). *Litologiya polezn. iskop.*, No.1, pp.75-92.

TIKHOMIROVA E.S. & CHERKASOVA E.V. (ТИХОМИРОВА Е.С., ЧЕРКАСОВА Е.В.), 1967: О распределении малых элементов в рудах Мангышлакского месторождения марганца (The distribution of minor elements in the ores of the Mangyshlak manganese deposit), *in* "Марганцевые месторождения СССР" *('The Manganese Deposits of the USSR')*, pp.258-273. Nedra Press, Moscow.

VARENTSOV I.M. (ВАРЕНЦОВ И.М.), 1962*a*: О геохимии Усинского марганцевого месторождения в Кузнецком Алатау (The geochemistry of the Usa manganese deposit in the Kuznets Alatau), *in* "Осадочные руды железа и марганца" (*'Sedimentary Ores of Iron and Manganese'*) *Trudy geol. Inst. Mosk., 70,* pp.28-64.

VARENTSOV I.M. (ВАРЕНЦОВ И.М.), 1962*b*: О главнейших марганценосных формациях (The principal manganese-bearing associations), *in* "Осадочные руды железа и марганца" (*'Sedimentary Ores of Iron and Manganese'*), *Trudy geol. Inst. Mosk., 70,* pp.119-173.

VARENTSOV I.M. (ВАРЕНЦОВ И.М.), 1964: К познанию условий образования Никопольского и других месторождений Южно-Украинского марганцерудного бассейна (An appreciation of the conditions of formation of the Nikopol' and other deposits in the Southern Urals manganese-ore basin). *Litologiya polezn. iskop.,* No.1, pp.25-39.

VARENTSOV I.M. *et al.* (ВАРЕНЦОВ И.М. и др.), 1967: Особенности распределения Ni, Co, Cu, V, Cr в рудах и вмещающих отложениях Южно-Украинского марганцерудного бассейна (Features of distribution of Ni, Co, Cu, V, and Cr, in the ores and surrounding deposits of the South Ukraine manganese-ore basin), *in* "Марганцевые месторождения СССР" (*'Manganese Deposits of the USSR'*), pp.179-198. Nauka Press, Moscow.

VERNADSKY V.I. (ВЕРНАДСКИЙ В.И.), 1954*a*: Геохимия марганца в связи с учением о полезных ископаемых (доклад на конференции, апрель, 1935 г. (The geochemistry of manganese in connexion with a study of mineral deposits (Conference proceedings, April 1935), *in* "Избранные сочинения" (*'Published Works'*), Vol.1, pp.528-542. Akad. Nauk SSSR Press, Moscow.

VERNADSKY V.I. (ВЕРНАДСКИЙ В.И.), 1954*b*: Очерки геохимии, история марганца (Review of the geochemistry and history of manganese) *in* "Избранные сочинения" (*'Published Works'*), Vol.1, pp.74-89. Akad. Nauk SSSR Press, Moscow.

DEPOSITS OF CHROMIUM

CLASSIFICATION OF DEPOSITS OF CHROMITE ORES

Of the small number of chromium-bearing minerals, only the chrome-spinels are important commercially and serve as a unique source of metallic chromium and the products of its chemical compounds. In the group of chrome-spinels with the general formula $(Mg,Fe)(Cr,Al,Fe)_2O_4$, the greatest interest centres around the following mineral species: mangochromite $(Mg,Fe)Cr_2O_4$; alumochromite $(Mg,Fe)(Cr,Al)_2O_4$; subferrichromite $(Mg,Fe)(Cr,Fe)_2O_4$; and to a lesser degree, subferrialumochromite $(Mg,Fe)(Cr,Fe,Al)_2O_4$.

In mining practice, the chrome ore, which consists of chrome-spinels and silicates, is given the selective term 'chromite' or 'chromite ore'. The deposits of chromites are spatially and genetically associated with complexes of ultramafic rocks and belong to the group of magmatic formations. The chromites are a definite petrographical variety amongst the ultramafic rocks and their facies, and are a component member of their differentiated series. Therefore, the conditions of formation of the chromites are directly connected with the formation of intrusive bodies of ultramafic rocks. The ultramafic massifs are distributed in the geosynclinal areas, in zones where the geosynclines and the platforms meet, and on the platforms. Such massifs are known in most of the geosynclinal areas of the world, the time of formation of which embraces the period from the pre-Palaeozoic to the Mesozoic inclusive. In these structures, they are arranged in the form of belts and are clearly connected with deep-seated regional faults. Under platform conditions, the ultramafic massifs are also associated with deep-seated faults (Wilson & Bateman, 1969).

The degree of saturation of the geosynclinal areas with ultramafic massifs is uneven. The area that they occupy in the various geosynclines comprises less than one percent and only in rare cases reaches 3 - 4% (Cuba). In the USSR, ultramafic massifs are most widely distributed in the Uralian folded region. The dimensions of the outcrops of individual massifs on the present surface are also extremely variable and vary from areas of less than 1 up to 2000 km^2. In the platform regions, the dimensions of the chrome-bearing massifs reach several tens of thousands of square kilometres.

In the folded regions, the ultramafic massifs are distributed in the eugeosynclines and are restricted to deep-seated fractures of the early phases of their development. The richest chrome-bearing intrusions are associated with long-developing deep-seated faults. The less chrome-rich ultramafic massifs occur in association with faults of 'short duration'.

The largest chromite deposits are included in massifs, the age of which is defined as Precambrian (India), Caledonian (Urals), Hercynian (Iran, Turkey, etc.), and Alpine (Yugoslavia, Albania, etc.). A comparison of the revealed reserves of chromites in ultramafic intrusions of various ages, demonstrates that the ancient structures of the folded regions are somewhat more chromiferous, possibly as a result of the deeper erosional exposure of these structures and the ultramafic massifs included in them.

Certain concentrations of chrome-spinels are always included in the rocks of the ultramafic petrographic associations: peridotite-orthopyroxenite--norite, peridotite, dunite-clinopyroxenite, gabbro-norite-harzburgite, picrite--dolerite, dunite-clinopyroxenite-ijolite, and kimberlite.

Of all the associations listed, only three are commercially chrome--bearing; these are the peridotite-orthopyroxenite-norite (on the platforms), the gabbro-norite-harzburgite and the peridotite (in the folded regions).   In the geosynclinal regions of the whole world, as in the USSR, the most widely distributed is the peridotite association, which combines the massifs of essentially harzburgitic composition with a certain proportion of dunites, lherzolites, wehrlites, pyroxenites, and troctolites.

Subassociations have been recognized on the basis of the presence and quantitative relationships between the above-listed rocks and the harzburgites: harzburgitic proper, dunite-harzburgitic, dunite-troctolite-harzburgitic, dunite--lherzolite-harzburgitic, and pyroxenite-harzburgitic.   The scale of the chromite mineralization and the quality of the ores corresponds to the petrographic composition of the rock assemblage forming the massifs.   Large concentrations of chromite-rich ores are associated with the most differentiated massifs, which include the dunite-troctolite-harzburgitic and dunite-harzburgitic subassociations. Owing to the high degree of serpentinization of the rocks, the recognition of individual petrographic varieties within the ultramafic massifs is not always possible.   Only in the most closely studied folded regions of the Urals, the Malyi Kavkaz (Minor Caucasus), and certain areas of Siberia, where large-scale mapping of the ultramafic massifs has been carried out, has it been possible to assign a particular massif to a definite subassociation, and thereby to make an assessment of its potential chrome content and the quality of the ores.

In the platform regions, the chromite deposits are associated with the harzburgite-orthopyroxenite-norite association.   This association is usually termed the stratiform or laminated type, since it consists of intrusive bodies, in which interlayering (pseudo-bedding) of the rocks of different composition is clearly manifested, with the predominant development of ultramafic types in the lower portion and gabbroid and more acid types in the middle and upper portions (Wager & Brown, 1968).   The deposits of chromites in them consist of layer-like segregations of relatively small thickness (from 0.1 up to 1.0 m), but of significant extent, reaching tens and even hundreds of kilometres.   In the USSR, no chromite deposits occurring under platform conditions are known, and the typical association has so far not been found.

According to Betekhtin's classification (1937) and Sokolov's additions (1948), four genetic types have been allocated for the chromite deposits.

1.   Early magmatic (protomagmatic) segregation deposits, developing in the early magmatic phase of formation of intrusions as a result of gravitational differentiation.   This genetic type includes the layer-like segregations of chromite in the stratiform layered intrusions of the platform areas (Bushveld, the Great Dyke of Rhodesia, Skaergaard, etc.).   In the USSR, commercial deposits of such type are so far not known.

2.   Late magmatic (hysteromagmatic) deposits, developed in the late phase of formation of intrusions from residual ore-silicate melts with the participation of volatile components.   In this genetic type, Sokolov (1948) has recognized a u t o m a g m a t i c (syngenetic with the surrounding rocks) and h e t e r o m a g m a t i c deposits (injected, and developed during the intru-

sion of residual ore-silicate melts from lower zones in the intrusive body). This genetic type includes all the principal, commercially important deposits of chromite ores in the massifs of the geosynclinal regions.

3. Hydrothermal deposits, formed as a result of solution and redeposition of chrome-spinels and chrome-bearing rock-forming minerals of the ultramafic rocks by hydrothermal solutions. Chromites of such origin form nests and veins. In the USSR, deposits of this type are not known. According to Varma (1971), such an origin applies to certain deposits in India.

4. Deposits of weathering (eluvial, deluvial, and littoral-marine deposits); they occur rarely and have no great value. The deluvial ore segregations have been worked in the USSR in the Saranovsk deposit. In Japan, a chromite concentrate is obtained from littoral-marine placers.

The classification presented has been constructed on the assumption that chromite concentrations develop in the intrusive chamber during its cooling. New data on the chromite-bearing intrusions, such as the established restriction of the intrusions to regional faults of deep-seated occurrence, and of the largest ore concentrations in them to zones adjacent to feeding channels, the clear spatial and genetic association between the ores of definite chemical composition and definite petrographic facies, and the best-quality ores with the deepest facies, along with new data on deep-seated geology, provide a basis for suggesting that the processes of separation of ore-silicate melts from the parent magmatic masses took place both under subcrustal conditions (upper mantle), and over the route of their advance along deep-seated faults in the upper part of the Earth's crust.

Factors determining the prospective chromite content of the ultramafic massifs of the geosynclinal regions are:

1) the assignment of the massif to the dunite-harzburgitic, dunite-troctolite-harzburgitic, and harzburgitic subassociations of the peridotite association, or to the gabbro-norite-harzburgitic association;

2) the restriction of the massif to a regional fault of long duration, separating structures of the first order;

3) the occurrence of the massif in Caledonian or Hercynian structures; older and younger intrusions are less promising;

4) the deep erosion of the intrusion. The dimensions and shape of the massif are of no small importance. In the large massifs, ore bodies of greater dimensions are localized (the Kempirsai massif), and only in detached apophyses from buried ultramafic bodies may mineralization sometimes be included in massifs of small extent (the Saranovsk massif).

In the Soviet Union, chromite deposits and ore-shows are known in many regions of the Urals, the Caucasus, Central and Eastern Siberia, in Kamchatka, Chukotka, Sakhalin, and a number of other places. However, the Urals still remain the principal source. In the Uralian folded region, more than 25 regions are known in which mining of chromite ore has taken place at various times. The most important are the Verblyuzh'egorsk, Gologorsk, Klyuchevsk, Ak-Karga, Orsk-Khalilovo, Alapaevo, Belorets, and a number of other ore regions. Mining of chromite has also been carried out in Transcaucasia in the Shorzha, Gei-Dara, Ipyag, and certain other ultramafic massifs.

All the chromite deposits of the USSR occur in ultramafic massifs, distributed in the folded regions, and they are not known in platform structures. An exception is the Zlatogorsk layered gabbro-peridotite massif, with its non--commercial mineralization, occurring in the Kokchetav block.

At the present time, the exploitation of deposits in the USSR is being carried out in two regions of the Urals, the Don group of chromite deposits, lying in the southeastern part of the Kempirsai massif, and in the Saranovsk deposit.

CHROMITE DEPOSITS, ASSOCIATED WITH THE PERIDOTITE ASSOCIATION

CHROMITE DEPOSITS, OCCURRING IN MASSIFS OF THE DUNITE-TROCTOLITE-HARZBURGITE SUBASSOCIATION

This subassociation includes the Don chromite deposits, which are unique in scale of mineralization, occurring in the southeastern part of the Kempirsai massif in the Southern Urals. Here also belong the chromite deposits of the Shorzha massif, lying in the ultramafic belt of the Minor Caucasus. Owing to the inadequate study of the ultramafic belts of the other geosynclinal regions of the USSR, this chromite-bearing subassociation is still not known in them.

Of the foreign deposits of chromite, this association includes certain deposits in Turkey (Guleman) and Cuba (the Majari-Baracoa).

DEPOSITS OF CHROMITES OF THE KEMPIRSAI MASSIF

The Kempirsai chromite massif lies in the Aktyubinsk district of the Kazakh SSR, on the southern extremity of the Urals, within the Central Urals uplift. Discoveries of chromite in this massif were first recorded by Karpov in 1920. In 1936, the geologist P. Dolgov discovered the actual outcrops of chromite. In 1937, a party under the direction of geologist A. Konev discovered the Dzharlybutak and Don deposits (Gigant, Sputnik, Almaz-Zhemchuzhina, etc.). Further geological research and scientific investigation work defined the huge scale of the deposits.

Geological position and shape of the massif

The Kempirsai ultramafic massif is located on the southern end of the Urals, within the Uraltau meganticlinorium, which consists of Precambrian and lower Palaeozoic sediments. To the east, the Uraltau meganticlinorium adjoins the Magnitogorsk synclinorium, being separated from it by a regional fracture. According to seismic-probe data, in the vicinity of the southern part of the Kempirsai massif, on the latitude of Khromtau, the thickness of the Earth's crust down to the Moho is 51 km, and the thickness of the 'basalt' layer has been determined here at 25 km. The Kempirsai massif extends in a submeridional direction for 82 km, in conformity with the direction of the fault zone. The width of the massif varies considerably. Thus, in the north it reaches 0.6 km, farther south, in the latitude of Batamshinsk village, it is 13 km, and in the southern part it is up to 31.6 km. The total area of the massif is 920 km^2 (Fig. 76).

Fig. 76.
Structural-petrographical diagram of the Palaeozoic basement of the Kempirsai ultramafic massif and its relationship to the Palaeozoic and Proterozoic rocks (compiled from information from the East Uralian Geological-Exploration Expedition).

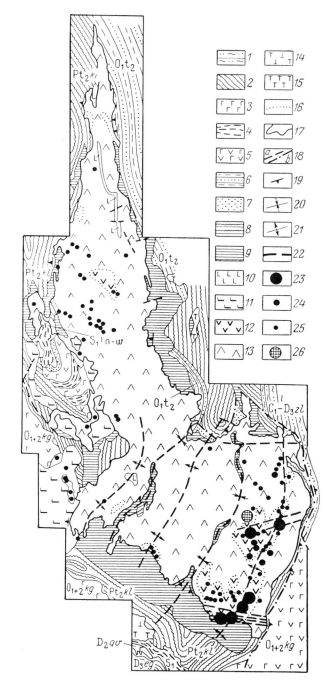

1) Lower Carboniferous - Upper Devonian sandstones, conglomerates, argillites, and siltstones, Zilair Group (C_1-D_3zl); 2) Upper Devonian cherty rocks, siltstones and crystalline limestones, Egida Group (D_3eg); 3) Middle Devonian (Givetian) coarse-pebbly and blocky conglomerates and diabase porphyrites, Aitpai Group (D_2gv); 4) Lower Silurian (Llandoverian-Wenlockian) diabases, Amygdaloidal basalts, and siliceous tuffs (S_1ln-w); 5) Lower and Middle Ordovician eruptives of basic and intermediate composition, tuff-sandstones, siliceous schists, conglomerates, and limestones, Kuagash Group ($O_{1+2}kg$); 6) Lower Ordovician (upper Tremadocian) clay-chert slates with seams of quartzites, arkosic sandstones and siliceous tuffites (O_1t_2); 7) Lower Ordovician (Tremadocian) argillites, sandstones, seams of tuffs and siltstones, Kidryasovsk Group (O_1tkd); 8) Upper Proterozoic sericite--chlorite-quartz, clay-quartz, chert-graphite, and other schists, lower Kayala subgroup of Kayala Group (Pt_2kl); 9) gabbro-amphibolites; 10) gabbros; 11) olivine and amphibolized gabbros; 12) serpentinites after dunites; 13) serpentinites after peridotites; 14) serpentinites after pyroxene dunites (see Fig. 77); 15) undifferentiated serpentinites; 16) boundaries of areas of maximum development of dunites; 17) boundary of Kempirsai ultramafic massif; 18) tectonic fractures (a - confirmed, b - assumed); 19) attitude of contact of ultramafic massif; 20) axes of anticlines; 21) axes of synclines; 22) outline of magma--feeding channel (see Fig. 77); 23-26) chromite deposits: 23) large, 24) medium, 25) small and ore-shows, 26) lean segregated ores.

In attitude, the northern and southern parts of the massif are substantially different. The northern part of the massif rests mainly conformably with the direction of the schistosity of the Upper Proterozoic sediments in the hanging wall, and with the upper Tremadocian deposits of the Ordovician, in the footwall. This part of the massif is a homoclinally dipping body, up to 2.5 km thick, with westerly dips of 40 to 60°. The eastern contact here is, in many cases, apparently tectonic. In the area where the homoclinal segregation merges with the southern part of the massif, the country rocks are Ordovician and Cambrian sediments.

The country rocks of the southern part of the massif are amphibolites. According to absolute-age data, these rocks are Cambrian in age. These same gabbro-amphibolites occur within Upper Proterozoic sediments. Tectonic contacts between the massif and Upper Devonian and Lower Carboniferous sediments are known. Within the southern part of the massif there are two belts of roof xenoliths, consisting of gabbro-amphibolites and Upper Proterozoic schists. One of them can be traced in the form of a discontinuous submeridional belt towards the west of the area of distribution of the commercial chromite deposits. The other has been developed in the area of the River Tassai and separates the Southwestern arch uplift in the massif from the Tagashasai uplift. From geological and geophysical data the massif, in its southern half, has the shape of a laccolith, occurring between Upper Proterozoic and lower Palaeozoic rocks. In the south, the contact surface of the hanging wall of the intrusion dips south and southeast at 30-50°, passing below the amphibolites. In the east, it has a vertical or steep easterly dip, but in a number of places it is cut by tectonic fractures.

From geological data (Nepomnyashchikh, 1950), the thickness of the massif in the southern part is 4 - 5 km. Segalovich (1971), who has reviewed all the geophysical information on the massif, has determined its thickness in the southeastern part at 16 km, and in the western part, up to 8 - 6 km.

Below the southeastern chromite-bearing portion of the massif, geophysical work has revealed a feeding channel, with a steep easterly dip, carrying it below the structure of the Magnitogorsk synclinorium. Below the northern, markedly elongated portion of the massif, a feeding channel has also been established which extends in a submeridional direction, and descends steeply towards the Magnitogorsk synclinorium.

From geophysical data and absolute-age determinations on a phlogopite from the contact-mineralized rocks, the time of the intrusion has been determined as late Caledonian (404 - 380 m.y.) (Pavlov *et al.*, 1968).

The upper surface of the intrusion is complicated by several arched uplifts and depressions. In the south of the massif, where its width reaches 30 - 32 km, there is clear evidence of three arched uplifts: the Southwestern, the Tagashasai, and the Southeastern (region of the Main ore field). The northern portion of the massif is about 10 km wide and is complicated by the Batamshinsk arched uplift, and farther west by the Taiketken uplift, which is isolated from it. The arches alternate with depressions in the roof of the massif, which have been filled with amphibolites and schists. The erosion section reveals various horizons in the intrusion. At many points on the contacts of the intrusive body there are zones of tectonic fractures, and faults and crumpling of the rocks are observed. However, in a number of places on the southern margin and in the north of the massif, contact mineralization of the country rocks is known. In these cases, for example, garnet, idocrase, pyroxenes, prehnite, and other minerals are developed after amphibolites.

The petrographic composition and structure of the massif

The following principal varieties of rocks are developed in the massif: lherzolites, amphibole peridotites, porphyritic harzburgites, banded harzburgites, wehrlites, dunites, enstatite dunites, sulphide-bearing dunites and olivinites, saussuritized and amphibolized troctolites, pyroxene-amphibole gabbros, anorthosites, chromite dunites, and chromite ores of different complexity.   All the above rocks have been serpentinized to a varying degree up to the formation of serpentinites, having lost the features of their primary composition.

The dyke (vein) rocks of ultramafic composition consist of bronzitites, actinolitites, hornblendites, sulphide-bearing diopsidites and websterites, dunites, and those of gabbroid composition comprise gabbro-diabases, beerbachites, olivine norites, and plagioclasites.   The Kempirsai massif consists mainly of peridotites.   All the remaining varieties, including the dunites, occur in markedly subordinate amount.

Peridotites

Harzburgites predominate amongst the peridotites of the massif.   They consist of different structural varieties, among which massive types with porphyritic aspect predominate.   Banded harzburgites are quite widely developed.

The massive harzburgites are restricted mainly to the less eroded peripheral portions of the massif with preserved remnants of the roof.   The banded harzburgites, as a rule, occur at places of maximum erosion of the massif, and also form 'windows' of various shapes and sizes in the field of the massive harzburgites.   Banded harzburgites are relatively widely distributed in the Southeastern, Batamshinsk, and Tagashasai arched uplifts.   In the inter-arch lows, mainly, massive porphyritic harzburgites are developed, which are characterized by the presence of large grains (3 - 8 mm) along with small forms of orthopyroxene (or bastite), which are relatively evenly distributed in the olivine (serpentine) matrix.   In the banded harzburgites there are preferentially oriented zones of increase and decrease in the amount of pyroxene grains, or linear (in plan) chains of pyroxene grains in a matrix of olivine (serpentine) grains.

The harzburgites involve:

1) primary minerals: olivine, orthopyroxene, and accessory chrome-spinel;

2) minor minerals: serpentine (lizardite, and rarely antigorite in the contact zones with the gabbro-diabases, bastite, and a low-polarizing gel-like serpentine), amphibole (tremolite), chlorite (clinochlore), brucite, powdery magnetite, talc, and very rarely phlogopite.

The quantitative mineral composition of the harzburgites is variable and depends on the fluctuating amounts of primary minerals and varying degree of serpentinization.   A quantitative grain-count of the minerals in a thin-section of banded serpentinized harzburgite has revealed the following amounts (in volume %): olivine 14.0; serpentine 67.8; orthopyroxene 2.5; bastite 7.3; tremolite 3.6; talc 1.9; chlorite 0.3; spinel 1.5; powdery ore mineral (magnetite) 1.1.   Assuming that the bastite, tremolite, talc, and chlorite have been formed from pyroxene, and the serpentine and magnetite, from olivine, the ratio of primary minerals was evidently as follows (in volume %):   olivine 82.9; orthopyroxene 15.6;   chrome-spinel 1.5.

A similar count for the massive porphyritic harzburgite also shows the following amounts of primary minerals (in volume %):   olivine 79.5;   orthopyroxene 19.0;   and chrome-spinel 1.5.

Table 5.   *Chemical Composition of Harzburgites (in wt %)*

| Components | Intensely serpentinized harzburgite | | | |
| | banded | | porphyritic | |
	analytical data	H_2O, S, and TiO_2	analytical data	after elimination of H_2O and S
SiO_2	35.86	43.07	37.94	43.62
TiO_2	0.02	–	tr.	–
Al_2O_3	1.20	1.44	0.77	0.88
Cr_2O_3	0.25	0.30	0.37	0.42
Fe_2O_3	5.93	7.12	5.63	6.47
FeO	1.28	1.52	1.86	2.14
MnO	–		0.04	0.04
MgO	38.65	46.44	39.38	45.26
NiO	0.08	0.10	0.27	0.31
CoO	0.01	0.01	0.01	0.01
CaO	0.00	–	0.40	0.46
Na_2O	0.00	–	0.26	0.30
K_2O	0.00	–	0.08	0.09
H_2O^+	15.22	–	12.24	–
H_2O^-	1.28	–	0.92	–
V_2O_5	0.00	–	–	–
S	0.13	–	0.05	–
Total	99.91	100.00	100.22	100.00

Analyses carried out in the Central Chemical Laboratory of Institute of Geology of Ore Deposits, Petrography, Mineralogy and Geochemistry (IGEM) by K. Sokova.

Clinopyroxene (diopside) occurs sporadically in the harzburgites, up to 1 - 3 vol. %, although this mineral (Pavlov *et al.*, 1968) has been developed during the post-magmatic phase and is connected with the dykes and veins of pyroxenites (diallagites and websterites).

The primary minerals have been placed in the following order: olivine — orthopyroxene — accessory chrome-spinel.   This applies both to the minerals of the large size grade (1 - 6 mm) and to the small grains, measured in fractions of a millimetre.   The texture of the rocks is hypidiomorphogranular.   Recrystallization textures frequently occur.

The chemical composition of the serpentinized harzburgites are presented in Table 5.

The dimensions of the olivine grains range from fractions of a millimetre up to 4 mm.   They possess rectilinear outlines with a short-prismatic habit; $\gamma - \alpha = 0.028 - 0.032$; $\gamma'' = 1.687$ (± 0.002); $\alpha' = 1.665$ (± 0.003), $2V = (+) 87 - 88°$.   The chemical composition of the olivine in the harzburgites is shown in Tables 6 and 7.

The orthopyroxene is often entirely bastitized.   Its grains are 2 - - 6 mm in cross-section; grains from 0.01 - 0.1 mm, characterized by wavy embayed outlines, are present in considerably lesser amount; $\gamma - \alpha = 0.010 -$ - 0.009, $2V = (+) 82 - 86°$, with $z:\gamma = 0$.   The chemical composition of the orthopyroxene from the harzburgites is given in Table 6.

Table 6.   *Chemical Composition of Olivines and Orthopyroxenes from Harzburgites (in wt %)*

Components	Olivine		Orthopyroxene		
	1	2	3	4	5
SiO_2	41.28	40.70	54.44	53.04	52.50
TiO_2	0.00	tr.	tr.	0.03	tr.
Al_2O_3	0.00	tr.	1.60	2.74	1.00
Fe_2O_3	0.06	0.00	0.92	0.98	0.83
FeO	7.22	8.52	4.84	4.75	5.16
MnO	0.12	0.14	0.14	0.12	0.14
MgO	50.12	50.18	35.09	34.90	34.95
CaO	0.00	tr.	0.89	1.10	2.50
H_2O^-	0.00	0.06	0.06	0.00	0.06
H_2O^+	0.27	0.38	0.94	1.60	2.22
NiO	0.28	0.25	0.08	n.d.	n.d.
CoO	n.d.	n.d.	n.d.	n.d.	n.d.
Cr_2O_3	0.01	0.00	0.70	n.d.	0.60
S	0.00	0.00	n.d.	n.d.	n.d.
Total	99.36	100.23	99.70	99.26	99.96

1. $(Mg_{1.848}, Fe_{0.145}, Ni_{0.006}, Mn_{0.001})_2SiO_4$; Spec.No.172-63, drill-hole 50, depth 849 m.

2. $(Mg_{1.819}, Fe_{0.174}, Ni_{0.004}, Mn_{0.003})_2SiO_4$; Spec.No.611-66, drill-hole 79, depth 410 m.

3. $(Mg_{1.888}, Fe_{0.112})_2Si_2O_6$; Spec.No. 172-63, drill-hole 50, depth 849 m.

4. $(Mg_{1.809}, Fe_{0.137}, Ca_{0.049}, Mn_{0.005})_2(Si_{1.941}, Al_{0.063}, Fe^{3+}_{0.014})_{2.18}O_6$; Spec.No. 610-66, drill-hole 106, depth 213 m.

5. $(Mg_{1.734}, Fe_{0.107}, Ca_{0.154}, Mn_{0.005})_2(Si_{1.989}, Fe_{0.014})_{2.003}O_6$; Spec.No.611- -66, drill-hole 79, depth 410 m.

Analyses carried out in the Central Chemical Laboratory of IGEM by Yu.Nesterova, V. Moleva and R. Teleshova.

Table 7. *Amount of FeO and Fayalite Component in Olivines from Harzburgites*

Components	Tagashasai arched uplift		Southeastern arched uplift (Main ore field)			
	1	2	3	4	5	6
FeO, wt %	8.99	9.16	7.09	7.60	7.88	9.23
Fa, mol %	9.10	9.27	7.12	7.65	7.93	9.35

1. Spec.No. 339-62, drill-hole 81, depth 1181 m; 2. Spec. No. 10-60, drill-hole 1, depth 837 m; 3. Spec.No. 190-63, drill-hole 50, depth 400m; 4. Spec. No. 262-66, drill-hole 137, depth 315 m; 5. Spec.No. 10-61, surface; 6. Spec. No. 588-66, drill-hole 233, depth 330 m.

Microchemical analyses carried out in Central Chemical Laboratory of IGEM by R. Teleshova.

The accessory chrome-spinel possesses grains of irregular shape with dimensions from fractions up to 2 mm.

The features of the composition of the accessory chrome-spinels from the harzburgites are shown in Table 8.

Table 8. *Chemical Composition of Accessory Chrome-Spinels from Harzburgites*

Components	Chrome-picotite			Alumochromite
	Spec.No.190-38	Spec.No.10-60	Spec.No.277-37	Spec.No.172-1-63
SiO_2	-	1.59	2.35	0.72
TiO_2	0.08	0.11	0.20	0.17
Al_2O_3	27.50	34.74	23.57	22.39
Cr_2O_3	40.89	29.84	40.40	44.46
Fe_2O_3	1.99	1.50	4.83	1.61
FeO	14.55	15.05	14.90	16.62
MnO	-	0.19	-	0.18
MgO	13.93	16.20	13.78	13.52
CaO	-	0.00	-	0.00
Na_2O	-	-	-	-
K_2O	-	-	-	-
H_2O^-	-	0.14	0.10	-
H_2O^+	-	0.58	0.36	-
V_2O_5	-	0.21	0.11	0.16
P_2O_5	-	0.06	-	-
NiO	-	0.14	-	0.08
CoO	-	0.17	-	-
Total	98.94	100.52	100.60	99.91

Analyses carried out in Central Chemical Laboratory of IGEM by O. Ostrogorskaya

From the chemical analytical data presented it is seen that the harzburgites of the massif are high-magnesia varieties of ultramafic rocks affected in significant degree by processes of serpentinization.   The chemical composition of the rock-forming minerals of the harzburgites is distinguished by its low iron content.   The amount of the fayalite component in the olivines varies from 7.1 to 9.3%.   The orthopyroxenes are also distinguished by an extremely low iron content with 6% $FeSiO_3$.   The accessory chrome-spinels of the harzburgites in the massif are characterized by extremely large amounts of aluminium oxide (22.4 - 34.7 wt %) and low chromium content (Cr_2O_3, 30 - 44.5 wt %).

Dunites and enstatite dunites

The non-pyroxene dunites and dunites with up to 10% orthopyroxene are widely developed in the Southeastern arched uplift, where all the largest deposits of chromites are distributed.   Here, they are involved in the ore-bearing zones and form individual fields and elongate belts.   Outcrops of less sizeable dimensions are known in the vicinity of the Batamshinsk uplift in the middle course of the River Kuagach and in Kempirsai gorge, and also in a number of other places.   Besides the independent fields, dunites occur amongst the harzburgites in the form of thin bands and elongate schlieren-like bodies, varying in dimensions.

A microscopic study of the dunites demonstrates a high degree of serpentinization.   The mineral composition of the typical dunites is as follows (volume %): olivine (relicts), not greater than 25; serpentine (lizardite),73 - 98; chrome-spinel, up to 1.5; ore dust in stringers of lizardite, 1.0.

Tremolite, talc, brucite, and chlorite occur sporadically.   The chemical composition of the dunites of the massif is given in Table 9.

The principal minerals of the dunites are characterized by the following features.

Olivine is colourless, and possesses a weakly defined cleavage; the primary dimensions of the grains are 1 - 5 mm. $\gamma - \alpha = 0.032 - 0.035$, $2V$ varies from +86 to +90° (data of more than 50 measurements); $\alpha = 1.656$ ($\pm$ 0.002), $\gamma = 1.688$.

Chemical analyses of olivines from dunites of various regions and depths in the massif are given in Tables 10 and 11.

Table 9.   *Chemical Composition of Strongly Serpentinized Dunites (in wt %)*

Components	Batamshinsk arched uplift, Kempirsai gorge	Southeastern arched uplift Spornoe deposit
SiO_2	33.76	33.90
TiO_2	tr.	0.00
Al_2O_3	0.76	0.36
Cr_2O_3	0.35	0.26
Fe_2O_3	7.27	5.19
FeO	0.69	0.71
MnO	0.05	0.06
MgO	39.52	39.62
NiO	0.29	0.28
CoO	0.01	0.01
CaO	0.21	2.25
Na_2O	0.20	0.00
K_2O	0.08	0.00
H_2O^+	15.04	15.72
H_2O^-	2.00	1.44
V_2O_5	–	0.00
S	0.05	0.14
Total	100.28	99.94

Analyses carried out in Central Chemical Laboratory of IGEM by P. Volkov and K. Sokova.

Table 10. *Chemical Composition of Olivines from Dunites (in wt %)*

Components	Batamshinsk arched uplift		Southeastern arched uplift (Main ore field)						
	1	2	3	4	5	6	7	8	9
SiO_2	40.66	40.30	41.00	41.00	40.50	41.16	41.22	40.78	41.10
TiO_2	0.10	0.05	tr.	tr.	tr.	0.00	tr.	0.00	tr.
Al_2O_3	0.05	0.70	0.20	tr.	tr.	0.00	tr.	0.00	tr.
Fe_2O_3	1.14	1.91	0.32	0.00	0.00	0.36	0.00	0.58	0.00
FeO	8.10	8.47	8.25	7.44	8.60	6.04	7.00	5.60	6.20
MnO	0.13	0.29	0.10	0.20	0.14	0.10	0.16	0.10	0.22
MgO	49.82	48.30	50.04	51.00	50.20	51.22	51.40	51.49	52.02
CaO	0.09	n.d.	0.00	0.00	0.00	0.00	tr.	0.00	tr.
H_2O^-	0.00	0.00	0.00	0.08	0.08	0.10	0.00	0.00	0.04
H_2O^+	0.12	0.30	0.26	0.36	0.43	0.09	0.24	0.87	0.38
NiO	n.d.	n.d.	0.15	0.23	0.23	0.38	0.24	0.42	0.32
CoO	n.d.	n.d.	n.d.	n.d.	n.d.	0.05	n.d.	0.00	n.d.
S	n.d.	n.d.	0.00	0.00	0.00	0.00	0.00	0.00	0.00
Cr_2O_3	tr.	tr.	tr.	tr.	tr.	0.02	tr.	0.007	tr.
Total	100.21	100.32	100.32	100.31	100.18	99.52	100.26	99.85	100.28

1. $(Mg_{1.840}, Fe_{0.157}, Mn_{0.003})_2 SiO_4$; Spec.No.317-68, drill-hole 6, depth 675 m.

2. $(Mg_{1.797}, Fe_{0.197}, Mn_{0.06})_2 SiO_4$; Spec.No.296-68, surface.

3. $(Mg_{1.830}, Fe_{0.165}, Ni_{0.003}, Mn_{0.002})_2 SiO_4$; Spec.No.595-66, drill-hole 59, depth 560 m.

4. $(Mg_{1.841}, Fe_{0.151}, Ni_{0.004}, Mn_{0.004})_2 SiO_4$; Spec.No.599-66, drill-hole 154, depth 515 m.

5. $(Mg_{1.818}, Fe_{0.176}, Ni_{0.004}, Mn_{0.002})_2 SiO_4$; Spec.No.40-66, drill-hole 167, depth 819 m.

6. $(Mg_{1.871}, Fe_{0.119}, Ni_{0.008}, Mn_{0.002})_2 SiO_4$; Spec.No.168-63, drill-hole 50, depth 868 m.

7. $(Mg_{1.851}, Fe_{0.141}, Ni_{0.005}, Mn_{0.003})_2 SiO_4$; Spec.No.127-66, drill-hole 145, depth 575 m.

8. $(Mg_{1.876}, Fe_{0.115}, Ni_{0.008}, Mn_{0.001})_2 SiO_4$; Spec.No.169-63, drill-hole 50, depth 866 m.

9. $(Mg_{1.864}, Fe_{0.126}, Ni_{0.006}, Mn_{0.004})_2 SiO_4$; Spec.No. 37-66, drill-hole 167, depth 872 m.

Analyses carried out in Central Chemical Laboratory of IGEM by V. Moleva and Yu. Nesterova, E.Lomeiko, and R. Teleshova.

Table 11.   *Amount of FeO and Fayalite Component in Olivines from Dunites*

Components	Analysis number										
	1	2	3	4	5	6	7	8	9	10	11
FeO, wt %	5.70	6.18	6.58	6,95	7.63	7.84	8.05	8.07	8.25	8.85	7.79
Fa, mol %	5.71	6.19	6.60	6,97	7.67	7.89	8.11	8.13	8.32	8.95	7.84

Microchemical analyses carried out in Central Chemical Laboratory of IGEM by R. Teleshova.

Table 12.   *Chemical Composition of Olivines from Enstatite and Sulphide--Bearing Dunites (in wt %)*

Components	Batamshinsk arched uplift	Tagashasai arched uplift		Southeastern arched uplift (Main ore field)			
	1	2	3	4	5	6	7
SiO_2	0.33	0.95	1.14	0.56	0.54	0.79	0.59
TiO_2	0.62	0.36	0.65	0.19	0.16	0.15	0.13
Al_2O_3	21.06	29.85	31.89	14.36	12.04	11.20	9.19
Cr_2O_3	38.10	31.10	31.26	52.26	55.43	57,45	59.91
Fe_2O_3	9.06	7.37	3.84	1.69	2.68	3.40	1.94
FeO	19.52	15.82	16.41	17.26	16.39	13.53	14.53
MnO	0.26	0.24	0.23	0.29	0.25	0.22	0.21
MgO	11.04	13.43	14.23	12.48	12.56	13.50	13.84
CaO	0.00	0.00	0.00	0.00	0.00	0.00	0.00
Na_2O	0.00	n.d.	0.08	n.d.	n.d.	n.d.	n.d.
K_2O	0.00	n.d.	0.05	n.d.	n.d.	n.d.	n.d.
H_2O^-	0.00	0.13	0.06	0.00	–	–	–
H_2O^+	0.06	0.27	0.42	0.15	0.10	0,24	0.19
V_2O_5	0.25	0.17	0.20	0.21	0.07	0,09	0.08
P_2O_5	n.d.	0.03	n.d.	0.04	–	0.00	n.d.
NiO	0.13	0.11	0.11	0.08	0.08	0.10	0.10
CoO	0.14	0.15	n.d.	0.15	n.d.	0.20	0.13
Total	100.57	99.98	100.57	99.74	100.30	100.87	100.84

1) subferrichromepicotite, Spec.No.317-68, drill-hole 6, depth 675 m;   2) chromepicotite, Spec.No. 1-60, surface;   3) chromepicotite, Spec.No.237-63, surface; 4) alumochromite, Spec.No.53-61, drill-hole 57, depth 224 m;   5) alumochromite, Spec.No.209-1-63, drill-hole 129, depth 255 m;   6) magnochromite, Spec.No.8-61, surface;   7) magnochromite, Spec.No.29-61, drill-hole 41, depth 576 m.

Analyses carried out in Central Chemical Laboratory of IGEM by O. Ostrogorskaya.

The accessory chrome-spinel consists of small idiomorphic crystals measuring from hundredths of a millimetre up to 1 - 2 mm, arranged in most cases along the junctions of two, three, and more olivine individuals.   The colour of the translucent chrome-spinels ranges from orange-red to reddish--brown.   Partially or completely altered chrome-spinels are not translucent. The composition of the accessory chrome-spinels from the dunites of the massif is given in Table 12.

The bulk of the serpentine in the dunites, as in the harzburgites, on the basis of X-ray structural analysis, belongs to lizardite.   This mineral forms a network of stringers, oriented in the most varied directions.   In the centres, formed by the stringers of the mesh, a lower-polarizing serpentine has been observed, also belonging to lizardite. $\gamma - \alpha$ varies from 0.009 - 0.011 for lizardite in the stringers and 0.003 - 0.004 in the mesh.   The lizardite is biaxial positive.

Enstatite dunites are distinguished from those described only in the presence of enstatite (or bastite), the amount of which varies from 1 to 10%. Their distribution in the rock is extremely uneven.   From optical data, the composition of the enstatite is similar in composition to that in the harzburgites.

The sulphide-bearing dunites have a considerable distribution only in the ore-bearing complexes of the Main ore field and are distributed near the ore chromite bodies.   They consist of olivine, serpentinized in varying degree, accessory chrome-spinel, powdery magnetite and sulphides.   The amount of sulphides in such dunites varies from fractions of a percent up to 5 - 10%.   Amongst the sulphides, pyrrhotite, mackinawite (tetragonal sulphide of iron - FeS), pentlandite, and rarely chalcopyrite and maucherite.   The chemical composition of the olivines from the enstatite and sulphide-bearing dunites is given in Table 13.

The chemical analytical data (see Tables 10, 11, and 13) demonstrate that the olivines from the dunites and enstatite dunites are distinguished by their extremely high magnesium content.   The amount of the fayalite component in them varies from 6.0 to 9.0%.   In this respect, the olivines from the dunites distributed in the northern part of the massif (the Batamshinsk uplift) are somewhat more iron-rich as compared with those from those in the same rocks in the Main ore field.   The sulphide-bearing dunites are distinguished by the maximum content of the fayalite component, being 9.2 - 11.2%.   In regard to the composition of the accessory chrome-spinels from the dunites (see Table 12), their composition is also dependent on the location of the rocks in the massif.   Thus, the chrome-spinels of the Batamshinsk arched uplift (north of the massif) belong to the subferrichromepicotite type, and are distinguished by their increased iron content (FeO 19.5, Fe_2O_3 9.0 wt %), low chromium content, and high aluminium content.   The chrome-spinels of the dunites from the Tagashasai arched uplift are chrome-picotites and are also low chromium-bearing.   The dunites of the Main ore field contain high-chromium chrome-spinels and belong to the mineral species alumochromite and magnochromite (amount of Cr_2O_3, 52.2 - 59.9 wt %).

Wehrlites, lherzolites, and amphibole peridotites

Insignificant occurrences of wehrlites have been recorded among the dunites of the Main ore field and in the Batamshinsk uplift, where they are associated with diopside vein formations.   These rocks apparently developed as a result of metasomatic replacement of the olivine of the dunites by diopside.

Table 13.   *Chemical Composition of Olivines from Enstatite and Sulphide-Bearing Dunites (in wt %)*

| Components | Enstatite dunite | Sulphide-bearing dunite | |
	1	2	3
SiO_2	40.67	39.34	39.40
TiO_2	0.00	–	–
Al_2O_3	0.00	–	–
Fe_2O_3	0.61	0.61	2.21
FeO	7.60	8.77	10.27
MnO	0.15	–	–
MgO	49.97	47.43	45.68
CaO	0.02	0.66	0.34
H_2O^-	0.00	–	–
H_2O^+	0.90	0.27	tr.
NiO	0.38	0.07	0.01
CoO	0.08	tr.	tr.
CuO	–	0.01	0.35
S	0.00	0.01	–
Cr_2O_3	0.02	–	–
Total	100.40	97.17	98.26

1. $(Mg_{1.830}, Fe_{0.159}, Ni_{0.008}, Mn_{0.003})_2SiO_4$;   Spec.No.87-63, drill-hole 104, depth 325 m.

2. $(Mg_{1.816}, Fe_{0.184})_2SiO_4$; Spec.No.8a, drill-hole 23.

3. $(Mg_{1.775}, Fe_{0.225})_2SiO_4$; Spec.No.9a, drill-hole 23.

Analyses carried out in the Central Chemical Laboratory of IGEM by Yu.Nesterova and V. Lupanova.

Clinopyroxene-poor lherzolites (up to 10%) are known both in the ore-bearing zones, and also among the fields of harzburgites.   Their nature is similar to that of the wehrlites.   They are not widely distributed.

Amphibole peridotites are mainly distributed in the peripheral parts of the massif, although they are known in the zones of tectonic disturbances in the harzburgites, at some distance from the contacts with the country rocks. The amphibole (tremolite) is of secondary origin.

Troctolites

Troctolites are known on the western margin of the massif, near Shandash and Shelekhta creeks, tributaries of the River Kokpekta, and in its southern part, on the right bank of the upper course of Dzharly-Butak Creek, in the depression between the Southeastern and Tagashasai arched uplifts.   On the western margin of the massif, they form a submeridional belt on the surface,

100 - 300 m wide, where gradual transitions from troctolites to harzburgites are observed.   Near Dzharly-Butak Creek, they form small schlieren up to 50 - - 80 m across and are distributed amongst porphyritic harzburgites near the contact zone with the amphibolites.   These rocks have a dark-green colour, with brownish-red weathering, in which altered plagioclase grains stand out in the form of streaks.   The amount of plagioclase varies markedly from a few up to 50%.   Under the microscope, the following composition of the rocks has been identified: the olivine has been serpentinized in varying degree, the plagio-clase is An_{78-80}, clinopyroxene is rare or absent, and there are chrome-spinels and sulphides (pyrrhotite, pentlandite, and chalcopyrite);   secondary minerals are serpentine (lizardite), saussurite, tremolite, prehnite, and magnetite.

From the chemical analytical data, the olivine in the troctolites belongs to the high-magnesium varieties with a fayalite content of 7.75 - 9.0%. The composition of the chrome-spinel in the troctolites is distinguished by its low chromium content and belongs to the ferruginous chrome-picotite variety (Table 14).

Table 14.

Chemical Composition of Accessory Ferruginous Chrome-Picotite from Troctolites (in wt %)

Components	Spec.No.12-60	Spec.No.6-60
SiO_2	0.93	3.43
TiO_2	0.44	0.70
Al_2O_3	25.82	25.01
Cr_2O_3	37.33	32.74
Fe_2O_3	4.29	6.12
FeO	19.10	21.60
MnO	0.24	0.29
MgO	11.85	9.26
CaO	0.00	0.00
Na_2O	–	–
K_2O	–	–
H_2O^-	–	0.00
H_2O^+	–	0.78
V_2O_5	0.16	0.08
P_2O_5	0.04	0.15
NiO	0.15	0.10
CoO	0.19	0.12
Total	100.54	100.38

Analyses carried out in the Central Chemical Laboratory of IGEM by O. Ostrogorskaya

Dyke rocks

Amongst the dyke rocks, the most widely developed are the gabbro-diabases.   Within the Southeastern uplift, near the chromite deposits, diallagites and websterites are known, often containing a segregation of sul-phides.   Dunite dykes of small thickness (up to 1 - 2 m) and thin veins, 1 - 20 cm thick, occur both in the ore chromite bodies, and beyond them in the country rocks near the ore bodies.   Dunite dyke formations have not been among the fields of harzburgites.

The Chromium Content of the Massif

Segregations of chromite ores are located in various sectors of the massif and at different depths from the surface.   The ore bodies are variable in size, and their component ores vary in quality.   The distribution of the chromite deposits and ore-shows depends on features of petrographic composition, the internal structure of the intrusion, and its depth of erosion (Pavlov, 1949).   Mineral-ization depends on the location of the arched uplifts in the intrusion, to which the individual ore fields are also restricted.   Within the intrusion, four ore fields have been recognized, in which about 160 deposits and ore-shows are concentrated.   In the northern part of the massif, within the Batamshinsk uplift, are the deposits of the Batamshinsk ore field.   In the southern part of the massif, in the Stepninsk, Southwestern, and Tagashasai uplifts, there are groups of small deposits and ore-shows.   Near the Southeastern arched uplift is the Main ore field.

The ore chromite bodies of the Stepninsk ore field are distributed in the zone of junction between the troctolites and the harzburgites and they are of small size.   Their extent does not exceed a few tens of metres, and their thickness is 3 m.   The ores are distinguished by their low quality owing to the paucity of chromium in the ore-forming chrome-spinel (Cr_2O_3 40%) with more than 30% Al_2O_3.

The deposits and ore-shows of the Batamshinsk, Southwestern, and Tagashasai ore fields occur amongst peridotites with an insignificant segregation of dunites.   The ore bodies are somewhat larger than those of the Stepninsk area, but their dimensions usually do not exceed 100 m along strike and 10 m in thickness.   Their component ores are also low in chromium and high in aluminium.   However, the amount of Cr_2O_3 in the chrome-spinel concentrate in some cases reaches 43 - 45 wt %.

All the commercial deposits of high-quality chromite ores have been concentrated in the Southeastern uplift within the Main ore field.   Aluminous ores, typical of all the other ore fields of the massif, are here absent.

The Main Ore Field

The chromite deposits and ore-shows within the Main ore field are located in two submeridional zones, the Western and the Eastern.   The zones have been recognized through the wide development of large dunite segregations in the peridotites and the concentrations of ore segregations in them.   Within the zones, a schlieren-banded complex is developed, consisting of dunites, enstatite dunites, and harzburgites with minor participation of wehrlites and lherzolites.   Sulphide-bearing dunites are typically present.

The Western ore-bearing zone extends for 22 km in a northnortheasterly direction.   Near the Millionnoe-Pervomaisk deposits, it merges with the Eastern ore-bearing zone (Fig. 77).   From south to north, the Western zone includes the following deposits: Millionnoe, Almaz-Zhemchuzhina, Pervomaisk, Nos 16, 29, 29a, 31, 39, Khrom-Tausk and Geofizicheskoe VII.   Farther northwards, we have the zones of Komsomol'sk No.4, Geofizicheskoe V and III, 20 years of the Kazakh SSR, Geofizicheskoe XII, Aleksandrovsk, Iyun'skoe, and a whole series of ore-shows. The ore bodies have a westerly dip of from 15 to 75°.

The Eastern ore-bearing zone has the same extent as the Western and is subparallel to it.   In the north, it has a tendency to merge with the Western zone.   From south to north, the following deposits are involved: No.21, Spornoe, Sputnik, Gigant, Geofizicheskoe II and VI, Voskhod, Solov'evskoe, 40 years of the Kazakh SSR, Molodezhnoe, Vkraplennoe, and a number of ore-shows. The ore bodies of the deposits here have an easterly dip at 0 - 50°.   In the southern part of this zone, in deposits No.21, Spornoe, Voskhod, and Molodezhnoe, a southerly pitch of the ore bodies has been recorded, and in the northern part,

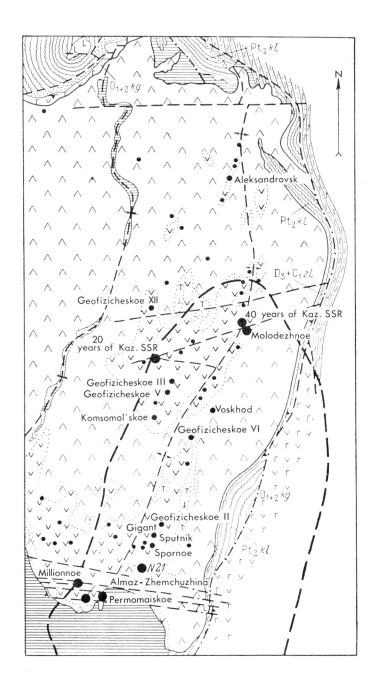

Fig. 77. Diagram of distribution of chromite mineralization in the Palaeozoic basement of the southeastern part of the Kempirsai Massif. For symbols see Fig. 76.

for example, in the Vkraplennoe deposit, the pitch is northerly.   The same tendencies have been recorded also for the ore bodies of the deposits of the Western ore-bearing zone.

The chromite deposits consist of a series of close, compressed and extended ore bodies (Fig. 78).   Vein-like ore bodies (the 20 years of the Kazakh SSR deposit) have been recorded.   Considerably less frequent are the schlieren-like segregations of chromites (the Geofizicheskoe XII deposit). The dimensions of the individual ore bodies vary within wide limits from a few tens of metres up to 1½ km along strike, with a thickness of a few up to 150 m.   The number of ore bodies, comprising a particular deposit, is varied. For example, the Molodezhnoe deposit consists of only one large body, the Almaz-Zhemchuzhina of five, the 40 years of the Kazakh SSR of 15, and the Millionnoe of 99.   The bodies are separated by segregations of dunites, pyroxene dunites, and less frequently, peridotites.

The ore segregations are elongated conformably with the overall strike of the ore-bearing zones, that is, submeridionally.   Exceptions are the Molodezhnoe, in which the ore body has a northeasterly strike (045°), and the Geofizicheskoe VI deposit, the ore body of which, close to equant in plan, is somewhat wider in cross-section (see Fig. 78).   Sometimes the ore bodies have been complicated by gentle folding.   Post-ore tectonics are extremely widely manifested. In all the deposits, the ore bodies are broken mainly by sublatitudinal fractures into individual blocks, and have sometimes been displaced for a distance of from a few tens up to 300 m.

In the ore-bearing complexes, the ore bodies of the deposits, as a rule, occur in dunites and only in rare cases, the compressed schlieren of the harzburgites converge towards their contacts, being separated by a thin selvedge. The contacts of the ore bodies with the country rocks are sharp, clear, and only in rare cases may thin zones (1 - 2 cm) be observed with a depleted segregation of chrome-spinel grains and decrease in their grainsize.

Each deposit of chromite ores is characterized not only by specific features of morphology of the ore bodies, but also by differences in their internal structure.   However, a number of common patterns have also been recorded.   Thus, in the peripheral parts of the ore bodies both in cross-section and in longitudinal section, a somewhat lowered density of segregation as compared with the internal parts has been recorded, significantly greater variation in structures and grainsize of the ore-forming chrome-spinels, the predominance of fine-grained and variable-grained varieties of the ores, and frequently the presence of segregations of nodular ores in them.

The internal parts of the ore bodies are distinguished by the high concentration of segregations of ore minerals, accumulations of massive, almost uniform ores, the marked constancy in the grain dimensions with a predominance of coarse-grained ores, and the maintenance of structures over significant intervals both along strike, and through the thickness.   In the coarse and thick ore bodies, a rough interlayering of the segregated and densely-segregated, medium- and coarse-grained ores has been noted, which forms an unusual rough banding, normally conformable with the attitude of the ore bodies.

The structures of the chromite ores are extremely varied.   Amongst them are segregated, uniform (massive), and nodular types (Fig. 79).   The segregated varieties of ores have been subdivided into densely-, medium-, and rarely--segregated types.   On the basis of grainsize of the chrome-spinels in the seg-

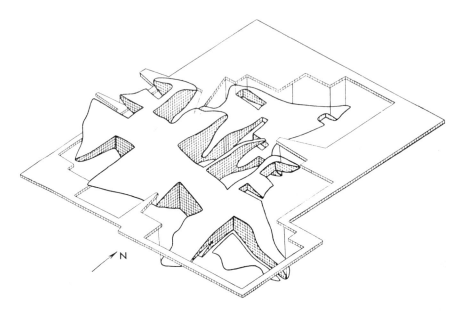

Fig. 78.   Block diagram of the Geofizicheskoe VI ore body (After G. Kravchenko).

regated ores, fine-grained (up to 1 mm), medium-grained (1 - 3 mm), and coarse-
-grained (more than 3 mm) types have been recognized.   The commercial high-grade
chromite ores of the southeastern part of the massif are characterized by densely-
-segregated varieties, in which up to 70 - 90% of chrome-spinels are present,
and 10 - 30% consist of gangue minerals.   On the basis of grain distribution
amongst the chrome-spinels in the ore, banded, schlieren-banded, evenly segregated, and patchy types are distinguished, with a eutaxitic and ataxitic aspect.
The most widely developed are the variable- and coarse-grained varieties of ores
with grainsizes of 1 - 2 and 3 - 4 mm respectively.

The large- and coarse-grained, densely-segregated ores occur mainly
along with massive and nodular types in the large ore segregations.   The massive
(uniform) ores contain insignificant traces of gangue minerals and consist of
medium- and coarse-grained chrome-spinels.   Such ores are of subordinate importance in the deposits of high-chromium ores and are more widely distributed
in the aluminous low-grade ores in the deposits of the Northern group (the Batamshinsk uplift).   The nodular ores are known in most of the deposits, although in
quantity they are markedly subordinate to the ores of segregated and massive types.
The size of the nodules is usually 1 - 2 cm across.   The nodules are often bean-
-shaped, that is, they are compressed and elongated in one direction; spheroidal
nodules are less frequent in occurrence.

The chemical composition of the raw chromite ores from the commercial
deposits of the Main ore field is liable to certain fluctuations (Table 15).
The amount of the principal components within the ore bodies varies according to
the distribution of ores of different structure within them.   Thus, for example,
in the Almaz-Zhemchuzhina deposit, the amount of Cr_2O_3 in the uniform chromite
ores is 58 - 59%, in the densely-segregated ores, 50 - 57%, in the medium-segregated ores 37 - 49%, and in the rarely-segregated ores 28 - 36%.   The average

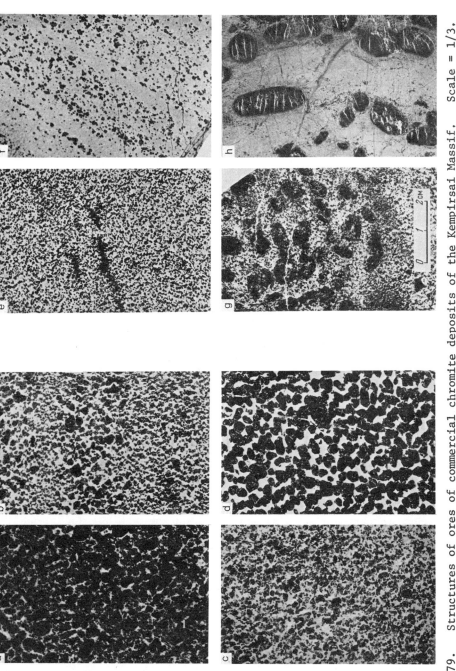

Fig. 79.   Structures of ores of commercial chromite deposits of the Kempirsai Massif.   Scale = 1/3.
a) densely-segregated variable-grained chromite ore, Molodezhnoe deposit, drill-hole 76/12, depth 525 m; *b*) banded ore with alternation of layers of different grain size, same as previously, depth 490 m; *c*) medium-segregated, medium-grained chromite ore, same as previously, drill-hole 75/11, depth 468 m; *d*) densely-segregated, coarse-grained ore with individual nodules, 40 years of Kazakh SSR deposit, drill hole 5, depth 151 m; *e*) fine-grained ore, drill-hole 149, depth 162 m; *f*) poorly-segregated streaky-banded ore, same as previously; *g*) ore of segregated-nodular construction, 40 years of Kazakh SSR deposit; *h*) ore with deformed nodules, same as previously, drill-hole 218, depth 62 m.

amount of this component in the ores of the deposits is 49.05 wt %. In the same way for SiO$_2$, the uniform ores contain 2 - 2.5%, and the poorly-segregated ores up to 18%. Its average amount in the deposit is 8.1%.

Table 15. *Amounts of Principal Components in Ores of Some Deposits of the Main Ore Field (in wt %)*

Deposit	Cr$_2$O$_3$	FeO	SiO$_2$	CaO	P
Almaz-Zhemchuzhina	49.05	12.5	8.1	0.42	0.002
Millionnoe	49.5	12.3	5.2	–	0.008
Pervomaisk	46.0	14.8	9.4	0.3	0.001
Geofizicheskoe VI	55.0	12.2	4.9	–	0.020
Komsomol'sk	49.6	11.6	7.8	–	0.012
Geofizicheskoe XII	10.5	8.4	29.2	0.23	0.001

The amount of iron (as FeO) varies within narrower limits, from 10 up to 15 wt %. In the massive and densely-segregated ores, its amount is greater, but with a decrease in segregation of the chrome-spinels in the ore, it decreases. The average amount[1] of FeO is 12.5%.

The primary composition of the chromite ores, which have not been affected by post-magmatic and hypergene alterations, is extremely simple. Such ores consist of two minerals, chrome-spinel and olivine. In some varieties, periclase is present, which has been enclosed in the chrome-spinels in the form of finely-dispersed grains, measurable in microns. However, the chromite ore of the Kempirsai Massif, which has undergone prolonged and intense post-magmatic alteration, always contains a certain quantity of secondary minerals. A complete list of minerals, identified in the chromite ores of this massif, numbers about thirty names. Olivine, chrome-diopside, chrome-actinolite, garnet (uvarovite), serpentine (lizardite and chrysotile), chrome chlorites (kämmererite and rhodochrome), fuchsite (?), brucite, magnetite, hematite, sulphides (pyrrhotite, mackinawite, pentlandite, pyrite, marcasite, chalcopyrite, and maucherite), hydroapatite and native copper; hypergene minerals are nontronite, cerolite, magnesite, hydromagnesite, artinite, opal, chalcedony, quartz, and iron oxides.

From this list of minerals, sulphides, native copper, and hydroapatite rarely occur in the chromite ores. In the ores, not affected by an ancient weathering crust, that is, at depths of more than 30 - 60 m from the surface, hypergene minerals only occur in the zones of post-ore tectonic disturbances. The best studied minerals, using chemical, X-ray — chemical, spectral, and X-ray — structural methods are those of the chrome-spinel group, the olivines, chrome-actinolites, chrome-chlorites, serpentines, and certain sulphides.

Below, only the most important minerals of the chromite ores of the deposit of the southeastern part of the massif are characterized.

---

[1] Iron is given in the ferrous form.

The ore-forming chrome-spinels of the Main ore field, consisting of magnochromite, are characterized by a large amount of chromium and comparatively small fluctuations in the amounts of the principal mineral-forming oxides. Nevertheless, regular changes in the ratios of chromium and aluminium, and magnesium and divalent iron do occur, according to the scale of the ore segregations and features of construction of the ores.   The chrome-spinels of the largest deposits are somewhat more chrome-rich as compared with those of the small deposits.   The amount of Cr_2O_3 in them is 60 - 64%, whereas in the chrome-spinels of the ores of the small deposits, it comprises 59 - 62%.   There is a wider variation in the magnesium index in the chrome-spinels (MgO:FeO, mol %), from 1.06 to 2.78.   In the large deposits, this index is about 2 and more, and in the small deposits and ore-shows, it is considerably less than 2 (Table 16).

Table 16.   *Chemical Composition of Ore-Forming and Accessory Chrome-Spinels of the Main Ore Field of the Kempirsai Massif (in wt %).*

Locality	Number of Analyses	Cr_2O_3		Al_2O_3		MgO:FeO, mol %	
		from - to	average	from - to	average	from - to	average
Ore-forming							
Large ore bodies, massive ores, and densely-segregated	26	60.90-64.59	62.90	7.67-10.19	8.50	1.82-2.78	2.28
Lean- and medium-segregated ores	21	60.09-64.03	62.28	5.77-11.22	9.04	1.38-2.17	1.81
Small deposits and ore-shows ...	7	59.31-61.94	60.68	8.56-12.16	10.11	1.40-2.26	1.66
Ore veinlets and segregation along banding of rocks.	7	57.05-62.48	59.57	7.78-12.85	9.86	1.06-1.51	1.23
Accessory							
Dunites	7	54.18-61.35	58.71	9.62-15.20	11.73	1.09-1.57	1.33
Pyroxene dunites.	2	48.46-50.38	49.42	18.42-19.52	18.97	0.98-0.99	0.985
Harzburgites	3	31.50-45.72	39.60	23.23-36.78	29.36	1.27-1.61	1.48

Differences in the composition of the chrome-spinels and ores of different structures are determined by the density of segregation of the ore grains, with increase in which the amount of Cr_2O_3 increases somewhat, and there is a marked increase in the magnesium index of the chrome-spinel (Pavlov *et al.*, 1969).

The chrome-spinels of the ores of the Main ore field are similar in composition to the accessory chrome-spinels from the peri-ore dunites and are distinguished sharply from the accessory chrome-spinels of the harzburgites, surrounding the ore bodies and the ore-bearing zones (Table 17).

Table 17.   *Chemical Composition and Certain Physical Parameters of the Chrome--Spinels from the Ores of the Main Ore Field*

Components	Almaz-Zhemchuzhina	Near Donsk	Drill-hole 79-S	West of Susanovka	Almaz-Zhemchuzhina	Geofizicheskoe VI	Almaz-Zhemchuzhina	20 years of Kaz. SSR deposit	40 years of Kaz.SSR deposit	
Specimen No.	131-60	10-58	402-61	1-61	135-60	135-60	154-60	700-60	88-64	704-62
SiO_2	2.33	2.97	0.86	1.20	0.48	0.73	0.59	0.26	0.23	0.26
TiO_2	0.13	0.17	0.21	0.16	0.16	0.21	0.10	0.13	0.26	0.31
Al_2O_3	10.20	11.73	9.33	8.35	9.28	8.36	8.78	7.87	8.83	8.40
Cr_2O_3	56.85	57.66	57.47	58.90	59.17	60.20	60.63	62.44	61.99	62.15
Fe_2O_3	2.34	1.90	2.61	1.19	2.92	3.68	1.89	2.04	2.21	2.96
FeO	17.00	11.25	14.65	17.78	15.01	9.49	11.70	10.86	11.62	10.64
MnO	0.20	0.15	0.18	0.24	0.20	0.14	0.17	0.15	0.18	0.17
MgO	11.02	14.09	13.25	10.62	12.92	15.82	15.50	15.69	14.94	15.25
CaO	0.00	0.00	0.00	0.00	0.00	0.54	0.00	0.00	0.00	0.00
Na_2O	–	–	–	–	–	0.12	0.05	0.05	0.04	0.00
K_2O	–	–	–	–	–	0.05	0.06	0.04	0.06	0.05
H_2O^-	0.00	0.00	1.10	0.71	0.20	0.00	0.00	0.00	0.00	0.04
H_2O^+	0.35	0.35				0.42	0.00	0.00	0.15	0.13
V_2O_5	0.01	0.05	0.04	0.04	0.05	0.08	0.09	0.09	0.10	0.10
P_2O_5	0.01	0.01	0.09	0.09	0.03	0.04	0.00	0.03	–	–
NiO	–	0.08	0.11	0.07	–	0.08	0.06	0.06	0.13	0.09
CoO	–	0.13	0.12	0.16	–	–	–	–	–	–
Total	100.44	100.54	100.02	99.51	100.42	99.96	99.62	99.71	100.74	100.55
a_0, Å	8.3082	8.2846	–	8.3108	–	–	8.3068	8.3116	–	–
R, prov. units	13.2	12.9	–	–	12.4	12.8	12.7	13.0	–	–
Density, g/cm^3	4.45	4.356	4.427	–	4.45	4.434	4.434	–	4.463	4.456

O l i v i n e   is rarely preserved in the chromite ores.   It has usually been replaced by lizardite and other minerals.   It most often occurs in the segregated and lean-segretated ores and in exceptionally rare cases in the monolithic blocks of densely-segregated and massive ores, occurring at great depths.

The olivines from the Almaz-Zhemchuzhina and 40 years of the Kazakh SSR — Molodezhnoe deposits (Table 18) possess a minimum amount of the fayalite component (3.0 – 6.3%), which is lower in comparison with those from the dunites (6 – 8.9%), and the harzburgites (7.1 – 9.3%).   The amount of fayalite component in the olivines of the ores increases with decrease in density of the ore segregation. In addition, the olivines from the roughly-segregated ores contain significantly more nickel (NiO 0.56 - 0.67 wt % in an olivine from the ore matrix and 0.15 - 0.42 wt % in an olivine from the dunites and harzburgites).

Table 18. *Chemical Analyses of Olivines from Chromite Ores of the 40 years of the Kazakh SSR — Molodezhnoe Deposit (in wt %)*

Components	Densely segregated ore		Dunites with streaky clots of chrome-spinels
	1	2	3
SiO_2	41.10	41.26	41.10
TiO_2	tr.	tr.	tr.
Al_2O_3	tr.	tr.	tr.
Fe_2O_3	0.00	0.00	0.00
FeO	3.00	3.40	6.20
MnO	0.03	0.03	0.22
MgO	54.18	54.20	52.02
NiO	0.67	0.56	0.32
CoO	–	0.01	–
CaO	0.00	0.00	tr.
H_2O^-	0.24	0.10	0.04
H_2O^+	0.90	0.60	0.38
Cr_2O_3	tr.	tr.	tr.
S	–	0.00	0.00
Total	100.12	100.16	100.28
Fe_2SiO_4, mol %	3.04	3.41	6.28

1. $(Mg_{1.926}, Fe_{0.061}, Ni_{0.013})_2SiO_4$; Spec.No.270-66, drill-hole 137, depth 506 m.

2. $(Mg_{1.922}, Fe_{0.068}, Ni_{0.010})_2SiO_4$; Spec.No.353-66, drill-hole 10, depth 490 m.

3. $(Mg_{1.864}, Fe_{0.126}, Ni_{0.006}, Mn_{0.004})_2SiO_4$; Spec.No.37-66, drill-hole 167, depth 872 m.

Analyses carried out in the Central Chemical Laboratory of IGEM by V. Moleva.

Chrome-actinolite and the chrome chlorites (kämmererite and rhodochrome) do not form large segregations. These minerals form branching veins and veinlets, from a few centimetres up to 50 cm thick, cutting the chromite ores. Near such veinlets, they are often present among the silicate component of the ores. Chrome-chlorite is usually developed after chrome-actinolite.

Chrome-actinolite is emerald-green in colour, with prismatic grains from 1 to 30 mm in size; $\gamma - \alpha = 0.022$, $2V = (-) 86$ to $(+) 88°$, $z : \gamma = 14 - 19°$; $\alpha = 1.623 (\pm 0.002)$; $\gamma = 1.646 (\pm 0.002)$. The chemical composition of the chrome-actinolites from ores in the Spornoe deposit is as follows (in wt %): SiO_2, 47.32; TiO_2, 0.29; Al_2O_3, 10.68; Fe_2O_3, 2.75; Cr_2O_3, 1.58; MgO, 19.96; FeO, 1.77; NiO, 0.08; CaO, 9.92; Na_2O, 1.74; K_2O, 0.20; H_2O^+, 1.76; H_2O^-, 1.60; S, 0.25; total, 99.96.

Chrome chlorite forms pseudohexagonal tabular prisms and amorphous foliated aggregates of pale-violet colour. It is colourless under the microscope; $\gamma = \alpha = 0.004 - 0.006$; $2V = (+) 40 - 42°$, $z:\gamma = 0$, $\alpha'' = 1.579$ (± 0.003) $\gamma' = 1.580$ (± 0.002).

The numerous chromite deposits distributed within the Main ore field, which are distinguishable in the shape of the ore bodies, their number, dimensions, and depth of occurrence, have a similar mineral composition with close quantitative relationships between the ore-forming minerals and similar chemical composition in them. All these deposits, grouped under the term 'Don', are distinguished by the high quality of the chromite ores. Some of them have been exploited by the Don chromite ore administration (Don GRP). Below, a description is given of the most typical deposits, occurring in the Western and Eastern ore-bearing zones, based on information from the Don GRP.

The 20 Years of the Kazakh SSR Deposit

This deposit occurs in the middle portion of the Western ore-bearing zone and is located 9 km north of Khromtau. The reserves of ores with a Cr_2O_3 content above 50 wt % on the basis of categories A + B + C exceed 25 million tonnes. Chromite mineralization of the deposit is represented in 60 bodies, and commercial reserves are present in 10 principal bodies. The ore bodies are elongated in a submeridional direction (010 - 015°) for a distance of 1.9 km, with a width in plan of from 40 m in the north up to 240 m in the central part. The dip of the ore bodies is westerly from 10 to 50° with a southerly slope of 5 - 10°. The maximum depth of occurrence of mineralization is 40 m in the north and 320 m in the south. The commercial ore bodies have a length along strike of from 80 to 1350 m, with a width in plan of from 33 up to 130 m and a cross-sectional thickness of from 3 - 5 up to 55m (Fig. 80).

The deposit has been broken by sublatitudinal faults into three blocks, displaced relatively to one another. The vertical displacement of the blocks reaches 80 m. In the northern part of the deposit the ores crop out on the eroded surface of the ultramafic rocks and are covered by Mesozoic-Cainozoic sediments. The immediate country rocks are serpentinized dunites and serpentinized after dunites. The contacts between the ore bodies and the dunites are sharp.

In the immediate vicinity of the contacts of the ore bodies, thin (1 - - 3 cm) veinlets of chromite are found in the dunites, oriented in various directions. Cases have been observed when a decrease in the grainsize of the chrome-spinels takes place in the contact zones of the ore bodies and a transition of the ores into segregated and lean-segregated varieties. The ore bodies in the principal mass consist of densely-segregated, medium- and coarse-grained chromite ores. Uniform ores occur in limited amounts. Medium- and rarely--segregated varieties of ores also play a subordinate role and are distributed in the peripheral portions of the ore bodies.

The upper horizons of the ore bodies down to depths of 20 - 25 m from the surface consist of friable varieties of ores, the origin of which has been controlled by the development of an ancient weathering crust. Below this level there are solid chromites. The mineral composition of the ores consists of chrome-spinel (magnochromite), serpentine (lizardite), and sometimes a relict olivine (forsterite), brucite, and a powdery magnetite. There are rare occur-

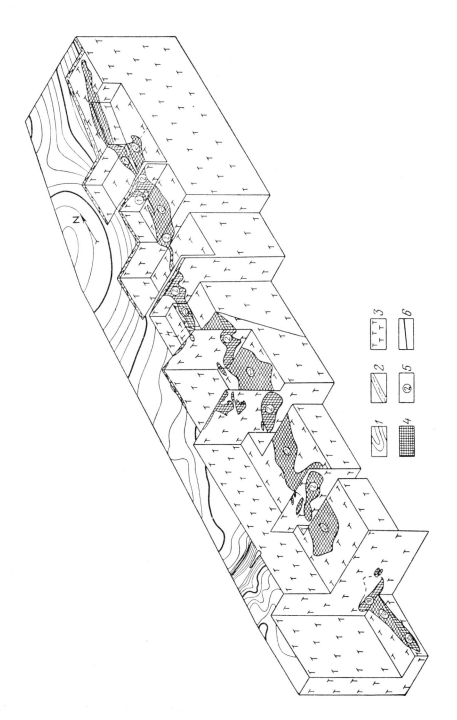

Fig. 80.   Block-diagram of ore segregations 1, 2, and 10 of the 20 years of the Kazakh SSR deposit
(After M. Shul'gin).
1) present surface;   2) Mesozoic-Cainozoic sediments;   3) peridotites;   4) chromite ores;
5) ore bodies;   6) tectonic disturbances.

rences of garnet (uvarovite), chrome-diopside, chrome-actinolite, chrome chlorites (kämmererite and rhodochrome), hydromagnesite, and sulphides (pyrrhotite, mackinawite, and pentlandite).   The hypergene minerals include magnesite, ankerite, calcite, quartz, chalcedony, and iron oxides.

The raw ores contain (in wt %): Cr_2O_3, 22.3 - 62.1; SiO_2, 1.3 - 21.7; Fe_2O_3, 9.2 - 15.1; CaO, 0.22 - 1.3; and P, 0.03.

The 40 Years of the Kazakh SSR — Molodezhnoe Deposit

This deposit is located 4 km eastnortheast of the 20 years of the Kazakh SSR deposit and is located in the middle portion of the Eastern ore-bearing zone. Exploratory drilling has established that the ore bodies of the 40 years of the Kazakh SSR and the Molodezhnoe deposits in the most southern portion at depth occur *en échelon* with each other, so that both deposits have been grouped into a single ore field and have been considered together.   The total reserves of the deposit comprise about 90 million tonnes of ore with a Cr_2O_3 content of about 50%.   In the ore field there are 23 ore bodies, replacing each other *en échelon*.

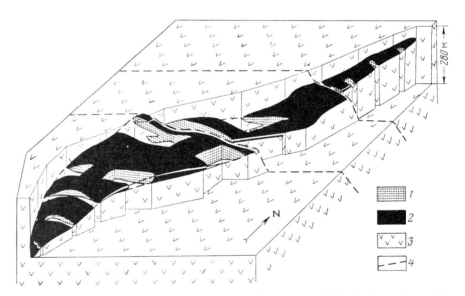

Fig. 81.   Block-diagram of buried ore body in the Molodezhnoe deposit.
1) ore body in section; 2) contact surfaces of ore body; 3) surrounding ultramafic rocks; 4) tectonic disturbances.

The ore bodies of the deposit strike submeridionally.   Part of the ore bodies have a gentle subhorizontal attitude, and another portion displays a tendency to dip westwards at 10 - 25°.   In the north, mineralization appears on the surface of the Palaeozoic basement, which is covered by thin Mesozoic - - Cainozoic sediments, and in the south mineralization occurs at depths of from 300 to 350 m.   The most extensive ore body No.4 extends for 934 m, with a width in plan of 200 m.   Its maximum cross-sectional thickness is 65 m.   The ore body

next in size, No.3, has a length of 664 m, a width of 282 m, and a thickness
of up to 85 m.   Other bodies along strike measure from 50 up to 420 m, with a
width in plan of 50 - 100m, and thicknesses of 5 - 40 m.   All the ore bodies
occur in serpentinized dunites, among which there are separate bodies of harzburgites, pyroxene dunites, and sulphide-bearing dunites (Fig. 82).

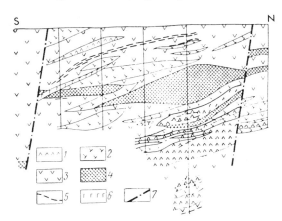

Fig. 82.
Longitudinal section through a sector
of the 40 years of the Kazakh SSR
deposit.

1) harzburgites;   2) enstatite
dunites;   3) dunites;   4) chromite
ore;   5) thin schlieren-banded
chromite segregations;   6) gabbro-
-diabases;   7) tectonic disturbances

        The Molodezhnoe deposit consists of a compact ore body, the largest
in the ore field under consideration.   Along strike, this body extends for
1540 m with a width in plan of from 200 to 300 m, and a maximum thickness of
140 m (Fig. 81).   The body does not crop out on the present surface, but occurs
at depths of 422 - 600 m and sinks evenly in a southerly direction.   This ore
body, elongated at 45°, dips in the northern part to the southeast at 25°;
towards the south it flattens out somewhat.

        The ore body is cut by tectonic fractures, mainly sublatitudinal in
direction and dipping steeply southwards and southwestwards at 70 - 85°.   The
throw varies from 8 up to 40 m.   The ore bodies are composed of both densely-
-segregated, almost uniform chromites, and in varying degree segregated varieties.

        The ores of the deposit contain (in wt %): Cr_2O_3, 19.3 - 58.95;
SiO_2, 0.3 - 30.0; Fe_2O_3, 9 - 14.5; CaO, trace - 0.84; and P, up to 0.005.
The amount of titanium, manganese, nickel, and other minor traces does not
exceed hundredths and a few tenths of a percent.   Complete chemical analyses
of the 'raw' chromite ores in their most typical representatives are given in
Table 19.

*The Batamshinsk Ore Field (Chromite Deposits and Ore-Shows of the Northern
Portion of the Massif).*

        In the northern portion of the massif, near Kuagach and the Kempirsai
and Istai-Keze gorges, over 30 small deposits and ore-shows have been counted.
Most of them are distributed in the axial part of the uplift, the crest of which
has undergone considerable erosion.   Fifteen kilometres north of Kempirsai
village, a separate deposit, the Buranovsk, is known, occurring in the area of
silicate-nickel deposition of the same name.   The Buranovsk deposit can be
traced in the shape of a vein-like body on the surface for 135 m, with a thickness of from 5 to 10 m.   It has been investigated only on the surface and can
be traced to depths of 50 - 60 m in a small number of drill-holes.

Table 19. *Chemical Composition of 'Raw' Chromite Ores from the 40 years of the Kazakh SSR — Molodezhnoe Deposit*

Components	Molodezhnoe deposit			40 years of Kazakh SSR deposit				
				Ore body 3			Ore body 4	Ore body 15
SiO_2	0.24	4.30	4.34	6.58	7.20	7.26	7.33	7.48
TiO_2	0.15	0.15	0.15	0.12	0.11	0.10	0.07	0.10
Al_2O_3	8.12	7.57	8.14	7.64	6.57	6.73	7.13	7.83
Cr_2O_3	58.95	54.93	50.28	49.32	48.68	43.33	48.48	47.21
Fe_2O_3	0.12	2.72	0.50	3.00	2.05	0.10	2.61	2.22
FeO	13.47	9.62	12.90	8.91	9.93	13.13	9.36	10.05
MnO	0.14	0.15	0.14	0.15	0.16	0.14	0.14	0.15
MgO	18.19	17.90	20.94	20.30	20.54	23.61	20.53	20.60
CaO	0.10	0.13	0.20	0.10	0.43	0.06	0.07	0.08
NiO	0.05	0.05	0.07	0.07	0.07	0.09	0.07	0.07
CoO	n.f.	n.f.	n.f.	n.f.	n.f.	n.f.	0.005	0.005
P	0.002	0.002	0.003	0.003	0,002	0.002	0.003	0.002
S	–	0.012	0.022	0.031	0.046	0.046	0.035	0.039
calcination loss	0.65	1.47	2.52	2.75	3.31	4.78	3.13	3.13
Total	100.18	99.0	100.20	98.97	99.09	99.37	98.96	98.96

Analyses carried out in laboratories of Kaz. Inst. Miner. Syr'ya and Aktyubinsk Group Geological Research Expedition

Chromite mineralization of the ore field under consideration is arranged mainly in fields of development of banded harzburgites, although massive varieties also occur amongst them.   The rocks directly surrounding the ore bodies are dunites, which form selvedges or thin bodies.   The chromite ores of the segregation are vein-like bodies or markedly compressed lenses and schlieren. On the surface, they extend from 30 up to 135 m with a thickness of from 1 up to 10 m.

The strike and dip of the ore bodies usually coincide with those of the layering in the harzburgites, but there are also deviations, when the strike of the ore body is at 45° and the dip at 35°, or the case when the ore schlieren at the southern end has a strike of 340°, and then, curving westwards, becomes sublatitudinal.   The Buranovsk deposit has a strictly meridional strike.   On the basis of the structures of the chromite ores in this field, three groups have been recognized (massive, densely-segregated, and segregated with which nodular varieties are often associated).

The ore bodies are composed in most cases of massive and densely-segregated ores, and in this respect, the massive types have been distributed within the bodies, and the densely-segregated types, along the periphery.   The contacts with the country rocks are sharp.

The chemical composition of the 'raw' ores from this region display a relatively low content of Cr_2O_3 (32 - 38 wt %) with a Cr_2O_3 : FeO ratio of 2.1 - 2.8.

The bulk composition of the ores of the Buranovo deposit is as follows (in wt %): SiO_2, 0.78; TiO_2, 0.21; Al_2O_3, 31.26; Cr_2O_3, 37.08; Fe_2O_3, 9.15; FeO, 6.10; MgO, 15.95; CaO, 0.00; H_2O^+, 0.15; total, 100.68; Cr_2O_3 : FeO = 2.6.   These data indicate that the ores of the region are suitable mainly for the refractory and chemical industries.

Table 20.   *Chemical Composition and Some Physical Parameters of the Ore-Forming Chrome-Spinels of the Batamshinsk, Tagashasai, Southwestern, and Stepninsk Ore Fields (in wt %)*

Components	Batamshinsk					Tagashasai			South-western	Step-ninsk
	1	2	3	4	5	6	7	8	9	10
SiO_2	0.37	1.75	1.97	0.91	0.78	1.47	0.39	1.66	0.73	4.50
TiO_2	0.43	0.23	0.34	0.25	0.21	0.52	0.43	–	0.35	0.28
Al_2O_3	21.84	22.98	26.20	32.29	31.26	32.82	25.75	25.12	23.78	32.89
Cr_2O_3	45.62	39.34	39.70	34.33	37.08	32.67	39.92	33.20	42.88	30.57
Fe_2O_3	3.96	3.90	2.67	2.70	9.15	4.20	3.53	9.04	6.80	2.51
FeO	12.23	14.11	11.80	10.98	6.10	11.60	12.80	10.05	10.18	13.40
MnO	0.17	0.23	0.08	0.15	0.00	0.13	0.19	–	0.16	0.15
MgO	15.30	15.62	15.62	17.71	15.95	16.62	16.34	14.76	15.36	14.99
CaO	0.00	–	0.00	–	0.00	0.10	0.00	–	0.00	0.00
Na_2O	0.08	–	–	–	–	0.17	0.10	–	0.12	–
K_2O	0.08	–	–	–	–	0.05	0.03	–	0.04	–
H_2O^-	0.00	0.25	0.19	0.14	–	0.00	0.28	–	0.00	0.00
H_2O^+		0.75	0.53	0.41	0.15	0.00	0.30	–	0.00	0.63
V_2O_5	0.17	0.13	0.13	0.09	–	0.07	0.16	–	0.11	0.06
P_2O_5	0.03	–	0.00	–	–	0.04	–	–	0.05	0.03
NiO	0.11	0.13	0.14	0.12	–	0.08	0.10	0.23	0.10	0.11
CoO	–	–	–	–	–	–	–	–	–	0.06
Total	100.39	99.42	99.37	100.08	100.68	100.54	100.32	94.06	100.66	100.18
a_0, Å	8.2547	–	–	–	–	8.2147	8.2365	–	8.2454	8.2051
R, prov. units	11.9	–	–	–	–	10.4	11.0	–	11.2	10.4
Density, g/cm^3	4.21	–	–	–	–	4.105	4.208	–	4.17	4.05

1) Spec.No. 289-60, southwest of Batamshinsk;   2) Spec.No. 186-37, Kempirsai Gorge, point 7;   3) Spec.No. 174-37, ore-show 13;   4) Spec.No. 260-37, ore-show 23;   5) Spec.No. 236-37, Buranovo deposit;   6) Spec.No. 7-60, Ivanovo ore-show;   7) Spec.No. 410-60, ore outcrop on R. Tagashasai;   8) Spec.No. 290-38, ore outcrop to west of R. Dzharly-Bulak;   9) Spec.No. 64-60, deposit 22;   10) Spec.No. 67-60, ore-show 17.

The mineral composition of the Batamshinsk ores is analogous to that of the above-described ores of the Main ore field (the Don deposits). However, the composition of their ore-forming chrome-spinels differs markedly. The chrome-spinels of the ores of the Batamshinsk ore field, as with the Tagashasai, Southwestern, and Stepninsk types, contain 22 - 32 % Al_2O_3 and only 20 - 40 and rarely 45% Cr_2O_3 (Table 20).

From the characteristics of the chrome occurrences in the Kempirsai Massif, it follows that the scale of mineralization and the quality of the ores, the shape and dimensions of the intrusion, the composition and degree of differentiation of the ultramafic rocks, the position of the feeding channel along which the magmatic and ore-silicate melts entered the system, and also the presence of arched structures in the massif favourable to the spread of mineralization, exerted an influence on the distribution of the chromite deposits, the scale of mineralization, and the quality of the ores. An important factor in the discovery of the deposits on the present surface of the intrusion was the degree of erosion, which revealed various petrographical complexes, bearing mineralization of various types. Thus, the distribution patterns of the chromite deposits have been subjected to three factors: magmatic, tectonic, and erosion. Differentiation of the ultramafic magma led to the formation of the ore concentrations, and the protectonic processes determined the distribution of these concentrations in the body of the intrusion. Erosion revealed the productive petrographical horizons of the intrusion.

The Deposits of the Shorzha Massif

The Shorzha chromite-bearing massif is located in the Krasnosel'sk region of the Armenian SSR on the northeastern coast of Lake Sevan, near Shorzha. The massif is involved in the Sevan-Amasi belt of ultramafic rocks, extending within the Armenian SSR in a southeasterly direction for 80 km and continuing farther into the territory of Azerbaidzhan. The massif includes more than ten small deposits and ore-shows of chromites.

The Shorzha Massif is located amongst Upper Cretaceous and Eocene sediments. The Upper Cretaceous consists of a thick sequence (up to 2000 m) of clastic rocks, with subordinate seams of tuffogenic formations and reefal limestones, assigned to the Turonian Stage. On this sequence rest medium- and thinly-layered limestones of the upper Senonian, up to 1500 m thick. The Eocene rocks rest conformably on the Upper Cretaceous deposits and consist of a sequence of shales. All the rocks have been crumpled into intense isoclinal folds, elongated in a sublatitudinal direction. The Shorzha Massif is restricted to one of the anticlinal folds.

In the northeastern portion of the massif, there are Jurassic tuffogenic rocks, which have been overthrust along a tectonic disturbance. The plane of the thrust has a northwesterly strike with a northeasterly dip at 35 - 40°.

The massif has been elongated latitudinally, and has an extent of about 3 km and a width in plan of from 100 to 700 m. In the western part, it has the shape of a narrow dyke-like layered segregation, dipping northwards at 75 - 80°, and is located amongst the limestones. Farther eastwards, the massif gradually expands, taking the shape of a laccolith-like body. The eastern portion of the massif is bounded by the line of thrust of the Jurassic sequence onto the Upper Cretaceous and Eocene sediments. According to K. Paffengol'ts, this thrust belongs to a pre-Eocene orogenic phase.

Fig. 83.   Diagram of the geological structure of the Shorzha peridotite massif (After A. Betekhtin)

1) dunites;   2) serpentinites;   3) peridotites;   4) pyoxenites;   5) troctolites;   6) Upper Cretaceous and Eocene limestones;   7) listwänites and dolomites;   8) porphyrites;   9) Jurassic tuffogenic rocks;   10) basalts;   11) alluvial deposits;   12) line of thrust;   13) chromite deposits and ore-shows.

According to Betekhtin (1937), and Abovyan (1961), the Shorzha Massif represents the apical portion of a deep-seated intrusive segregation, possessing the features of a laccolithic body.   The time of injection is assigned to the Late Eocene.   Knipper & Kostanyan (1964) and Morkovkina & Arutyunyan (1971) believe that the Shorzha Massif (like the other ultramafic massifs of the Malyi Kavkaz) has a secondary attitude, resulting from protrusion of ultramafic rocks formed earlier than the country rocks.   The massif consists mainly of harzburgites, amongst which irregular lensoid bodies of dunites have been recognized, being interstratified with harzburgites, and elongated conformably with the strike of the massif (Fig. 83).   Lherzolites and wehrlites are known in insignificant amount.   In the eastern portion of the massif, troctolites are distributed in the form of a northwesterly-striking belt, and these pass southwards gradually into harzburgites.   Elongate anorthosite schlieren occur in the troctolites.

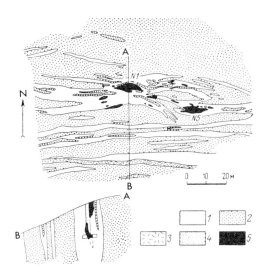

Fig. 84.
Diagram of the geological structure of the region of deposits 1 and 5 (After A. Betekhtin).

1) dunites;   2) peridotites;   3) basalts;   4) serpentinites;   5) chromite ironstones.

No patterns have been identified in the distribution of the dunite segregations amongst the harzburgites (Fig. 84).   The shape and attitude of the dunite schlieren and the elongated belts are in the main concordant with the overall shape of the massif.   Amongst the dyke series of rocks there are pyroxenites (websterites), diabases, diabase porphyrites, and diorite porphyrites, from 0.5 to 10 m thick over an extent of 10 - 60 m.

All the ultramafic rocks of the massif have undergone serpentinization in varying degree.   In the zone of contacts with the country rocks there are in a number of places serpentinites, in which the relict texture of the primary rock is absent.   The serpentine in the serpentinized peridotites and dunites consists of lizardite and antigorite.   Antigorite predominates markedly in the serpentinites proper.

In the contact zones between the ultramafic rocks and the limestones, there is a widespread development of listwänites.   Sometimes carbonatized, sheared and brecciated serpentinites occur here, which indicates the tectonic nature of the contacts in such sectors.

Little-altered varieties of harzburgites are rare in the massif.
Their primary composition, ignoring the degree of serpentinization, is evidently as follows: 70 - 80% olivine, about 30 - 20% orthopyroxene, and 1 - 2% accessory chrome-spinel.   Clinopyroxene (diopside) occurs sporadically.

The chrome-spinels form rare grains measuring from fractions of a
millimetre up to 1.5 mm and xenomorphs with respect to olivine.   Grains of
idiomorphic outline, enclosed in pyroxene, occur rarely.

The dunites are usually considerably serpentinized.   The olivine
has mainly been replaced by lizardite and less frequently by antigorite, which
has been developed both after lizardite and after relicts of olivine.   The
olivine in the unaltered dunites consists of idiomorphic grains measuring 1.5
- 2.0 mm.   On the basis of optical data ($2V$ = (+) 86 - 88°, $\gamma - \alpha$ = 0.034 -
0.035), it has been assigned to the high-magnesian varieties, with 7 - 10% of
the fayalite molecule.   The chrome-spinels are present in the form of idiomorphic or rounded grains measuring 0.5 - 1.5 mm.

The troctolites have gradual transitions into the peridotites and are
unique representatives of the gabbroid rocks, associated with them genetically.
The principal rock-forming minerals are plagioclase, comprising from 50 to 70%
of the rock, and olivine, 30 - 50%.   The secondary minerals comprise epidote,
sericite, prehnite, serpentine, hornblende, and rarely clinopyroxene.   The
accessory ore minerals are chrome-spinel, magnetite, and rarely sulphides.   On
optical data, the olivine is a high-magnesian type with 8 - 10% of the fayalite
component.

All the chromite ore bodies are restricted to the segregations of
dunites and, as a rule, they occur conformably with the attitude of the dunites.
Having a predominant sublatitudinal strike, the ore bodies possess a steep
northerly or southerly dip with a clearly defined pitch to the east (Abovyan,
1961).   The ore segregations consist of bodies of lens- and nest-like shape.
The nest-like bodies have been elongated vertically and converge towards column-
-like segregations.   The dimensions of the ore bodies are extremely insignificant.   The nest-like bodies are 3 - 4 m across, the lens-like bodies are up to
25 - 30 m in length, with a thickness of 1 - 4 m, and the columnar forms (with
lensoid cross-section) can be traced for 30 - 35 m downslope with maximum cross-
-sectional dimensions of 10 and 3 m.

The relationships between the ore bodies and the surrounding dunites
are varied.   Gradual transitions are known from the segregated chromite ores
into the surrounding dunites.   The ore bodies with segregated or massive ores
often possess sharp boundaries with the surrounding dunites also.   Definite
veins of chromites, cutting the dunites, have also been observed.   Amongst
the structures of the ores, massive (uniform), densely-segregated, banded, and
evenly-segregated varieties can be distinguished.   Nodular chromite ores are
also known (deposit 3).   There is a predominance of massive ores, forming
independent ore bodies, and they frequently occur amongst the segregated ores
in the form of small sectors.   The grainsize of the chrome-spinels in the ores
varies widely from 1 up to 10 mm.   The important minerals of the ores are also
serpentine (lizardite and antigorite) and sometimes olivine.   In subordinate
amount are uvarovite, kämmererite, chrome-diopside, calcite, brucite, and rarely
native copper, sulphides (pyrrhotite and pentlandite), and magnetite.

In chemical composition, the chromite ores of the Shorzha Massif are
varied.   Thus, in the ores of deposit 2, distributed in the northern contact

portion of the intrusion, the content of Cr_2O_3 is 32 - 35 wt %, and the ore-
-forming chrome-spinel contains about 40 wt % of Cr_2O_3, with about 25% of Al_2O_3,
and total iron calculated as FeO, about 16%.   The ores of the other deposits,
for example, deposits 27 and 28, which occur in the inner portion of the massif,
are considerably richer in chromium and are less aluminous.   The amount of Cr_2O_3
in such ores is more than 45 wt %, alumina 10 - 14%, and FeO about 16 - 20%.

The ore-forming chrome-spinels of the deposits of the Shorzha Massif
(Table 21) have been separated into two groups (high-chromium (Cr_2O_3, 46 - 53%;
Al_2O_3, 8 - 14%), and low-chromium with a large amount of alumina (Cr_2O_3, about
40%; and Al_2O_3, about 23%).   The high-alumina ores concentrate towards the
fields of widespread development of the peridotites, whereas the high-chromium
ores occur in the fields of maximum development of dunites.

Table 21. *Chemical Composition of Chrome-Spinels from Ores of the Deposits of*
the Shorzha Massif (in wt %)

Components	Deposit 1	Deposit 2	Segregated zone gallery 10	Ore body 27 gallery 4	Ore body 28 gallery 10
SiO	1.76	2.30	1.64	3.16	n.d.
Al_2O_3	7.68	22.70	12.27	14.46	8.90
Cr_2O_3	53.49	38.87	46.15	48.84	50.14
Fe_2O_3	6.94	4.97	4.61	4.46	4.97
FeO	14.07	11.33	15.01	16.53	12.65
MgO	10.87	13.40	10.50	12.12	12.22
CaO	nil	1.34	0.50	–	0.38
MnO	0.22	–	–	–	–
calcination loss	–	n.d.	n.d.	n.d.	n.d.
Total	95.03	94.91	90.68	99.57	89.26

Analysis of chrome-spinel from deposit 1 from Betekhtin (1937), remainder from
Abovyan (1961).

The Shorzha Massif is distinguished by the significant differentiation
of the ultramafic rocks, from massive and segregated chromites through dunites
and harzburgites to rocks of the gabbroid series (troctolites and anorthosites).
With its small dimensions (1.5 km^2), the massif also contains ore bodies of the
primary composition of their component ore-forming chrome-spinels according to
their distribution in the fields of harzburgites or in the dunite segregations.

CHROMITE DEPOSITS, OCCURRING IN MASSIFS OF THE DUNITE-HARZBURGITE SUBASSOCIATION

This subassociation includes the following chromite-bearing massifs:
Khalilovo and Akkarga in the Southern Urals, Klyuchevsk in the Middle Urals,
Voikar-Syn'ya in the Northern Urals, the Geidarinsk Massif in Transcaucasia,

and the Ospa and Khara-Nura massifs in the East Sayan, etc.   According to
Petraschek (1957) and Hiessleitner (1951, 1952), a significant number of chromite-bearing massifs in Turkey, Iran, Albania, Yugoslavia, and other countries
possess an analogous petrographical composition.

The Klyuchevsk Deposits

The Klyuchevsk chromite deposits occur in an ultramafic massif of the
same name, located on the eastern slopes of the Northern Urals, 50 km southeast
of Sverdlovsk.

The Klyuchevsk Massif is involved in the Alapaevo-Techen belt of
ultramafic rocks (Malakhov, 1966), to which, besides the Alapaevo and Techen
massifs, are also assigned the Ostaninsk, Rezhevsk, Bazhenovo, and Kazakbaevo
massifs.   The belt is located on the western limb of the Alapaevo-Breda depression.   The Klyuchevsk Massif is restricted to one of the folds which complicate the Aramil'sk synclinorium.   The massif, about 80 km^2 in area, is
mushroom-like in plan.   It extends in a submeridional direction for 17 km.
In the southern portion, its width reaches 8 km, and to the north it decreases
to 1 - 3 km.

The ultramafic rocks occur in Lower Silurian metamorphosed volcanogenic-sedimentary sequences.   According to Yu. Solov'ev, the stratigraphical
section of the Palaeozoic sediments of the Klyuchevsk Massif area comprises
Upper Ordovician graphite-quartz slates, and Lower Silurian mica-quartzite slates,
phyllites, gneisses, and marbles.   In the Upper Silurian deposits, metamorphosed
eruptives of basic composition, tuff-sandstones, and amphibolites are present
along with carbonaceous-chert, quartz-sericite, and greenschists.   The Lower
Devonian deposits consist of basaltic porphyrites and tuffs with seams of cherty
limestones.   The Ordovician deposits are 500 m thick, the Lower Silurian from
0 to 800 m, the Upper Silurian, 200 - 500 m, and the Devonian deposits 500 m.

The rocks, surrounding the massif, have been crumpled into eastward-
-overturned isoclinal folds, the axes of which are subparallel to the elongation
of the massif.   The eastern limbs of the folds are steeper, and the western
limbs gentler.

The western and eastern contacts between the ultramafic rocks and the
surrounding deposits have, on the basis of geophysical data, a westerly dip.
In this respect, the western contact is simpler than the eastern.   The southern contact is tectonic and dips steeply northwards.   The shape of the massif,
occurring in a synclinal structure, is lopolith-like.   On geophysical data,
the greatest thickness (4 - 5 km) has been identified in the southern portion
of the massif, where the presence of a magma-feeding channel is assumed to have
been present.

The time of intrusion of the ultramafic rocks, according to P. Sobolev,
is Middle - Late Devonian.   According to Pronin (1948), the age of the Klyuchevsk
Massif, like all the remaining massifs of the belt, is Carboniferous.   Within
the area of the Klyuchevsk Massif, rocks of ultramafic, gabbroid, and granitoid
complexes have been developed and also a number of metamorphogenic formations.
More than half of the area is covered by peridotites of harzburgite composition
with an original content of orthopyroxene of up to 25 - 30%.   The dunites and
apo-dunitic serpentinites form bands more than 1.5 km wide along the southern
and southeastern margins of the massif, and also crop out in the form of patches

elongated meridionally in the field of peridotites.   The areas of individual outcrops of dunites decrease from south to north, where they do not exceed 0.5 km^2 and are arranged predominantly along the eastern edge of the massif.   On the whole about one third of the area of the massif is occupied by dunites.

The pyroxenites and gabbros, which are younger with respect to the complex mentioned, are most widely developed in the west of the massif, where their outcrops form vast fields.   The pyroxenites consist mainly of fresh coarse--grained diallage, and they contain accessory magnetite (rarely chrome-spinels), and sometimes they have been converted to talc and antigorite.   In the south of the massif in the area of development of predominant dunites and the apo-dunitic serpentinites, dyke-like pyroxenite bodies are distributed, extending for more than 300 m, and there are thin veinlets of pyroxenite.

Along the contacts between the pyroxenites and the dunites, wehrlites are frequently developed, which are characterized by relative freshness of the pyroxenes along with intense metamorphism of the chrome-spinels and decrease in the amount of clinopyroxenes in the direction of the dunites.   Telegin (1967) considers the wehrlites to be the product of metasomatic effects of the pyroxenites on the dunites.   Sometimes the wehrlites occur in the form of individual segregatations outside any visible contact with the pyroxenites.

The dyke rocks of the granitoid complex (bimica granites) are concentrated on the northeastern margin of the massif.   Granite-gneisses and veined alaskite granites have been developed along its southeastern border.

The rocks of the dunite-harzburgite complex have been subjected to processes of serpentinization (lizardite and antigorite), chloritization, amphibolization, carbonatization, and graphitization, and under hypergene conditions, to silicification and nontronitization.   Metamorphic transformations have been most strongly manifested in the zones of the tectonic disturbances.   At the contact between the ultramafic rocks and the surrounding deposits (eastern contact) and along the boundary between the serpentinites and the veined granites and lamprophyres, there are serpentine-carbonate, talc-carbonate, and talc-chlorite rocks, which were formed as a result of the effects of an acid magma on the ultramafic rocks.   In the south of the massif, the belts of these rocks, several hundreds of metres long and up to 50 m thick, have a latitudinal, northwesterly, and southwesterly strike.   They are also widely developed within the ore fields.

In the area of the Klyuchevsk Massif, about 200 small deposits and ore-shows are known, located both in the field of dunites and amongst the peridotites also.   In the belt of dunites, bordering the massif on the south, there are schlieren-like and vein-like, often subparallel bodies of chromite ores. Adjacent elongated ore bodies form the ore-bearing zones.   Individual, most productive sectors of these zones are recognized as independent bodies.   The largest of them are the Revda, Pervomaisk, Kozlovsk, and Samokhvalovo.

The ore-bearing zone of the Revda deposits, located on the left bank of the River Iset' opposite the mouth of the River Syserta, extends for up to 1.5 km.   The strike of the zone is sublatitudinal, and the dip is northwards at 65 -- 90°.   Within the zone, from west to east, five deposits can be recognized, Revda I - V, and a number of ore-shows.   In the west, the zone thins out, and in the east it passes into the zone of the Pervomaisk deposit.

From the surface, the deposits have been exposed in exploratory trenches. The old working quarry, the largest in the Klyuchevsk deposits, extends for 175 m, with a width of 40 m and a depth of up to 15 m.   A sector of the deposit has been drilled to depths of 300 m.   Within the ore field on the surface and in the drill cores, bodies have been discovered which consist of ores of the segregated type with variable (from 1 to 16 m) thickness of ore layers and varying density of segregation.   Such bodies are separated by seams of non-ore dunites (Fig. 85).

The interlayering of the thin ore layers with varying density of segregation creates the banded structure of the ores.   The banding, as a rule, conforms with the spatial orientation of the ore bodies themselves.   In the ore belts themselves, a symmetrical structure sometimes appears.   From the dunite towards the middle of the belt there is a gradual increase in segregation up to the formation of almost uniform ore.   The chromite concentrations often have the form of streaks.   The streaks are either rectinlinear or they curve, branch, or replace one another *en échelon*.   In addition to the chrome--spinels in the ores, there are relicts of olivine and replacing serpentine, chlorite, brucite, talc, and also magnetite, sulphides, iron hydroxides, and carbonates (magnesite, calcite, and breunnerite).

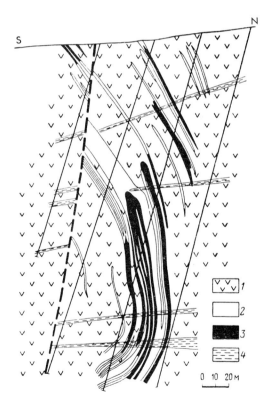

Fig. 85.
Geological section through a sector of the Revda deposit (After A. Sharypov).

1) dunites and apo-dunitic serpentinites;   2) dunites with increased segregation of chrome-spinels and 10 wt % of Cr_2O_3;   3) chromite ore with 10 – 16 wt % Cr_2O_3;   4) talc--carbonate rocks.

The ores of the Revda deposits, like those of the remaining deposits of the Klyuchevsk Massif, have been subdivided into rarely-segregated types with from 10 to 30% of ore grains, medium-segregated, with up to 50% of chrome-spinels, and massive types, containing more than 90% of ore grains.   Consequently, the massive ores contain 40 - 55% Cr_2O_3, the densely-segregated, 30 - 45%, the medium--segregated 15 - 30%, and the remainder, less than 15%.   All these types occur among the ores of the Revda deposits, although there is a marked predominance of ores with 8 - 10% Cr_2O_3.   In individual sectors it may increase to 36%.

At the same time, the ore-forming chrome-spinels are distinguished by the large content of chromium.   They contain (in wt %): Cr_2O_3, 52 - 59; Al_2O_3, 8 - 15; FeO, 14 - 25; Fe_2O_3, 3 - 9; MgO, 9 - 13.   Manganese, nickel, cobalt, vanadium, and titanium are present as isomorphous additives.   At the present time, the deposit has been worked to a depth of 15 m, and mineralization can be traced to depths of 300 m without any evidence of thinning out.

The zone of the Pervomaisk deposits is located to the east and south-east of the Revda deposits and they are separated quite arbitrarily.   The zone extends for about 1 km, with a width in different parts of 300 - 500 m.   Mineralization has been recorded to a depth of 250 m.   The zone has been exposed in trenches and in drill-holes.   During the period of working, a quarry and shaft were installed with cross-cuts to a depth of 30 m.

Within the ore-bearing zones, non-ore layers alternate with belts or layers enriched in chromite segregations.   With denser segregation, the boundaries of the ore segregations and the surrounding dunites are quite sharp, and when the segregation is exhausted, transitions from ores to dunites may be completely gradual.

The shape of the bodies which consist of interlayered non-ore seams and sectors of concentrated chromite segregation, is most commonly layer-like, and sometimes the bodies have the shape of lenses and nests.

The ore bodies are 5 - 20 m thick, up to 350 m long, and like the Revda bodies they are separated by apo-dunitic serpentinites.   The strike of most of the ore bodies, which crop out on the surface, is northeasterly, whereas the dip directions are quite variable and are determined by the fold shapes of the ore bodies (Fig. 86).   Bending of the ore belts occurs both along the strike and along the dip.   Drilling has established that the ore-bearing horizon, with its folded structure, in general slopes down towards the southwest.

The ore bodies do not have clear boundaries with the dunites, and are formed of segregated banded ores mainly with a small amount of chrome-spinels. Thus, sampling at various intervals from ten drill-holes located on profiles X and XI of the Pervomaisk sector, demonstrates that more than half of their total thickness is composed of ores with less than 10% of chrome-spinels, and a further fifth part includes ores with 10 - 20% of chromite component, and only a few with more than 1% occur in the fraction of the ores, in which chrome-spinels predominate over gangue minerals.   The total thickness of the ore intervals accounted for, comprises 680 m.   The average amount of Cr_2O_3 in the ores of the richest central portion of the Pervomaisk sector lies within 13 - 18%, although the chrome-spinels in chemical composition are analogous to those of the Revda deposits.   The deposit is characterized by a widespread manifestation of carbonatization of the serpentinites up to the development of carbonate rocks proper. Their segregations have the shape of veins and layers, up to 20 m thick, and they are arranged subparallel and independently of the orientation of the ore bodies.

Fig. 86.
Geological section through the region of the Pervomaisk deposit (After A. Sharypov).

1) dunites and apo-dunitic serpentinites; 2) dunites with increased segregation of chrome-spinels and 5 - 10 wt % of Cr_2O_3; 3) chromite ore with more than 10 wt % Cr_2O_3; 4) talc-carbonate rocks.

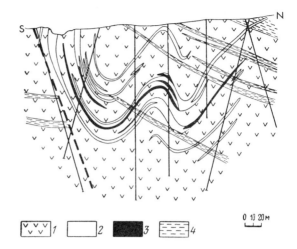

The geological structure of the remaining deposits, the country rocks of which are dunites, is similar to that described for the Revda and Pervomaisk deposits.   In them there is also a widespread development of ores of the segregated and streaky-banded type, and along with them there are ores with orbicular, orbicular-banded, nodular, lenticular-mesh, and brecciated structures.

The mineralization, associated spatially and genetically with the peridotites, is distinguished on the basis of shape of the ore segregations, structural features, and chemistry.   The ore bodies here are less elongate and more compact, and consist of nests, stocks, or short lenses, measurable in a few tens of metres along the long axis, with a length to thickness ratio of about 1 : 10.

Whereas in the south of the massif, in the dunite field, there is a predominance of ores of the segregated type with variable and normally low density of segregation, here the predominant structure is massive, large- and coarse-grained.   The massive ores either form the ore bodies entirely, or are located in the axial parts of the ore bodies.

The chrome-spinels of the ore bodies, occurring amongst the peridotites, are distinguished by the large amount of alumina (Al_2O_3, 20 - 25%) and the low content of chromium (Cr_2O_3, less than 40%).   In the matrix of the ores, there is an extraordinarily large distribution of chrome-bearing chlorites; chlorite also forms selvedges around the ore bodies.

CHROMITE DEPOSITS, OCCURRING IN MASSIFS OF THE DUNITE-LHERZOLITE-HARZBURGITE SUBASSOCIATION

This subassociation includes the chromites of the ultramafic massifs of the Kraka (Northern, Middle, and Southern), occurring on the western slope of the Southern Urals, and the Nurali Massif, lying on its eastern slope.   Of the foreign massifs, this subassociation may include the weakly chromite-bearing Konyukhe Massif in Yugoslavia.

No large deposits have been discovered amongst them.

CHROMITE DEPOSITS, DISTRIBUTED IN THE HARZBURGITE SUBASSOCIATION PROPER

This subassociation includes such chromite-bearing massifs of the Urals, as the Verblyuzh'egorsk, Verkhne-Ufalei, Itkul', Uspensk, Kulikovsk, and a number of others, and also certain chromite-bearing massifs in Yugoslavia, Greece, Turkey, Iran, Pakistan, and a number of other countries.

The Verblyuzh'egorsk Deposits

The deposits of the Verblyuzh'e Mountains belong to those where mining of the chromite ores came to an end with the expansion of the Kempirsai mines. At the present time, the deposits have no economic importance, but are of interest as a type of mineralization associated with the harzburgite subassociation. The Verbluzh'egorsk deposits occur in the south of the Chelyabinsk district not far from Kartaly station, being located on the eastern slope of the Urals within the Verblyuzh'egorsk ultramafic massif (sometimes called the Poltava Massif).

The massif is located in sedimentary-metamorphic rocks of Palaeozoic age, consisting of argillaceous, carbonaceous-argillaceous, mica-chlorite, argillaceous-chlorite-micaceous, or graphitic schists with seams of arkosic, quartzose, and ferruginous sandstones and coals, belonging to the Carboniferous System. On the northern side, the ultramafic rocks are in contact with the granite massif of the Dzhabyk-Karagai forest, which occupies an area of several hundred square kilometres.

The Verblyuzh'egorsk Massif has an area of about 50 km^2, is 16 km long with a width in the southern part of 5.5 km, it extends submeridionally and is convex in arcuate fashion towards the east (Fig. 87).

The massif is composed almost exclusively of harzburgites. The rocks have undergone intense alteration both during the process of autometamorphism, and under the influence of solutions, associated with the intrusion of granites. Rocks, unaffected by secondary changes, have not been observed in the massif, and they have all to some degree been altered by the processes of serpentinization, carbonatization, talcification, and silicification. Near the present surface, the rocks have been intensely altered by weathering processes.

In the rocks, encountered in the deepest drill-holes, one may observe relicts of the primary rock-forming minerals and from them establish the primary texture of the rocks. Assignment to peridotite is usually established on the presence of bastite (often converted to talc) and the typical shapes of the chrome-spinel grains, and the texture of the rock in the serpentinized peridotites, is emphasized by the segregation of a powdery magnetite along the boundaries of the serpentinized olivine grains, or on the simultaneous extinction of their relicts. However, in this respect, we must take account that processes of deserpentinization are quite widely manifested in the massif, resulting in the appearance of large, newly-formed olivine grains, which are developed, and do not conform with the primary texture of the rock, embracing grains of chrome-spinel and magnetite 'mesh'. The new-formed olivine is characterized by a clearly displayed cleavage.

Fig. 87.
Diagrammatic geological map of the
Verblyuzh'egorsk ultramafic massif.

1) mica-chert-chlorite and chert-clay
slates; 2) peridotites; 3) gabbros;
4) granites and granite-gneisses; 5)
granites; 6) chromite deposits.

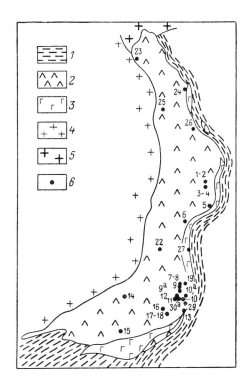

In individual sectors of the massif, in areas of development of
chromite mineralizaticn, there are serpentinized rocks with amounts of chrome-
spinels in excess of the accessory grade, separated in the form of streaky
accumulations, and with no visible relics of bastite.   These rocks are provisionally assigned to the dunites.   There are also small sectors of peridotites containing clinopyroxene (lherzolites ?).   Amongst the vein rocks, there
are metagabbros, granodiorites, plagioclasites, aplitophyres, amphibolites, and
chloritized actinolites.

Within the massif there are 32 chromite deposits, but the distribution
of mineralization is uneven and 18 of them are concentrated in the southeastern
portion over an area of less then 3 km^2, known as the Main ore field (Fig. 88).

The deposits in the Main ore field are arranged in two ore-bearing
zones.   The Western ore-bearing zone, with northeasterly strike, includes
deposits 7 - 9, 11, 12, 17, 18, 30, and 30a.   All the deposits of this zone
are arranged parallel to a dyke of microdiorite, and the strike of the ore
bodies coincides with that of the zone itself.   Deposits 19, 10, 10a, 20, and
29 have been assigned to the Eastern zone.   An interesting feature of this
zone is the fact that the ore bodies of the deposits included in it have a
latitudinal or northwesterly strike, whereas the zone itself strikes submeridionally.

The ore bodies dip at from 30 up to 80°;   in the Western ore-bearing
zone, the dip is predominantly easterly, and in the Eastern zone, southwesterly.

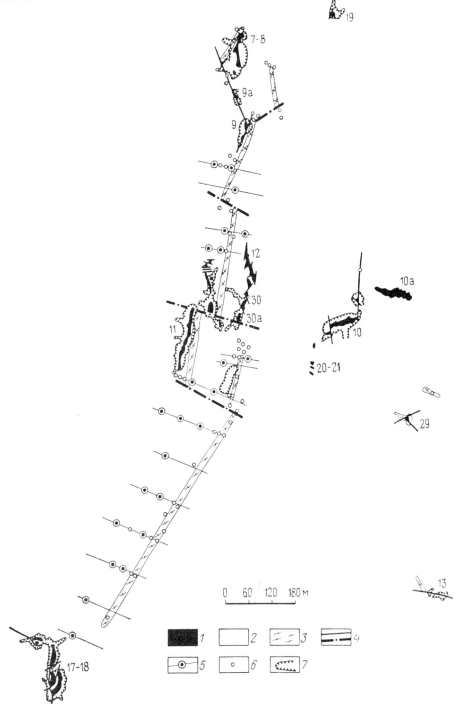

Fig. 88. Diagram of distribution of chromite ore bodies in the Main ore field of the Verblyuzh'egorsk Massif.

1) chromite ore; 2) peridotites; 3) microdiorites; 4) tectonic disturbances; 5) drill-holes through exploration profiles; 6) mining operations; 7) boundaries of working quarries. Figures denote deposit numbers.

The ore bodies are like simple lenses in shape, and less frequently have a
vein-like form.   They have been broken by post-ore disturbances into a series
of blocks, displaced relatively to one another.   The ore bodies are from a few
tens of metres up to 100 - 150 m long, with the most extensive ore body No. 30
having a length of 245 m.   The thicknesses of the ore bodies are 1 - 16 m,
and the greatest thicknesses correspond to the most extensive ore bodies.
Increase in thickness is usually observed in the middle portions of the ore
bodies, which also adds to the lensoid shape of the bodies (Fig. 89).

The chromite ores are characterized by massive, segregated, and
banded structures.   The ores of the massive type do not display directional
features in the arrangement of the ore and non-ore grains and their aggregates.
They contain up to 90% of chrome-spinel.   Amongst the silicates, there is a
wide distribution of chlorite.   Such ores completely predominate in the deposits of the Western ore-bearing belt.

In the segregated chromite ores, the chrome-spinels comprise up to
half of the volume of the rock and are enclosed in a matrix of chlorite, antigorite, and carbonates.   The ore contains geikielite, jefferisite, and antigorite, and uvarovite has been found also.   The ores of the segregated type
markedly predominate in the deposits of the Eastern ore-bearing zone.

Table 22.   *Chemical Analyses of Uniform Chromite Ores and Chromite Concentrates*
from Segregated Ores of the Verblyuzh'egorsk Deposits (in wt %)

Components	1	2	3	4	5	6	7
SiO_2	5.62	4.74	7.25	7.11	2.28	3.93	7.79
TiO_2	0.32	0.46	0.30	0.23	–	–	0.15
Al_2O_3	7.48	9.29	5.67	4.66	22.54	22.21	14.86
Cr_2O_3	47.03	47.54	46.05	47.15	41.20	37.59	40.31
Fe_2O_3	14.47	14.94	13.10	14.37	–	–	3.76
FeO	7.67	6.01	8.55	7.95	14.61	16.11	11.50
MnO	0.47	0.34	0.66	0.37	–	–	0.22
MgO	13.41	13.96	14.38	14.26	18.03	18.56	17.14
CaO	0.00	0.00	0.00	0.00	–	–	0.00
H_2O^-	0.10	0.04	0.09	0.07	–	–	–
H_2O^+	2.24	1.90	2.86	2.71	–	–	2.52
Total	98.81	99.22	98.91	98.88	98.66	98.40	98.25

1-4) uniform chromite ore, deposit 11, shaft, depth 37m;   5-7) chrome-spinel
concentrates from segregated ore;   5) deposits 17-18;   6) deposit 19;   7) deposit 21. Analyses 1-4 and 7 quoted from Kashin (1937);   5-6 from Sokolov *et al.*
(1936).

Banded chromite ores in the deposits of the Verblyuzh'e Mountains
occur rarely.   The chrome-spinels of some of the Verblyuzh'egorsk deposits
have been substantially metamorphosed.   Especially intense metamorphism shows

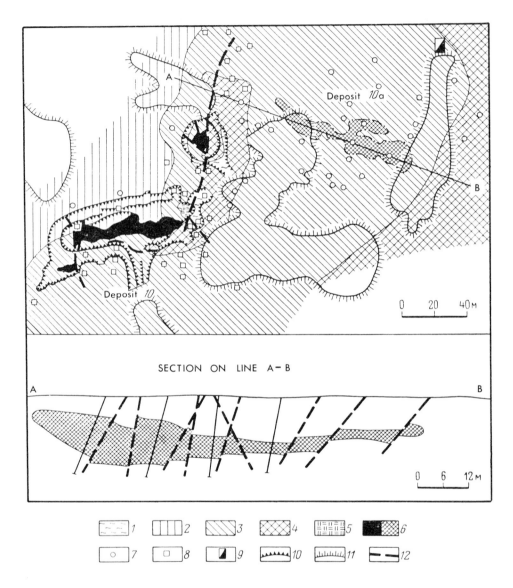

Fig. 89. Diagrammatic geological map of the area of deposits 10 and 10a in the Verblyuzh'egorsk Massif (After N. Kasatkin)

1) loams; 2) apo-peridotitic serpentinites; 3) carbonatized apo--peridotitic serpentinites; 4) carbonatized and silicified apo--peridotitic serpentinites; 5) limonite-schuchartite rocks; 6) chromite ore; *a*) near-surface ore bodies, *b*) buried ore body; 7) drill-holes; 8) prospecting pits, open pits, and trenches; 9) shaft; 10) quarry boundaries; 11) boundaries of waste dumps; 12) tectonic disturbances.

up in deposits 11, 12, 9, 30, and 10, where massive ores mainly predominate. However, there are deposits of massive ores, in which the chrome-spinels have been quite weakly metamorphosed (deposits 14 and 15).   Metamorphism of the chrome-spinels in the segregated ores is also insignificant.

The processes of metamorphism have led to the alteration of the chemical composition of the chrome-spinels, since they have been accompanied by oxidation of the iron and the removal of aluminium and magnesium (Sokolov, 1936).

Two types of ores are recognized on the basis of composition: massive, with a Cr_2O_3 content of up to 50% and Al_2O_3, 4 - 9%, and densely--segregated ores.   The chrome-spinel concentrate of these ores contains 37 - 41% Cr_2O_3 and up to 22% Al_2O_3 (Table 22).   However, the composition of the ores and the ore--forming chrome-spinels is distinguished not only by the different degree of their metamorphism.   It is determined apparently, by the different concentration of the components in the ore-silicate melts, which formed the deposits of the Western and Eastern ore-bearing zones.

CHROMITE DEPOSITS, ASSOCIATED WITH THE GABBRO-NORITE-HARZBURGITE ASSOCIATION

This association includes two chromite-bearing massifs, the Northern and the Southern Saranovsk, and also the Tesovsk Massif with its chromite ore--shows, which occur in the outer ultramafic belt of the western slope of the Middle Urals.   In the other folded regions of the USSR, this association has still not been discovered.   There are also no examples amongst the foreign chromite-bearing massifs.   However, there are insufficient grounds for grouping this association with that of the stratiform platform intrusions (harzburgite-orthopyroxenite-norite).   The twofold composition of the massifs of the present association (gabbroids and peridotites) indicates their formation from a more acid magma.

The Saranovsk Deposit

The Main Saranovsk deposit of chromites lies on the western slope of the Urals in the Chusovo region of the Perm' district and is located 100 km east of the regional centre.

The Saranovsk chromite-bearing, gabbro-peridotite massif, of very small dimensions, possesses quite large reserves of chromite ores.   The area of outcrop of the ultramafic rocks is 0.22 km^2 in all, with an extent of 1700 - 1800 m and an average width of about 200 m.   The maximum width of the massif in the middle portion, including the gabbroids, is about 400 m.   The strike is submeridional (000 - 345°).   Towards the north, the massif gradually sinks and in the buried portion can be traced for 500 m.   At 1 km south of the Northern Saranovsk Massif, a second ultramafic massif (the Southern) crops out on the surface, and it has somewhat smaller dimensions and a submeridional strike also. In it are the chromite bodies of former mines (the Bisersk, Lyubushkin, and the Bol'shoi Pester'), and also a number of ore bodies, discovered during exploration in recent times.   Nowadays, these deposits are not worked, and the only one being exploited is the Main Saranovsk mine in the Northern Massif.

The Saranovsk Massif is located on the western limb of the Central Urals uplift and lies in the outer belt of ultramafics in the Urals. Its origin must be associated with the Caledonian activation of submeridional faults within the Proterozoic basement and is associated with the arcogenic structures of the western slope of the Urals (Smirnov, 1959, 1969).

The zone, where this and some other ultramafic massifs are located, lies outside the true eugeosynclinal structures of the Urals, consists of miogeosynclinal formations, and is characterized by the presence of a rigid basement. Its position is thus reflected in the features of construction and composition of the rocks and ores of the Saranovsk Massif.

Near the Saranovsk Massif, metamorphic micaceous, quartz-mica-chlorite slates and phyllites are developed, with subordinate sequences of quartzites, limestones, and dolomites. The immediately surrounding rocks are quartz-mica schists. Their age is believed to be Late Proterozoic.

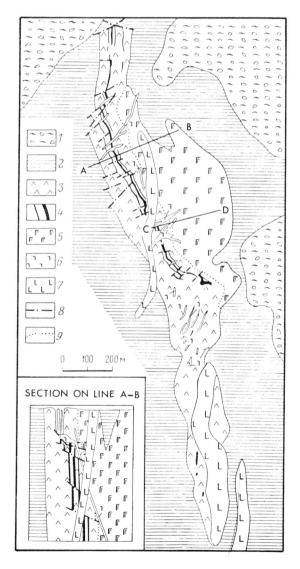

Fig. 90.
Diagrammatic geological map of the Northern Saranovsk chromite--bearing massif (compiled from information from the Saranovsk ore administration)

1) deluvial cobbly-clay sediments; 2) quartz-sericite, quartz-sericite-chlorite, epidote- and actinolite-chlorite schists, etc.; 3) apo-peridotitic serpentinites; 4) chromite ore bodies; 5) gabbro--norites; 6) metamorphosed diabases; 7) gabbro-diabases; 8) tectonic disturbances; 9) boundary between peridotites and gabbro--norites.

The massif is located in the area of the southern periclinal closure of the anticline, also termed the Saranovsk.   It is evident that the massif is restricted to one of the tectonic disturbances, which passes in the eastern limb of the Saranovsk anticline near its axial zone.   The contacts of the ultramafic massif are conformable with the country rocks (Fig. 90).

In shape, the massif is a conformable homoclinal body, dipping steeply eastwards.   Geological-exploration drilling data have established that both the contacts of the massif and the chromite mineralization included in it at depths of 300 - 400 m, from a steeply-dipping attitude (80 - 85°) gradually pass into more gentle dips down to 40 - 30° (Fig. 91).   With the contacts of the massif conformable with the country rocks, such an arrangement of the foot-wall of the intrusion indicates that the steep limb of the anticline eastwards along the dip passes into a synclinal structure, which at depth probably accomm-odates the main body of the intrusion.   It is evident that erosion has exposed only the upper western portion of the massif, possibly representing an apophysis from a phacolite-like intrusive body, the bulk of the rocks of which are located to the east of its outcrop on the surface, lying in a synclinal depression.   To the east of the area of outcrop on the surface of the ultramafic rocks and gabb-roids, the syncline is apparently cut by a tectonic fracture.

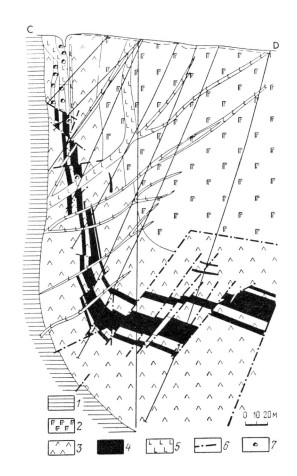

Fig. 91.
East-west geological section through the Main Saranovsk deposit along the line CD (see Fig. 90).

1) schists;   2) gabbros and gabbro-norites;   3) peridotites; 4) densely-segregated and mass-ive chromite ores;   5) vein rocks: gabbro-diabases, por-phyrites, etc.;   6) tectonic disturbances;   7) subsurface mine workings

The peridotites and gabbro-norites form a single intrusive body, which is cut by gabbro-diabase dykes.   The drill-cores exhibit an extremely sharp transition from the gabbro-norites into peridotites, but no traces of the effects of one rock upon the other have been discovered.   This suggests an unusual, markedly layered intrusion, to the lower (peridotitic) portion of which all the chromite mineralization has been restricted.

The ore-bearing zone proper has a width in plan of 40 - 45 m and includes three subparallel vein-like ore bodies, the Western, Central, and Eastern;   it is located approximately in the axial part of the ultramafic belt.

Its position in the ultramafic belt is somewhat asymmetrical.   In the north, it approaches the western (lower) contact of the massif with the surrounding rocks over 25 - 30 m and lags behind the transition zone from the ultramafics to the gabbroids over 60 - 70 m.   In the southern part of the massif, the distance of the ore-bearing zone from the surrounding slates is 120 - 140 m, and from the gabbroids, about 20 m.   The western edge of the massif of the zone of contact between the ultramafics and the surrounding rocks as far as the ore-bearing zone consists entirely of serpentinized peridotites.   The peridotites of the massif consist of serpentinites, in most cases with persistant relict textures.   The peridotites are distinguished by the varying ratios of their earlier olivine, pyroxene, and chrome-spinel components.

In individual sectors, especially in the zone of the western contact, the amount of pyroxenes falls to 7 - 10%.   The quantity of accessory chrome-spinels increases markedly as one approaches the ore-bearing zone.

The peridotites are often coarse-grained.   They have a poikilitic, poikilo-ophitic, and pegmatoid structure.   The poikilitic peridotites are characterized by quite constant grainsize of the olivines (2 - 4 mm).   The olivine grains are usually rounded, and small idiomorphic grains of chrome-spinel are often enclosed in the larger grains of orthopyroxene or in their aggregates.

The pegmatoid peridotites have this same structure, but a coarser and impersistent grainsize, the dimensions of the grains of the olivine replaced by serpentine measuring from 1 - 2 up to 10 - 30 mm and more.   A clinopyroxene is sometimes found, replaced by chlorite and tremolite.

On the eastern side of the ore-bearing zone there are apo-peridotitic serpentinites with poikilitic and poikilo-ophitic texture;   they are 20 - 60 m thick (Table 23).   In these rocks, the large (1 - 2 cm) grains of orthopyroxene (now antigorite, bastite, and talcified bastite) enclose rounded idiomorphic grains of completely serpentinized olivine measuring from 1 to 3 mm and more across.   Large grains (1 - 2 cm) of clinopyroxene of irregular shape also occur rarely, replaced by a felted-acicular colourless amphibole (tremolite) and a fine-flaky chlorite.   In a number of drill-holes amongst the poikilitic peridotites of the Eastern zone, segregations of vari-grained pegmatoid peridotites have also been found.

The transitions from the provisionally recognized ore-bearing zone into the surrounding peridotites of the Eastern and Western belts are gradual. The peri-ore peridotites are characterized by an intense saturation by segregations of small grains of chrome-spinels.   The amount of chrome-spinels in the rock varies considerably: from 3 - 5 up to 20 - 25% and more.   It is quite characteristic that the idiomorphic grains of chrome-spinels, measuring from fractions of a millimetre up to 2 - 3 mm, are distributed between large (1 - 4 cm) idiomorphic grains of serpentinized olivine and have been enclosed in the grains of ortho- and clinopyroxene. The pyroxenes are subjected to the crystalline forms of the olivine.

The peridotites, distributed between the main ore bodies, often have a medium-grained fabric, and the olivine grains in dimensions and shape are the same as in the poikilitic peridotites. In the interstices between the olivines are pyroxenes and chrome-spinels. The amount of chrome-spinels in the inter-granular space may reach 90 - 95% and then in such cases, a segregated chromite ore develops with a pseudo-sideronitic texture. Near the ore bodies, such 'ore peridotites' are frequently found along with non-ore types.

Table 23. *Chemical Composition of Apo-Peridotitic Serpentinites from the Saranovsk Massif (in wt %)*

Components	1	2	3	4	5	6	7	8
SiO_2	34.16	34.78	35.36	35.95	36.90	37.50	36.15	36.72
TiO_2	0.11	0.15	0.15	0.17	0.21	0.08	0.17	0.14
Al_2O_3	3.16	2.15	1.93	5.20	2.89	3.68	2.83	1.11
Cr_2O_3	2.58	1.05	0.93	1.00	0.67	0.55	0.54	0.35
Fe_2O_3	10.41	11.95	9.91	5.71	3.48	9.03	6.07	11.10
FeO	1.80	–	2.16	6.25	6.01	2.56	7.20	4.96
MnO	0.10	0.02	0.08	0.10	0.06	0.13	4.15	0.21
MgO	35.15	36.18	35.94	32.17	35.30	33.60	31.68	31.84
CaO	nil	nil	nil	1.35	1.61	0.88	0.31	2.78
NiO	0.29	0.20	0.46	0.28	0.33	0.32	–	–
Calcination loss	11.27	12.20	11.54	12.20	12.40	11.50	10.00	10.07
H_2O^-	0.85	1.48	1.30	–	–	0.70	0.49	0.88
Total	99.88	100.16	99.76	100.38	99.86	100.53	99.59	100.16
Reference	Saranovsk Ore Administration	I. Zimin		I. Idkin *et al.*	I. Zimin and I. Idkin	Kh. Rumyantseva and Yu. Smirnov	Yu. Smirnov	

Commercial ore concentrations consist of three subparallel vein-like bodies. The Central body extends for 1200 m, with a thickness of 10 m; the Western, 910 m and 5 m; and the Eastern, 1100 m and 3 - 3.5 m. Exploratory drilling at depth has revealed accompanying lens-like bodies of small thickness (up to 1.5 m) and impersistent along the strike, located both below and above the principal ore bodies. In construction, the ores are similar to that of the principal ore bodies.

All three vein-like bodies consist of almost massive chromite ores, although they always contain a small quantity of gangue minerals. The most chrome-spinel rich are the ores of the Central Ore body. The ores possess a vari-grained, and in bulk a medium-grained fabric, with the grainsize of the chrome-spinels varying from fractions of a millimetre up to 2 - 3 mm. The typical ores have a banded fabric with the banding conformable with the contacts.

The Western vein-like ore body is separated from the Central body by a belt of country rocks with a persistent thickness of 6 - 8 m, and the Eastern body is separated from the Central body, by a band of somewhat smaller thickness. The ore bodies in the visible sections nowhere merge either along the strike or along the dip, if we disregard late tectonic approaches. Post-ore tectonic disturbances have broken the ore bodies into a system of blocks, displaced relatively to one another, with throws of up to several tens of metres and less.

To the east the ultramafic rocks are markedly replaced by extremely texturally uniform, medium-grained leucocratic rocks of gabbroic composition. Macroscopically, they are light-grey, medium-grained rocks with subordinate amounts of dark-coloured minerals (30 - 35%).

The primary minerals have not been preserved in the rocks. The feldspars have been replaced completely by a zoisite-clinozoisite aggregate with chlorite. The dark-coloured minerals have also been completely replaced by chlorite and amphibole of the actinolite-tremolite series. The chlorite is a finely-lamellar clinochlore. Serpentine (antigorite) sometimes occurs in insignificant amount. Intense metamorphic transformation of the gabbroid rocks has led to the absence of accessory ore minerals in them, if we disregard the isolated, and very rare chrome-spinel grains. These occurrences of chrome-spinels are extremely interesting. Since the contact between the two rock complexes (the ultramafics and the gabbroids) bears no traces of mutual interaction (chilled facies are absent, and the rocks in their junction zone are medium- and coarse--grained), and accessory chrome-spinels have been found in the gabbro-norites, an affinity is assumed between the peridotites and the gabbroids.

The ore bodies, and the ultramafic and gabbroid rocks have been cut by gabbro-diabase dykes and veins of strongly chloritized diabase-porphyrites and more acid rocks.

Among the uniform and very densely-segregated chromite ores from the vein-like ore bodies, three most clearly defined varieties have been distinguished macroscopically: 1) massive, with a very insignificant amount of silicate, which has been distributed evenly through the ore matrix (a very weakly defined banding or linearity is sometimes seen); 2) very densely-segregated chromite ores with evenly dispersed silicate grains clearly elongated in one direction; and 3) densely-segregated chromite ores, with a chain-like distribution of elongated silicate grains, forming 'short lines' in one section and very fine layers, with a 'single-grain' thickness across. In this way, the elongate grainlets of silicates are arranged either end-to-end, or are separated from each other by chrome--spinel grains.

The ore consists of chrome-spinel (*ca* 95%) and silicate, consisting of serpentine, developed after pyroxene and rarely after olivine. Later minerals, developing after serpentine and formed during the process of hydrothermal mineralization, associated with the subsequent injection of more acid melts, such as talc, chlorite, carbonates, and less frequently amphibole, have not been described here.

The chrome-spinels are characterized by isomorphism of the grains and a clearly defined porphyritic fabric, depending on the presence of two families of chrome-spinel grains (large, from 1 to 3 mm in size, and small, the grains of which are measured in fractions of a millimetre (0.2 - 0.7)).   Both types possess rectilinear outlines and are idiomorphic, although amongst the large family there are sometimes grains with smoothed angles between the faces.

The segregated and lean-segregated chromite ores, distributed near the vein-like ore bodies and between them, macroscopically have extremely variable structural patterns, which are determined by variations between the silicate and ore components and the grainsize of the chrome-spinels and the silicates.   Silicates predominate sharply over chrome-spinels.   A common characteristic feature of the segregated ores is the presence of a clearly-defined pseudosideronitic fabric.

In all the varieties of segregated ores, there are chrome-spinels of two sizes: a large family (1 - 3 mm) and a small family (0.1 - 0.5 mm).   The presence of two clearly defined families of chrome-spinel grains in the ores indicates a change in the physicochemical conditions of crystallization of the ore melt, apparently dependent on its displacement during intrusion from deep--seated zones into higher levels.

The order of crystallization of the minerals in the rocks proceeds according to a somewhat varied scheme.   In the peridotites, crystallization begins with the precipitation of olivine, after which accessory chrome-spinel and pyroxene follow, and in the massive almost uniform chromite ores, the chrome-spinel commences the crystallization, and pyroxene ends it.   This indicates that crystallization proceeds from an already segregated portion of the magmatic melt of different composition.

The chemical composition of the 'raw' chromite ores of the Main Saranovsk deposit is shown in Table 24, where average values have been given for the three principal ore bodies (Western, Central, and Eastern), and also the composition of the most representative ores of the massive and segregated types. Owing to the low content of chromium and the high iron index the ores of the deposit are unsuitable for the production of ferrochrome and have been used as a refractory and chemical raw material.

Table 24. *Chemical Composition of the 'Raw' Ores of the Saranovsk Deposit (in wt %)*

Components	Western ore body	Central ore body	Eastern ore body	Massive ores from Central ore body	Segregated ores in aureole of ore bodies
Cr_2O_3	35.86	38.24	33.17	38.81	34.72
Al_2O_3	15.11	18.80	20.35	14.86	15.84
FeO	17.68	18.53	20.05	18.34	18.60
MgO	16.85	15.16	16.20	12.16	18.03
CaO	0.24	1.36	–	–	–
SiO_2	6.25	5.00	5.14	3.04	5.68
Calcination loss	1.63	1.96	2,98	2.58	5.66

In the chromite ores of the Saranovsk deposit, owing to intense hydrothermal influences, the normal minerals such as chrome-spinel, lizardite, antigorite, bastite, chlorites, actinolite, and magnetite are frequently accompanied
by phlogopite, albite, talc, quartz, calcite, magnesite, hornblende, pyrite,
chalcopyrite, pyrrhotite, and sporadic apatite.   Relicts of primary olivine and
pyroxenes are absent.   The chrome-spinels in individual sectors of the ore bodies have been quite intensely metamorphosed, but their less altered varieties
are also present.

The chrome-spinels of the ores are distinguished by their constant
composition (Table 25).   The amounts of the individual oxides vary as follows
(in wt %): Cr_2O_3, 44 - 47; Al_2O_3, 18 - 19; Fe_2O_3, 5 - 7; FeO, 14 - 17; MgO,
12 - 13.   A somewhat increased content of titanium and vanadium is typical.

The accessory chrome-spinels are markedly different from the ore-forming types both on the basis of the principal spinel-forming oxides, and in the
amount of trace-elements.   The most marked differences have been observed in
the amounts of di- and trivalent iron, magnesium, and aluminium.   In comparison
with the ore-forming chrome-spinels, the accessory types are more ferriferous,
and contain less magnesium and aluminium.   Along with the general increased
amount of titanium (as opposed to the chrome-spinels of the dunite-peridotite
association), the accessory chrome-spinels in the Saranovsk Massif are richer
in titanium (TiO_2, 0.54 - 1.7 wt %) as compared with those of the ore-forming
types (0.36 - 0.55 wt %).

The ore-forming chrome-spinels of the large and small families, occurring together in the ores, are completely identical in composition.

A comparison of the features of the petrographical and chemical composition of the rocks, the ores in them, and the component chrome-spinels of the
ores, and also the order of crystallization of the ultramafic rocks and chromite
ores in the Saranovsk Massif and in other chromite-bearing massifs of the Urals,
. reveal marked differences between them.

A comparison between the composition of the harzburgites of the Saranovsk Massif and the average composition of those of the eastern slopes of the
Urals, based on Malakhov's data (1967), shows that whereas the average FeO content in the harzburgites of the eastern slope of the Urals based on 553 analyses
is 8.04 wt %, in the Saranovsk apo-harzburgites, it is 12.9%.   The harzburgites
of the Saranovsk Massif contain (in wt %): Al_2O_3, 3.95; Cr_2O_3, 0.68; TiO_2,
0.24;   and CaO, 0.78;   whereas the harzburgites of other chromite-bearing massifs
contain: Al_2O_3, 2.06; Cr_2O_3, 0.40; TiO_2, 0.06; and CaO, 0.34.   Consequently,
the composition of the harzburgites of the Saranovsk Massif is substantially
different from that of the harzburgites of the peridotite association proper.
The differences are also observed in the composition of the ore-forming chrome-
-spinels.   Thus, in those from the Saranovsk deposit, the content of iron and
titanium is markedly greater than in those from the deposits occurring in the
peridotite association proper, where Fe_2O_3 is normally 2 - 4 wt %, FeO 9 - 13%,
and TiO_2 is present in traces up to 0.2 wt %.   This again emphasizes the fact
that the parent ultramafic magmatic melt, which formed the Saranovsk Massif,
differed in its increased iron content, which was also reflected in the compositions of the chromite ores.

The variable grainsize of the peridotites, and sometimes their porphyritic aspect, are reflexions of the fact that nucleation of the crystals of
rock-forming olivine and pyroxene took place over a major interval of time,

Table 25. *Chemical Composition of the Chrome-Spinels of the Saranovsk Massif (in wt %)*

Components	Analysis number												
	1	2	3	4	5	6	7	8	9	10	11	12	13
	Accessory				Ore-forming								
SiO_2	1.46	4.28	1.63	1.70	0.73	0.88	1.12	0.76	0.59	0.99	1.54	1.15	1.18
TiO_2	1.71	1.20	0.54	0.97	0.51	0.55	0.37	0.36	0.50	0.39	0.38	0.53	0.41
Al_2O_3	11.54	10.46	14.53	13.12	18.32	16.29	17.39	18.39	18.43	18.02	17.39	17.93	17.38
Cr_2O_3	40.21	29.09	37.92	37.25	46.22	45.68	45.76	44.59	43.30	44.72	44.72	45.41	45.46
Fe_2O_3	2.16	15.45	13.36	13.23	5.01	5.08	6.04	5.77	7.56	6.11	6.16	5.91	5.83
FeO	33.63	28.37	21.76	25.15	15.24	17.69	14.88	17.27	16.68	16.05	16.64	14.97	15.33
MnO	0.44	0.69	0.26	0.43	0.19	0.21	0.16	0.22	0.21	0.24	0.21	0.22	0.27
MgO	7.58	8.18	9.42	7.34	13.54	12.09	13.83	12.40	12.44	13.19	12.69	13.54	13.51
CaO	0.00	0.00	0.00	0.00	0.00	0.00	0.00	0.00	0.00	0.00	0.00	0.00	0.00
Na_2O	n.d.	n.d.	0.00	0.00	n.d.	0.00	0.00	0.00	0.00	0.00	n.d.	n.d.	n.d.
K_2O	n.d.	n.d.	0.01	0.00	n.d.	0.00	0.00	0.00	0.00	0.00	n.d.	n.d.	n.d.
H_2O^-	0.00	0.19	0.06	0.00	0.00	0.00	0.00	0.00	0.00	0.00	0.06	0.00	0.00
H_2O^+	0.50	1.21	0.40	0.53	0.20	0.56	0.49	0.30	0.23	0.30	0.29	0.41	0.39
NiO	0.20	0.18	0.15	0.13	n.d.	0.15	0.18	0.18	0.15	0.13	0.16	0.16	0.15
CoO	0.18	0.19	0.10	0.11	n.d.	0.05	0.05	0.07	0.07	0.05	0.10	0.08	0.09
V_2O_5	0.28	0.28	0.16	0.23	0.17	0.18	0.16	0.16	0.22	0.16	0.14	0.15	0.15
Total	99.89	99.77	100.30	100.23	100.13	100.31	100.43	100.47	100.38	100.35	100.48	100.46	100.15
Composition of chrome-spinels after recalcutation to pure material (wt %)													
Al_2O_3	12.43	14.54	15.97	14.04	18.92	17.67	18.54	19.44	19.41	18.45	18.26	18.16	18.27
Cr_2O_3	43.33	40.25	39.84	41.11	44.75	47.39	47.26	45.89	44.11	47.07	46.83	47.32	47.57
Fe_2O_3	14.14	14.52	14.28	14.44	5.18	6.47	6.44	6.10	7.62	6.56	6.53	6.17	6.20
FeO	23.20	24.21	21.63	24.22	14.88	16.73	14.50	16.54	16.56	15.00	16.36	14.85	15.09
MgO	6.90	6.48	8.28	6.19	13.27	11.74	13.26	12.03	12.30	12.92	12.02	13.05	12.87

1-4) accessory chrome-spinels, separated from: 1) coarse-grained pegmatoid apo-peridotitic serpentinite, quarry near ore body, Spec.No. 18-68; 2) apo-peridotitic serpentinite with poikilitic texture, drill-hole 421, profile XIII, depth 220 m, Spec.No. 147-68; 3) vari-grained apo-peridotitic serpentinite from peridotites, separating the Western and Central ore bodies, horizon 280 m, Spec.No. 82-68; 4) apo-peridotitic serpentinite, eastern contact of Eastern ore body, horizon 280 m, Spec.No. 92-69.

5-13) ore-forming chrome-spinels from: 5) almost uniform chromite ore from central part of Western ore body, shaft, horizon 280 m, Spec.No. 78-68; 6) densely-segregated, almost massive chromite ore from Central ore body, 50 m from western contact, as previously, Spec.No. 83-68; 7) almost massive chromite ore from Central ore body, 6 m from western contact, as previously, Spec.No. 85-68; 8) massive chromite ore from eastern contact of Central ore body, as previously, Spec.No. 89-68; 9) densely-segregated chromite ore from central part of Eastern ore body, as previously, Spec.No. 89-68; 10) chromite segregations in coarse-grained apo-peridotitic serpentinites, southern part of quarry, south of Central ore body, Spec.No. 22-68; 11) densely-segregated mesh-like chromite ore, drill-hole 425, profile XXX, depth 273 m, Spec.No. 176-68; 12) densely-segregated chromite ore of vari-grained fabric with patches of lean-segregated fine-grained ore, large grains measuring 1 - 3 mm from ore concentrate, Spec. No. 5-k-68; 13) small grains, less than 0.1 - 0.2 mm in size from same sample, Spec.No. 5-m-68.

Analyses carried out in Central Chemical Laboratory of IGEM AN SSSR by O. Ostrogorskaya.

during the various phases of crystallization of the magmatic melt.   This is
indicated by the conjoint occurrence of olivine grains, measuring more than a
centimetre in size and grains of 0.5 - 2 mm, and also the presence of pyroxenes
with and without ingrowths of olivine.

The shape of the ore bodies, the nature of the contacts, and the composition and the textural-structural features of the chromite ores demonstrate
that the ore melt with chrome-spinel crystals nucleated in it, was distributed
in the peridotites under the influence of tectonic forces.

The subparallel nature of the ore bodies and the maintenance of thicknesses over significant distances both along strike and dip, and the clearly
manifested linearity and banding of the fabrics in the chromite ores, indicate
injection and displacement of the crystallizing ore melt along opening tectonic
fractures in the already consolidated peridotites.

The occurrence of the Saranovsk Massif in the area of a miogeosyncline,
and not a eugeosyncline, which is typical of the chromite-bearing massifs of the
Urals and other fold regions, the clear spatial and genetic association with
gabbros and gabbro-norites, the specific chemical composition of the rocks, ores,
and ore-forming chrome-spinels, and also the unusual trend in crystallization,
provide grounds for establishing a different associational assignment to the
complex of rocks of the Saranovsk Massif and other chromite-bearing massifs of
the Urals, and also for placing the Saranovsk Massif in the gabbro-peridotite
association.

From the characters noted, the Saranovsk Massif seemingly occupies an
intermediate position between the associations of the platform and geosynclinal
regions.

REFERENCES

ABOVYAN S.B.   (АБОВЯН С.Б.), 1961: Геология и полезные ископаемые
северо-восточного побережья озера Севан *(The Geology and Mineral Resources of the Northeastern Coast of Lake Sevan)*.   Akad. Nauk armen. SSR Press,
Yerevan.

BETEKHTIN A.G.   (БЕТЕХТИН А.Г.), 1937: Шоржинский хромитоносный
перидотитовый массив (в Закавказье) и генезис месторождений хроми-
стого железняка вообще (The Shorzhinsk chromite-bearing peridotite massif
(in Transcaucasia) and the origin of deposits of chrome ironstone in general),
in "Хромиты СССР" *('Chromites of the USSR')*, Vol. 1, pp.7-156.   Akad. Nauk
SSSR Press, Moscow.

HIESSLEITNER, G., 1951/1952: Serpentin- und Chromerz-Geologie der Balkanhalbinsel
und eines Teiles von Kleinasien.   *Jb. geol. Bundesanst. Wien, Sonderband* 1, T.
1-2.

KNIPPER A.L. & KOSTANYAN Yu.L.   (КНИППЕР А.Л., КОСТАНЯН Ю.Л.), 1964:
Возраст гипербазитов северовосточного побережья озера Севан (The age
of the ultramafic rocks of the northeastern coast of Lake Sevan).   *Izv. Akad.
Nauk SSSR, ser. geol.*, No.10, pp.67-79.

MALAKHOV A.E.   (МАЛАХОВ А.Е.), 1967:   Материалы по вопросам геологии и полезным ископаемым Урала *(Information on Problems of the Geology and Mineral Resources of the Urals)*.   Gosgeoltekhizdat, Sverdlovsk.

MALAKHOV I.A.   (МАЛАХОВ И.А.), 1966:   Петрохимия ультрабазитов Урала (The petrochemistry of the ultramafic rocks of the Urals).   *Trudy Inst. Geol. ural'. Fil. Akad. Nauk SSSR, 79.*

MORKOVKINA V.F. & ARUTYUNYAN G.S.   (МОРКОВКИНА В.Ф., АРУТЮНЯН Г.С.), 1971:   О радиологическом возрасте гипербазитов Севанского хребта (Армения)   (The radiometric age of the ultramafic rocks of the Sevan Range (Armenia)).   *Izv. Akad. Nauk SSSR, ser. geol.*, No.11, pp.133-137.

NEPOMNYASHCHIKH A.A.   (НЕПОМНЯЩИХ А.А.), 1950:   О форме и размерах Кемпирсайского ультраосновного массива (The shape and dimensions of the Kempirsai ultramafic massif).   *Dokl. Akad. Nauk SSSR, 73*, pp.1275-1277.

PAVLOV N.V.   (ПАВЛОВ Н.В.), 1949:   Химический состав хромшпинелидов в связи с петрографическим составом ультраосновных интрузивов (The chemical composition of the chrome-spinels in connexion with the petrographical composition of ultramafic intrusives).   *Trudy Inst. geol. Nauk Mosk, ser. rudn. Mestorozh.*, No.13, vyp. *103.*

PAVLOV N.V., CHUPRYNINA I.I. & OSTROGORSKAYA O.P.   (ПАВЛОВ Н.В., ЧУПРЫНИНА И.И., ОСТРОГОРСКАЯ О.П.), 1969:   О составах сосуществующих оливинов и хромшпинелидов из пород и руд дунит-гарцбургитовой формации (на примере Кемпирсайского массива) (The compositions of coexisting olivines and chrome-spinels from the rocks and ores of the dunite-harzburgite association) (as exemplified by the Kempirsai Massif). *Geologiya rudn. Mestorozh.*, No.2. pp.17-29.

PAVLOV N.V., KRAVCHENKO G.G. & CHUPRYNINA I.I.   (ПАВЛОВ Н.В., КРАВЧЕНКО Г.Г., ЧУПРЫНИНА И.И.), 1968:   Хромиты Кемпирсайского плутона *(The Chromites of the Kempirsai Pluton)*.   Nauka Press, Moscow.

PETRASCHEK, W., 1957:   Die genetischen Typen der Chromlagerstätten und ihre Aufsuchung.   *Erzmetall.*, Bd 10, h.6.

PRONIN A.A. (ПРОНИН А.А.), 1948:   Морские фации турнейского яруса на восточном склоне Урала   (Marine facies of the Tournaisian Stage on the eastern slope of the Urals).   *Dokl. Akad. Nauk SSSR, 62*, pp.389-391.

SEGALOVICH V.I.   (СЕГАЛОВИЧ В.И.), 1971:   О строении Кемпирсайского ультраосновного массива (The structure of the Kempirsai ultramafic massif). *Dokl. Akad. Nauk SSSR, 198*, pp.178-181.

SMIRNOV Yu.D.   (СМИРНОВ Ю.Д.), 1959:   Малые интрузии основных и ультраосновных пород алмазоносных районов западного склона Среднего Урала   (The minor intrusions of basic and ultramafic rocks of the diamond-bearing regions of the western slopes of the Central Urals).   *Inf. Sb. vses. geol. Inst.*, No.16, pp.75-85.

SMIRNOV Yu.D.   (СМИРНОВ Ю.Д.), 1969:   Закономерности размещения некоторых магматических формаций в полициклической складчатой области (на примере западного склона Урала) (Distribution patterns of some magmatic associations in a polycyclic folded region (as exemplified by the western slope of the Urals)).   *Materialy IV vses. petrogr. Soveshch.*, pp.153--155.   Baku.

SOKOLOV G.A. (СОКОЛОВ Г.А.), 1948: Хромиты Урала, их состав, условия кристаллизации и закономерности распространения (The chromites of the Urals, their composition, conditions of crystallization, and distribution patterns). *Trudy Inst. geol. Nauk Mosk., 97, Ser. rudn. Mestorozh., 12.*

SOKOLOV G.A. *et al.* (СОКОЛОВ Г.А. и др.), 1936: Геохимические исследования на горе Верблюжей; Труды южноуральской экспедиции (Geochemical investigations on Mt Verblyuzhei; the work of the Southern Urals Expedition). *Akad. Nauk SSSR: Sovet po izucheniyu proizvoditel'nykh sil i Lomonosovsk. Inst.,* S. Ural'skaya, 5.

VARMA, O.P., 1971: Some reflections on the genesis of chromite; a study in reference to Indian deposits (abstr.). *Proc. Indian Sci. Congr., Assoc.,* 58th Sess., pt 3, pp.308-309.

WAGER, L.R. & BROWN, G.M., 1968: *Layered Igneous Rocks.* Oliver & Boyd, Edinburgh.

WILSON, H.D.B. & BATEMAN, A.M. (Eds), 1969: Magmatic ore deposits; a symposium. *Econ. Geol. Monogr., 4.*

DEPOSITS OF TITANIUM

THE CLASSIFICATION OF TITANIUM DEPOSITS

Of the large number of titanium minerals, only ilmenite, rutile, anatase, leucoxene, and loparite in part are of commercial importance. On the
basis of the genetic features of the deposits, they are divided into magmatic,
weathering crust, sedimentary, volcanogenic-sedimentary, and metamorphogenic.
Of fundamental importance in the production of titanium at the present time are
the ores of the magmatic deposits and the placers. In the USSR, the main raw-
-material sources for the production of titanium are the ilmenite-magnetite
ores of the magmatic deposits and the littoral-marine types (ilmenite, rutile,
and zircon), and also the alluvial and eluvial-deluvial ilmenite placers.

In the capitalist countries (USA, Australia, Canada, Norway, and
Finland) and the developing countries in 1969, 61.6% of ilmenite concentrate
and titanium slag was obtained from ores of magmatic deposits, and 27.2% of
ilmenite and 11.2% rutile concentrates from placers. In all, these countries
in 1969 produced 3.13 million tonnes of ilmenite and 400,000 tonnes of rutile
concentrates. In 1971, 3.76 million tonnes of titanium concentrates were
obtained.

The titaniferous ores of the commercial deposits belong to different
genetic types (Table 26). The amount of TiO_2 in the richest ilmenite-magnetite
ores of the magmatic deposits reaches 23%, and in the ilmenite-hematite ores
there are on average 23%. The placer deposits may contain titanium in considerably lesser amounts. Placers are regarded as commercial that contain one or
more percent of the ore mineral (Alekseevsky, 1970). In Australia, beach
sands with 0.5% of ore minerals are exploited. Easy enrichment of the beach
sands and substantial reserves enable their exploitation to be carried out on
a large scale. There is practical interest in the ores of the metamorphogenic
deposits, for example, the ilmenite-magnetite ores in amphibolites (Otanmäki,
Finland). The ancient metamorphosed placers (the Yagersk deposit in Timan)
deserve attention, and also the rutile deposits in schists and gneisses (Harford,
USA; Plumo-Hidalgo, Mexico), and also deposits of anatase (Minas Gerais, Brazil).

The prospective deposits for titanium also include certain other endogenic and exogenic types, from the ores of which titanium has not yet been extracted, mainly because of the absence of economically suitable technological methods. They include: 1) magmatic titanomagnetite deposits in gabbro-diabases;
2) magmatic titanomagnetite and ilmenite-titanomagnetite deposits in pyroxenites,
peridotites, hornblendites, olivinites, and gabbros; 3) magmatic (and metasomatic) perovskite-titanomagnetite deposits in pyroxenites and olivinites; 4)
hydrothermal ore-shows of rutile and ilmenite in various complexes of igneous
rocks; 5) rutile-bearing eclogites and amphibolites; 6) deposits of sphene;
and 7) deposits of bauxites.

Table 26. *Genetic types of Commercial Deposits of Titanium*

Genetic type of deposit	Titaniferous ores	Amount of TiO_2 in ore, %	Principal titanium minerals	Examples of deposits
Magmatic	Ilmenite-magnetite and ilmenite-titano-magnetite in gabbros	12 - 18 in uniform ores	Ilmenite	Kusinsk, Kopansk, Medvedevsk (Urals)
	Ilmenite-magnetite in gabbro-anorthosites	7 - 23	"	Tehawus (USA), Telness (Norway)
	Ilmenite-hematite in anorthosites	average 32	"	Lac-Tio, Puyjelom (Canada)
	Loparite in lujavrites, foyaites, and urtites	38.3 - 41.0 (in loparite)	Loparite	Lovozero Massif (Kola Peninsula)
Exogenic	Weathering crust	3 - 20	Ilmenite, rutile	Ukraine, Kazakhstan
	Eluvial-deluvial placers	3 - 25	Ilmenite	Ukraine
	Alluvial placers	0.5 - 20	"	Ukraine, Urals
	Littoral-marine (ancient and modern) placers	0.5 - 35	Ilmenite, leucoxene, rutile	Ukraine, shores of Baltic and Azov seas, Australia, India, Sri Lanka, Sierra Leone
Volcanogenic-sedimentary	Ilmenite in tuffs, tuffites, and tuff-sandstones	$\sim$ 20 (in magnetic fraction)	Ilmenite	Yastrebovsk horizon (Voronezh district)
	Sandstones (sometimes petroliferous)	$\leqslant$ 8 - 10	Leucoxene, ilmenite	Yagersk (Timan); Bashkiria; Agartsa (Armenia)
	Ilmenite-magnetite in amphibolites	average 12.2	Ilmenite	Otanmäki (Finland)

Malyshev (1957) has also pointed out other ore-shows and deposits of titanium. However, they are all characterized either by insignificant reserves, or by small amounts of TiO_2.

The largest magmatic deposits of titanium were formed during Precambrian time. They include the ilmenite-magnetite and ilmenite-hematite deposits of the anorthosite association (USA, Canada, and Norway). In the USSR, Lower Proterozoic intrusions of the gabbro-anorthosite association are known on the Kola Peninsula: the Tsaginsk Massif, and the Keiv and Kolmozero-Voron'e intrusions, containing an ilmenite-titanomagnetite mineralization. Considerably smaller ilmenite-magnetite and ilmenite-titanomagnetite deposits were formed during the Palaeozoic Era. In the USSR they are represented by the deposits of the Kusinsk Group (Urals), restricted to a Cambrian gabbro intrusion. In Transbaikalia, the Palaeozoic gabbro-anorthosite massif is spatially and genetically associated with the Kruchinina deposit of titanomagnetite and ilmenite. Deposits of loparite were formed during the Palaeozoic Era in the nepheline syenites of the Kola Peninsula.

In contrast to the magmatic deposits, the formation of most of the commercial titanium placers took place during the Mesozoic and Cainozoic Eras. In the USSR, the leading role amongst them has been played by the buried littoral-marine beach sands of the Ukraine. Abroad, the most important are the modern beach sands (Australia, India, etc.).

The principal metamorphic deposits of rutile in schists, gneisses, amphibolites, and ecologites were formed during the Precambrian. Amongst the metamorphosed deposits there are Upper Proterozoic sandstones with rutile (the Bashkirian uplift), Palaeozoic sandstones, enriched in leucoxene (the Yagersk deposit in Timan), and Eocene sandstones, containing titanomagnetite and ilmenite (the Agartsinsk deposit in Armenia). Thus, the greater part of the commercial magmatic and metamorphogenic deposits of titanium were formed during the Precambrian and the Palaeozoic, and the placer deposits, during the Mesozoic and Cainozoic eras.

MAGMATIC DEPOSITS

The magmatic deposits of titanium may be spatially and genetically associated with ultramafic and basic rocks of the normal and alkaline series. However, all the principal commercial deposits, from the ores of which titanium concentrates are at present obtained, are restricted to intrusive rocks of the normal series, mainly gabbros and anorthosites. The massifs, which consist of these rocks, are mainly distributed on the crystalline shields and ancient platforms, and also in their surroundings. The Precambrian ore-bearing intrusions of basic composition have an especially significant development on the Baltic and Canadian Shields.

In the USSR, the greatest practical importance attaches to the ilmenite-magnetite and ilmenite-titanomagnetite magmatic deposits in gabbros. On the western slopes of the Urals in the region of the miogeosyncline, the Kusinsk gabbro intrusion was formed during Cambrian time, carrying an ilmenite-magnetite and titanomagnetite mineralization. The quite numerous deposits of titanomagnetite formed in the region of the Uralian eugeosyncline during the Palaeozoic in ultramafic rocks (the Gusevogorsk, Kachkanar, etc.), as a rule, contain little ilmenite.

The Palaeozoic activated zones of the ancient platforms are associated with the formation of multi-phase plutons, consisting of agpaitic nepheline syenites and ijolite-urtites, bearing a loparite mineralization (the Kola Peninsula). In the zones of activation of the ancient platforms, massifs of the alkaline--ultramafic association, enriched in titanomagnetite and perovskite, were also formed during the Palaeozoic. However, the practical importance of the magmatic deposits of titanium, associated with rocks of the alkaline series, is at present small.

The commercial and prospective magmatic deposits of titanium of the geosynclinal and orogenic regions are restricted to the following associations (Kuznetsov, 1964): 1) gabbro-diorite-diabase (Kusinsk, Kopansk, Medvedevsk, and Matkal, Urals); 2) gabbro-pyroxenite-dunite (Gusevogorsk, Kachkanar, Urals; the Lysansk group of deposits, East Sayan); 3) gabbro-syenite (Yelet'-Ozero, Kola Peninsula).

On the crystalline shields and ancient platforms (activated sectors), the magmatic deposits of titanium are associated with the following associations: 1) differentiated gabbroic and noritic intrusions (Bushveld Complex, South Africa); 2) gabbro-diabases of the migmatite association of the amphibolite facies and their associated anatectites (Pudozhgorsk and Koikar, Karelia); 3) anorthosite (Kruchinina, Transbaikalia; Roseland and Tahawus in the USA; Lac Tio in Canada); 4) the association of intrusions of ultramafic-alkaline rocks of the central type (Kovdor and other deposits of the Kola Peninsula); 5) the associations of intrusions of agpaitic nepheline syenites of the central type (Lovozero Massif, Kola Peninsula).

The most significant commercial deposits of titanium are restricted to the large massifs of the anorthosite association. The area of individual massifs may reach several hundreds and even thousands of square kilometres. Examples are the huge gabbro-anorthosite massifs of the Adirondack Mountains (USA), within which there are the ilmenite-magnetite deposits of Sanford Hill, Ore Mountain, and others. The reserves of rich ores in the individual deposits reach 100 million tonnes and more. In the USSR, the Volyn labradorite massif in the Ukraine, characterized, however, by a lean ilmenite mineralization, is on a significant scale.

This same association also includes the smaller ore-bearing massifs of gabbros and anorthosites (up to several tens of square kilometres). For example, in Transbaikalia, such a massif is associated with the Kruchinina ilmenite-titanomagnetite deposit.

Massifs of the gabbro-diorite-diabase association are usually small in area (up to a few square kilometres), but rich ilmenite-magnetite deposits may be restricted to them (the Kusinsk, etc.).

The magmatic deposits of titanium, spatially and genetically associated with basic and ultramafic rocks of the normal series, were formed as a result of differentiation of a basaltic magma (Malyshev, 1957; Sokolov, 1957; Lebedev, 1963, 1965; Godlevsky, 1968; Bogatikov *et al.*, 1970; Karyakin, 1970; etc.). The titanium ores in the complexly differentiated gabbro-anorthosite massifs are, as a rule, restricted to melanocratic rocks or to the zone of transition from melanocratic rocks to more leucocratic types. Lebedev (1963), on the basis of the Chineisk gabbro-anorthosite massif (Transbaikalia), has noted that '... the thin lens- or layer-like bodies form ultramafic differentiates of the pyroxenite

and peridotite type, which are mainly associated with titanium mineralization'
(p.114). The formation of ilmenite-magnetite, ilmenite-hematite, and ilmenite-
-titanomagnetite ores most probably occurred as a result of intra-cameral differentiation of a basic magma.

The ore bodies have various shapes. There are layer-, lens-, and
vein-like bodies of uniform or segregated ores. The uniform ores show gradual
transitions through segregated ores into the surrounding rocks. Normally, the
ore bodies lie conformably with the banding of the basic-ultramafic massifs.

Zavaritsky, on the basis of method of formation, has recognized segregation and fusive ores in the magmatic deposits. The fusive ores are more
typical of the commercial deposits of titanium. The most valuable massive ores
are those that have been subjected to metamorphism. Under its influence, high-
-quality, easily-enriched ores have been formed, consisting of an aggregate of
individual grains of ilmenite and magnetite.

In some deposits, the ilmenite contains small inclusions of hematite
(Lac Tio, Canada) or magnetite (Matkal, Urals), which cannot be separated by
present-day methods into an independent concentrate. There is also a wide
distribution of titanomagnetite, containing minute inclusions of ilmenite (breakdown of solid-solution texture), for example, in the Pudozhgorsk, Koikar (Karelia),
and many other deposits (Borisenko *et al.*, 1968; Yudin & Zak, 1970). Although
the content of titanium in the titanomagnetite is 13.6%, its extraction from
concentrates of this mineral is not at present being carried out.

The most important magmatic deposits of titanium in the USSR are:
Kusinsk, Kopansk, Medvedevsk, Matkal (Urals); Pudozhgorsk (Karelia); Yelet'-
-Ozero (Kola Peninsula); Kruchinina (Transbaikalia); and Lysansk and Kedransk
(East Sayan). Abroad, they are: Lac Tio, Puy-Jelon, Mills, etc. (Canada),
Tahawus etc. (USA), and Telness etc. (Norway).

THE KUSINSK GROUP OF ILMENITE-MAGNETITE AND ILMENITE-TITANOMAGNETITE DEPOSITS

The deposits of this group (Kusinsk, Medvedevsk, Kopansk, Matkal, etc.)
occur in the Southern Urals, and they are restricted to four gabbro massifs of
the same names, elongated in the form of a discontinuous chain with northeasterly strike (Fig. 92). Their overall extent is 76 km, with a width of up to
2 km. It is likely that these massifs are portions of the one Kusinsk intrusion. According to Myasnikov (1959), and Shteinberg *et al.* (1959), it is a
layered body (dipping southeastwards at 35 - 60°), lying conformably with the
surrounding sedimentary rocks. Its position is determined by a large fault
between two varied rock sequences of late Proterozoic age. In the west (footwall of the intrusion), they consist of limestones, and dolomites of the Satka
Group and phyllites, quartzites, and conglomerates of the Bakal Group. In the
hanging wall are the granite-gneisses of the Kuvash Group and granites of the
Ryabinovsk Massif, and also quartzites of the Kuvash and Zigal'ga groups.

The basic rocks of the Kusinsk intrusion consist of normal gabbros,
amphibolized gabbros, and gabbro-amphibolites. The metamorphic grade of the
rocks decreases southwards (Table 27).

The absolute age of the gabbroic rocks of the Kusinsk intrusion,
according to Ovchinnikov *et al.* (1967), is about 1300 m.y. The intrusion
was most probably formed during the Baikalian phase of folding.

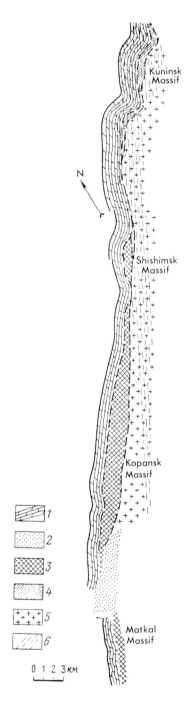

1

2

3

4

5

6

0 1 2 3 KM

The basic rocks of the massifs are characterized by a pseudostratification, caused by an alternation of melanocratic and leucocratic varieties. There are ilmenite-magnetite, ilmenite, and titanomagnetite ores lying conformably with the banding of the gabbros. In the Kusinsk deposit, there is a widespread distribution of uniform ilmenite-magnetite ores, and in the remaining deposits, there is a predominance of segregated titanomagnetite, magnetite-ilmenite, and ilmenite ores, which have a typical sideronitic texture.

The ore bodies in the deposits of the Kusinsk group mainly have a layer- and lens-like shape. In the section of the Matkal deposit (according to V. Novikov) it is seen that the layer-like accumulations of segregated ores are restricted only to the mesocratic and melanocratic gabbros and have a common dip with them (Fig. 93). The transitions from the ore sectors into the non-ore parts are gradual.

In the Kusinsk deposit, the vein- and lens-like bodies of uniform ilmenite-magnetite ores lie conformably with the banding of the surrounding gabbro-amphibolites. In mineral composition, these ores consist of aggregates of grains (0.2 - 0.5 mm) of magnetite (60 - 80%) and ilmenite (20 - 40%), and in addition there are: spinel, hematite, pyrite, chalcopyrite, and pyrrhotite, and the gangue mineral, chlorite. According to chemical analytical data, the average amounts in the uniform ores are (in wt %): Fe, 53 (FeO, 27.6; Fe_2O_3, 48.2); TiO_2, 14.21; V_2O_5, 0.65; S, 0.122; P, 0.01; SiO_2, 3.18; Al_2O_3, 2.98; Cr_2O_3, 0.67; CaO, 3.33; MnO, 0.50; MgO, 0.005; K_2O and Na_2O, 0.12; H_2O, 0.25 (Malyshev, 1957). In contrast to other deposits, the uniform ores of the Kusinsk deposit contain a considerable quantity of independent grains of ilmenite. The magnetite of this deposit is almost devoid of titanium ($\leqslant 0.n\%$). On the other hand, in the Medvedevsk, Kopansk, and Matkal deposits, the titanomagnetite is, as a rule, enriched in titanium (up to 13.4% TiO_2); laminar inclusions of ilmenite are usually observed in it.

Fig. 92. Diagram of the South Uralian band of basic rocks (After Myasnikov, 1959).

1) Proterozoic shale-carbonate rocks (Satka Group); 2) Proterozoic quartzites; 3) gabbros; 4) amphibolites; 5) granite-gneisses; 6) zone of crumpling.

Table 27.   *Characteristics of the Ores of the Kusinsk, Medvedevsk, Kopansk, and Matkal Deposits*

Deposit	Country rocks	Shape of ore bodies and types of ores	Principal ore minerals
Kusinsk	Gabbro-amphibolites	Vein- or lens-like bodies of uniform ores, up to 10 - 12 m thick. Segregated ores of minor importance	Magnetite and ilmenite
Medvedevsk	Amphibolized and saussuritized gabbros	Mainly lens-like bodies of segregated ores. Rarely thin (up to 2 m) vein-like bodies of uniform ores	Ilmenite, magnetoilmenite, and titanomagnetite
Kopansk	Gabbros	Zones of segregated ores and thin vein-like segregations of uniform ores	Titanomagnetite, ilmenite, and magnetoilmenite.
Matkal	Amphibolized gabbros	Layered bodies of segregated ores with fine lens-like bodies of uniform ores	Titanomagnetite, magnetoilmenite, and ilmenite

In the uniform ores of the Matkal deposit, titanomagnetite is mainly present, and the amount of ilmenite grains does not exceed 3 - 5% (Karpova & Martynova, 1970).   It is typical that the magnetite-ilmenite of these deposits usually displays small inclusions of magnetite (breakdown texture of solid solutions).   In the ilmenite grains of the Kopansk deposit the quantity of magnetite inclusions reaches 7.3% (Karpova & Burova, 1970).

The amount of TiO_2 (Table 28) in the ilmenite of the Kopansk and other magmatic deposits is similar to the theoretical value (52.66% TiO_2). The various forms of occurrence of ilmenite in the ores depends to a significant degree on their metamorphism.   In the most metamorphosed ores of the Kusinsk deposit, as noted by Myasnikov (1959), the laminar inclusions of ilmenite have been precipitated in the form of independent grains, and the magnetite has been cleared of titanium.

In commercial respects, the metamorphosed ores have the greatest value, because during their enrichment a high-grade ilmenite concentrate is obtained (Sysolyatin, 1963).

In addition to the Kusinsk group of deposits, we must note the Gusevo-gorsk deposit of titanomagnetite, described in the chapter on 'Vanadium'.   In

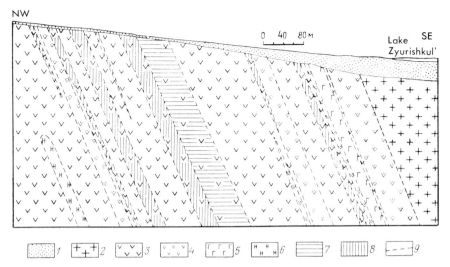

Fig. 93. Diagrammatic geological section through the Matkal deposit

1) eluvial-deluvial sediments; 2) plagioclasites and cataclased
granites; 3) meso- and melanocratic gabbros; 4) leucocratic gabbros,
and anorthosites; 5) pegmatoid gabbro; 6) mesocratic gabbro-norites;
7) segregated ilmenite ores; 9) assumed contacts

Table 28. *Chemical Composition of Ilmenite from Magnetite Deposits (in wt %)*

Components	Analysis number					
	1	2	3	4	5	6
SiO_2	0.02	0.35	0.53	0.45	tr.	2.80
TiO_2	49.29	49.87	47.39	47.90	51.23	44.00
Al_2O_3	1.95	0.50	0.51	0.21	0.30	3.00
Fe_2O_3	6.68	4.53	6.45	6.73	5.11	9.70
FeO	37.60	43.23	42.63	42.72	40.15	32.32
MnO	0.25	1.16	1.18	2.21	–	2.60
MgO	3.74	0.47	0.38	0.15	2.66	3.96
CaO	–	–	0.39	–	–	1.20
V_2O_5	0.16	0.14	0.07	–	0.56	–
Nb_2O_5	–	0.04	0.05	–	–	–
Total	99.69	100.29	99.58	100.37	100.01	100.04 including H_2O^+ 0.46
Density, g/cm³	4.65	4.71	4.77	4.77	4.68	
Analyst	P.Nissen-baum	T. Burova			V.Moleva	V.Arkhangel'-skaya

1) ilmenite-titanomagnetite ore, Kopansk Massif; 2) ore gabbro, same locality;
3) fine-grained ore gabbro, same locality; 4) the same, Matkal Massif; 5) ilmenite-magnetite ore, Kusinsk Massif. Samples 1-5 after Karpova & Burova (1970);
6) ore pyroxenite of Gusevogorsk Massif (H_2O^+, 0.46), after L. Borisenko and E.
Uskov.

the ores of this deposit, ilmenite in the form of individual grains, with equant
or irregular shape or in the form of small laminar inclusions (breakdown textures of solid solutions), usually occur in clear association with titanomagnetite.
The dimensions of the individual ilmenite grains are from 0.1 X 0.1 up to 0.5 X
X 0.5 mm, rarely up to 1 mm.   Their amount in the ore pyroxenites may reach
3 - 4% (Borisenko, 1966).

The work carried out in the enrichment plant of the Kachkanar Mining
and Concentration Combine has shown the possibility of obtaining ilmenite as a
by-product from the dump tailings, using the gravity-magnetic method.   This
method separates into a concentrate only the ilmenite, precipitated in the form
of individual grains.   The laminar inclusions of ilmenite, present in the titanomagnetite, are not selected.   Thus, it seems possible to use the titanium-poor
iron ores of the magmatic deposits for obtaining an ilmenite concentrate (1.2 -
- 1.4% TiO_2 in the Gusevogorsk deposit).

EXOGENIC DEPOSITS

Deposits of this type are found in marine and continental sediments,
chief amongst which are the placers and weathering crusts.   Volcanogenic-
-sedimentary formations are involved in this same group.   The ore minerals
of titanium in the exogenic deposits consist of ilmenite and leucoxene, and
less frequently, rutile and anatase.

PLACER DEPOSITS

The principal commercial type of titanium raw-material is the littoral-marine, normally complex titano-zircon placers (beach sands) (ilmenite-
-rutile-zircon).   Of lesser significance are the continental (alluvial and
eluvial-deluvial) placers of ilmenite and the weathering crusts on crystalline
rocks (basic rocks, etc.).   Placers in the Soviet Union include more than half
of the reserves of titanium in the principal commercial types of its deposits
(Gurvich & Alekseevsky, 1967).

In the USSR, the most important are the ancient (or buried) placers:
Neogene, Palaeogene, Mesozoic, and Paleozoic.   They are distributed in the
various regions of the Russian Platform, Western and Eastern Siberia, Kazakhstan,
and Transbaikalia.   The present-day beach sands of the Baltic and Azov Seas,
occurring in the USSR, occupy a subordinate position with respect to the
ancient littoral-marine placers.   On the other hand, abroad (Australia, etc.),
the predominant role in the production of titanium concentrate is played by
present-day complex beach sands.

The formation of the bulk of the commercial placers of the USSR took
place on the platforms, where beach sands were formed mainly under conditions
of transgressing seas, and continental placers (in lakes, deltas, and river
valleys, and in gullies) were formed under conditions of regressing seas.   The
formation of the titanium placer deposits of different types is clearly associated with the denudation and accumulative processes, which took place during
the changing conditions of transgression and regression of the sea.

Palaeogeography exerted a considerable influence on their formation,
that is, the relief, climate, nature of the soils and weathering crusts, and

the facies conditions of sedimentation.   As noted by Malyshev (1957), one of
the principal conditions for the formation of the large titanium placers was
chemical weathering, which contributed to the liberation of ilmenite and other
resistant accessory minerals from the crystalline rocks.   The most important
source of ilmenite is the metamorphic para-schists (Mikhalev & Korobova, 1972).

The principal rocks, forming placers of various types, are sands;
among these there is a predominance of quartzose varieties.   In the alluvial
and eluvial-deluvial placers, secondary kaolins and kaolin clays play a substantial role.

The principal ore mineral (ilmenite) occurs for the most part in the
form of grains, having various degrees of rounding.   Less frequently, there
are laminar crystals or their fragments.   The ilmenite is black, and in the
leucoxenized varieties, dark-brown.   Alteration of the mineral is associated
with oxidation and hydration processes.   In comparison with the ilmenite of
the magmatic deposits, the ilmenite of the placers contains less iron, and more
titanium (Table 29).   Its density decreases from 4.7 to 3.7 g/cm^3.

Table 29.   *Characteristics of the Titanium Minerals of the Ukrainian Placers*
          After M. Veklich *et al.* (1965)

| Mineral | Amount | | | Density g/cm^3 | Colour of mineral |
	TiO_2	Fe_2O_3 + + FeO	H_2O		
Ilmenite	up to 50	49–50	–	> 4.7	Black
Hydrated ilmenite (phase I)	51–60	29–49	0.01–12	4.35–4.7	"
Arizonite (phase II)	60–70	11–34	11–23	4.0–4.35	Reddish-brown, dark brown, rarely almost black
Leucoxene (phase III)	70–96.4	0.77–16.0	0–18	3.71–4.0	Reddish-brown, yellowish-brown, steel-grey to almost white
Rutile (terrigenous)	96–99.3	0.1–1.90	0–0.5	4.20–4.26	Black, dark-red, and orange- -yellow.

(Products of alteration of ilmenite — bracketing Hydrated ilmenite, Arizonite, Leucoxene)

The titanium-zircon littoral-marine placers have usually been formed
as a result of decomposition of rocks of different composition, and repeated
redeposition of earlier formed placers.   Any clear association with the sources
of supply is absent from them.   In the Ukraine, for example, the placers were
apparently formed as a result of the breakdown of the crystalline rocks of the
Ukrainian Shield and partially of the Voronezh Massif.

        In composition, the littoral-marine placers are usually oligomictic
(quartzose).   Their commercial interest lies in the ilmenite, rutile, leucoxene, and zircon, and also kyanite, xenotime, anatase, malacon, and certain
others.   The principal rock-forming mineral is quartz; of lesser importance
is kaolinite.   The sands are usually very fine- and fine-grained.

        These placers are characterized by layer- or lens-like ore segrega-
tions, the thickness of which reaches tens of metres, and an extent of several
tens of kilometres, with a width of up to thousands of metres (Fig. 94).   They
rarely have strictly defined lithological boundaries.   However, as noted by
Alekseevsky (1970), the ore minerals in the members which consist of interbedded
sands and clays, are restricted mainly to the sandy layers.   Therefore, a
layered distribution of the ore component is typical in the ore bodies, and
this is usually persistent in plan.

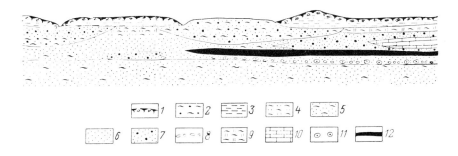

Fig. 94.   Typical section of productive sediments of the Cenomanian of the
           central regions of the Russian Platform (After Gurvich, Kazarinov
           & Khmara, 1964).

           1) soil-vegetation layer;   2) loams;   3) clays;   4) sandy clays;
           5) argillaceous sands;   6) very fine- and fine-grained sands;   7)
           medium- and coarse-grained sands;   8) gravels;   9) micaceous clays;
           10) limestones;   11) phosphorite nodules;   12) productive layer.

        The commercial amount of ilmenite and rutile in the placers is from
tens to hundreds of kilogrammes per cubic metre.   In the Ukraine, the most
important littoral-marine placers are restricted to the Poltavian and Sarmatian
sediments, for example, the Samotkan deposit.   The ore-bearing sands of both
ages have a similar mineral composition: quartz, ilmenite, rutile, leucoxene,
anatase, zircon, kyanite, sillimanite, staurolite, tourmaline, andalusite,
spinel, chromite, apatite, corundum, limonite, carbonates, kaolinite, micas,
manganese oxides, garnet, feldspars, and pyrite.   The ratio of ilmenite to
rutile ranges from 3 to 8 (Zherdeva & Abulevich, 1960).

        The ore minerals in the Sarmatian sands are spread throughout the
entire section, but in the Poltavian section, they are only in the upper part
of the layer.   The grainsize of the ilmenite is 0.07 - 0.25 mm, and rutile,
0.07 - 0.15 mm.   The average amount of TiO_2 in the ilmenite is 67.7%, which
indicates its substantial leucoxenization.   According to Zherdeva & Abulevich
(1960), the Samotkan placer was formed as a result of erosion and redeposition
of sands of earlier age.

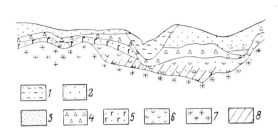

Fig. 95.
Section through sedimentary rocks on the Korosten granites (After P. Khatuntseva)

1) sand-clay sediments;   2) sands;
3) very fine-grained kaolinitic sands;
4) flints;   5) glauconitic sands;
6) sands of varying grainsize with lenses of secondary kaolins;   7) weathering crust of Korosten granites;
8) ilmenite placers

The continental ilmenite placers are mainly distributed in Lower Cretaceous, Palaeogene, and less frequently in Quaternary deposits, often directly on a weathering crust.   As a rule, their connexion with the source of supply has usually been identified.   Thus, the alluvial placers lie approximately 5 - 30 km from it.   In the Ukraine, many deposits of this type (Fig. 95) are associated with basic, and also with acid rocks of the Korosten Complex, in the weathering crusts of which there are relatively large grains of ilmenite ($\leqslant$ 1 mm).

In composition, the continental placers are usually polymict (quartz, feldspars, and kaolinite).   The grainsize of the ilmenite is 0.1 - 0.25 mm and more.   Roundness, especially in the eluvial-deluvial placers, is feeble.

The ore bodies of the alluvial placers are usually ribbon-like and sinuous segregations in shape, restricted to river valleys.   The ore minerals have been deposited in the lower horizons, that is, in the most coarsely-clastic material, consisting of coarse-grained sand, gravel, or small-pebble material. The commercial content of ilmenite in the alluvial placers varies from a few tens to several hundreds of kilogrammes per cubic metre, sometimes reaching 2000 kg/m^3. However, the average amount of ilmenite is considerably less, for example, in the Irsha alluvial placer, it is about 20 kg/m^3 (Malyshev, 1957).

DEPOSITS OF THE WEATHERING CRUST OR RESIDUAL DEPOSITS

These deposits may contain significant amounts of ilmenite, and less frequently, rutile.   Modern and buried titaniferous weathering crusts have been found.   They have been formed on rocks of varying composition, and they may be ortho-eluvial and para-alluvial.   The formation of the crusts is clearly associated with the process of chemical weathering, which takes place with particular intensity under humid climatic conditions.   Against a background of substantial removal of sodium, potassium, calcium, and magnesium, and the formation of clay minerals of the kaolinite and other groups, accumulation of the more resistant accessory minerals takes place in the crust, including ilmenite and rutile.   In the weathering crusts, formed on the crystalline rocks, rounding of the ore-mineral grains is absent.   Well-preserved crystal shapes are quite often observed.

The thickness of the weathering crusts may reach several tens of metres. For example, the crust preserved on the crystalline rocks of the Ukrainian Shield has a thickness of up to 50 m, comprising on average 5 - 20 m.   Its formation occurred mainly at the end of the Triassic, the beginning and middle of the Jurassic, and partly during the Early Cretaceous (Veklich *et al.*, 1965).

The titanium-bearing crusts usually have a kaolinite profile. The amount of ilmenite in them may reach several hundreds, and that of rutile several tens of kilogrammes per cubic metre. In the crusts developed on rocks of one petrographic composition, the amount of titanium minerals is quite evenly spread. The mineral composition of the solid rocks may substantially influence the quantity of ore minerals in the crust. For example, in the Kundybaevo deposit (Kazakhstan), the weathering crusts, formed on metamorphic rocks, contain up to 180 kg/m^3 of ilmenite and up to 73.8 kg/m^3 of rutile (Niyazov, 1970). In this respect, the ore bodies of the deposit, which have a lensoid and layer-like shape, extend strictly along the strike of the solid rocks. A change in the composition of the accessory minerals in the metamorphic rocks is accompanied by their quantitative and qualitative change in the weathering crust.

The Volyn' gabbro-anorthosite massif is characterized by a crust, enriched only in ilmenite (*ca* 300 - 500 kg/m^3), which is associated with the absence of rutile in the main gabbroid rocks (Malyshev, 1957).

One of the features of the weathering crusts, formed on basic crystalline rocks, is the predominant accumulation of ilmenite, containing little chromium. This substantially increases the quality of the ilmenite concentrate obtained.

SEDIMENTARY-VOLCANOGENIC DEPOSITS

The deposits of this group have been discovered relatively recently (Blinov *et al.*, 1963). Typical representatives are the titaniferous volcanogenic-sedimentary formations of the Yastrebovsk horizon in the south of the Voronezh district (middle course of the River Don from Losevo village to Boguchar). This region consists of sedimentary and volcanogenic-sedimentary rocks of Palaeozoic age and sedimentary rocks of Mesozoic-Cainozoic age, resting with angular unconformity on the Precambrian basement. In structural respects, the region lies in the southeasterly portion of the Voronezh Anteclise.

The deposits of the Yastrebovsk horizon (Devonian) have a thickness of a few metres up to 35 metres (near Nizhnii Mamon village). The depth of this horizon is on average 50 - 70 m. The overall extent of the volcanogenic--sedimentary rocks is approximately 100 km, with a width of 20 - 40 km. The principal strike coincides with the Losevo-Mamon zone of faults, with which, according to Blinov *et al.* (1963), intense volcanic activity has been associated.

The largest amount of ilmenite is restricted to the coarsely-clastic tuffs, tuffites, and tuff-sandstones, in which effusive fragments consist mainly of rocks of basic composition. The terrigenous material in the tuffs accounts for up to 10%, and in the tuff-sandstones, up to 90%. The matrix is a magnesian--iron chlorite.

The coarsely-clastic varieties of the tuffogenic rocks are most enriched in ilmenite (sometimes up to 50 volume %). Fragments of ilmenite crystals or corroded grains predominate, and less frequently, idiomorphic crystals. In the coarsely- and medium-clastic tuffogenic rocks, the dimensions of the ilmenite grains are on average 0.25 - 0.3 mm, and in the very finely clastic rocks, less than 0.1 mm. The density of the ilmenite is 4.25, and its colour black or brownish-black. There is less leucoxenized ilmenite than unaltered material. The amount of ilmenite markedly decreases with increase of terrigenous material in the sequence.

It is likely that the formation of the volcanogenic rocks, which have been enriched in ilmenite, took place in a shallow-water marine basin and was the consequence of submarine eruptive activity.

METAMORPHOGENIC DEPOSITS

This group combines the metamorphosed and metamorphic deposits of titanium.

METAMORPHOSED DEPOSITS

Deposits of this group, according to Malyshev (1957), are formations of different origin, which during the course of geological time have undergone various changes along with the surrounding rocks. For example, they include the Upper Proterozoic metamorphosed placers within the Bashkirian uplift. One of the deposits is restricted to the sandstones of the Zil'merdak Group, where layers (up to 2.5 m thick) have been found, enriched in ilmenite (up to 250 - - 400 kg/tonne) and zircon (up to 30 kg/tonne).

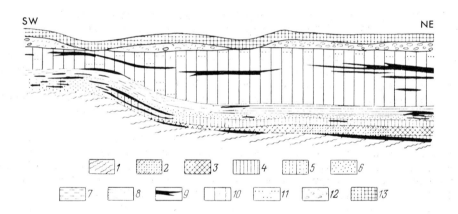

Fig. 96. Diagrammatic section through productive layer (constructed on restored surface of pre-Eifelian relief (After A. Sushon, 1962).

1) metamorphic schists; 2) leucoxene-quartz sandstones of varied grainsize with significant amount of coarse- and large-grained material; 3) leucoxene-quartz sandstones of varying grainsize; 4) predominantly fine-grained leucoxene-quartz sandstones with seams and lenses of varied grainsize; 5) predominantly fine-grained leucoxene-quartz sandstones, poorly sorted, with individual gravel-sized and large grains of quartz, with lenses of sandstones of varied grainsize; 6) fine-grained leucoxene-quartz sandstones; 7) interstratified siltstones and fine-grained leucoxene-quartz sandstones; 8) interstratified silt-stones and argillites with lenses of fine-grained sandstones; 9) argillites; 10) fine-grained quartzose sandstones; 11) seams of coarser-grained sandstones with individual gravel-sized grains of quartz; 12) polymict and leucoxene-quartz conglomerates, and coarse- and large-grained sandstones; 13) super-seam argillites.

A somewhat unusual representative of this type is the Yagersk deposit of leucoxene (Southern Timan), restricted to the petroliferous sandstones of the Eifelian and Givetian stages. This deposit is a buried Devonian placer (Kalyuzhnyi, 1960). The main ore segregation, termed the 'third layer', has been divided into lower and upper ore horizons. The lower ore horizon consists mainly of coarse- and large-grained quartzose sandstones with seams of siltstones and argillites. The upper ore horizon consists of polymict conglomerates and quartzose sandstones of varying grainsize. A productive horizon rests on Riphean metamorphic schists (Fig. 96). According to Shvetsova (1971), 80% on average of the heavy fraction of the Yagersk deposit consists of semi-rounded grains of leucoxene, having dimensions of 0.2 - 1.5 mm. They are grey or yellowish-grey in colour. Rounding of the grains is insignificant. The primary shape of the crystals is usually preserved. However, unaltered ilmenite has been found in the form of isolated grains. In rich samples of the ore-bearing sandstone there is 8 - 10% TiO_2 (Malyshev, 1957). It is evident that the deposit was formed during the Devonian as a result of the erosion of weathering crusts of Riphean metamorphic crusts.

The high-quality ilmenite-magnetite uniform ores and segregated ilmenite ores may have been formed during the regional metamorphism of originally magmatic ores. An example of commercial deposits of this type is the Otanmäki deposit in Finland, which is restricted to amphibolites that were formed through the metamorphism of an ore-bearing gabbro. Rich ores of this deposit contain up to 19%, and on average 12.2% TiO_2.

METAMORPHIC DEPOSITS

Deposits of this group are restricted to ancient schists, gneisses, eclogites, and amphibolites. They have been formed as a result of metamorphism of intrusive, eruptive, and sedimentary rocks, enriched in titanium. The principal ore mineral is rutile. The richest deposits of this type are known in the USA (Harvord) and in Mexico (Plumo-Hidalgo). The ores of the Harvord deposit consist of Precambrian chlorite schists, enriched in rutile (up to 20%). In the Precambrian gneisses of the Plumo-Hidalgo deposit, the amount of rutile reaches 25%.

Deposits of this type known in the USSR are significantly poorer in titanium. In the Kuznechikha deposit (Middle Urals), the amphibolites contain about 1.5% rutile, and in the eclogites of the Shuba deposit, the amount of rutile reaches 4.5%. Rutile ore-shows are also known in other metamorphic sequences in the USSR. However, technological difficulties prevent the use of the metamorphic rutile deposits at present.

Thus, the number of deposits with suitable reserves, and most satisfactory for enrichment, include the placer ores. Amongst them in the USSR, the leading place is occupied by the ancient littoral-marine placers of ilmenite and rutile (and also zircon). An important place is occupied by the leucoxene sandstones, containing petroleum, and also the ilmenite ores of the volcanogenic-sedimentary deposits.

The greatest practical interest in the magmatic deposits, on the basis of technological conditions, concerns those ores in which ilmenite is present in the form of separate grains (*e.g.* ilmenite-magnetite ores). However, the deposits of titanomagnetite, characterized by the greatest reserves of

Table 30.   *Amounts of Scandium and Vanadium in Ilmenite (in g/tonne)*

Type of deposit	Scandium			Vanadium		
	Number of Samples	Amount		Number of Samples	Amount	
		from - to	average		from - to	average
Ilmenite-magnetite in gabbros and gabbro-amphibolites	3	64 - 90	77	6	800 - 1200	1000
Titanomagnetite in pyroxenites	4	75 - 100	88.1	2	280 - 580	430
Weathering crust on basic rocks	6	50 - 70	59.6	2	200 - 450	325
Alluvial placers	9	52 - 73	60.6	8	168 - 448	291
Nearshore-marine placers (ancient)	27	26 - 77	43.3	20	84 - 880	364
Nearshore-marine placers (modern)	5	23.5 - 35	27.5	3	71 - 560	252

Note: Scandium determined by Yu. Sotskov by neutron-activation method; vanadium determined by L. Yalanskaya and L. Maslenkova by quantitative spectral-analysis method.

titanium, may also be regarded as present as a possible raw-material source for the extraction of ilmenite (Borisenko *et al.*, 1973). The value of the ilmenite concentrates, obtained from the ores of the magmatic deposits and the exogenic deposits, is increased by the presence of trace-elements, primarily scandium (Table 30). Its maximum concentrations (up to 100 g/tonne Sc) have been identified in an ilmenite from gabbroic and basic rocks. The possibility of the side-production of scandium during the process of reworking ilmenite concentrates has recently been demonstrated (Takezhanov, 1972). Considerable amounts of reworked material from these concentrates suggest that vanadium could be extracted during their treatment. The possibility of the side-production of vanadium during the production of TiO_2 has been demonstrated by Kurumchin & Yatsenko (1967).

REFERENCES

ALEKSEEVSKY K.M. (АЛЕКСЕЕВСКИЙ К.М.), 1970: О титан-циркониевых россыпях и их положении среди других типов месторождений (The titano-zircon placers and their position amongst other types of deposits), *in* "Металлогения осадочных и осадочно-метаморфических пород" (*'The Metallogenesis of the Sedimentary and Sedimentary-Metamorphic Rocks'*), pp.210-222. Nauka Press, Moscow.

BLINOV V.A., DYUBYUK K.A., KUZ'MINA L.S. & ODOKII B.N. (БЛИНОВ В.А., ДЮБЮК К.А., КУЗЬМИНА Л.С., ОДОКИЙ Б.Н.), 1963: О концентрации титана в вулканогенно-осадочных образованиях ястребовского горизонта на юге Воронежской области (The concentration of titanium in volcanogenic-sedimentary formations of the Yastrebovo horizon in the south of the Voronezh district). *Geologiya rudn. Mestorozh.*, *5*, No.1, pp.109–113.

BORISENKO L.F. (БОРИСЕНКО Л.Ф.), 1966: Редкие и малые элементы в гипербазитах Урала *(Rare and Minor Elements in the Ultramafic Rocks of the Urals)*.   Nauka Press, Moscow.

BORISENKO L.F., LEBEDEVA S.I. & SERDOBOVA L.I. (БОРИСЕНКО Л.Ф., ЛЕБЕДЕВА С.И., СЕРДОБОВА Л.И.), 1968: О титаномагнетите и магнетите из железорудных месторождений различного генезиса (The titanomagnetite and magnetite from iron-ore deposits of different origin). *Geologiya rudn. Mestorozh.*, *10*, No.4, pp.40–53.

BORISENKO L.F. *et al.* (БОРИСЕНКО Л.Ф. и др.), 1973: Ильменит бедных титаномагнетитовых руд и возможности его извлечения (The ilmenite of lean titanomagnetite ores and the possibility of extracting it), *in* "Исследования в области рудной минералогии" *('Investigations in the Field of Ore Mineralogy')*, pp.32–38.   Nauka Press, Moscow.

GODLEVSKY M.N. (ГОДЛЕВСКИЙ М.Н.), 1968: Магматические месторождения (Magmatic deposits), *in* "Генезис эндогенных рудных месторождений" *('The Origin of Endogenic Ore Deposits')*, pp.7–83.   Nedra Press, Moscow.

GURVICH S.I. & ALEKSEEVSKY K.M. (ГУРВИЧ С.И., АЛЕКСЕЕВСКИЙ К.М.), 1967: Титаноциркониевые россыпи (Titano-zircon placers), *in* "Успехи в изучении главнейших осадочных полезных ископаемых в СССР" *('Successes in the Study of the Principal Sedimentary Mineral Resources in the USSR')*, pp.226–238.

GURVICH S.I., KAZARINOV L.N. & KHMARA N.V. (ГУРВИЧ С.И., КАЗАРИНОВ Л.Н., ХМАРА Н.В.), 1964: Древние редкометально-титановые россыпи, методы их поисков и оценки *(Ancient Rare-Metal — Titanium Placers, and Methods of Prospecting for and Assessing Them)*.   Nedra Press, Moscow.

KALYUZHNYI V.A. (КАЛЮЖНЫЙ В.А.), 1960: Тиман — новая провинция россыпей (Timan, a new province of placers) *in* "Закономерности размещения полезных ископаемых. 4. Россыпи" *('Distribution Patterns of Mineral Deposits. 4. Placers')*, pp.117–125.   Moscow.

KARPOVA O.V. & BUROVA T.A. (КАРПОВА О.В., БУРОВА Т.А.), 1970: Ильменит из основных пород Копанского массива (Ilmenite from basic rocks in the Kopansk massif), *in* "Минералы базитов в связи с вопросами петрогенезиса" *('The Minerals of Basic Rocks in Relation to Problems of Petrogenesis')*, pp.216––231.   Nauka Press, Moscow.

KARPOVA O.V. & MARTYNOVA A.F. (КАРПОВА О.В., МАРТЫНОВА А.Ф.), 1970: Некоторые особенности распределения элементов-примесей в титаномагнетитах Маткальского массива (Some features in the distribution of trace-elements in the titanomagnetites of the Matkal massif).   *Izv. Akad. Nauk SSSR, ser. geol.*, No.1, pp.84–91.

KARYAKIN A.E. (КАРЯКИН А.Е.), 1970: Структруы рудных полей магматических месторождений *(The Structures of the Ore Fields of Magmatic Deposits)*. Nedra Press, Moscow.

KRUMCHIN Kh.A. & YATSENKO A.P. (КРУМЧИН Х.А., ЯТСЕНКО А.Р.), 1967: Комплексное использование медно-ванад-ных кеков титанового производства (Utilization of copper-vanadium residue from titanium production). *Tsvet. Metall. 40, No.4, pp.72-74.*

LEBEDEV A.P. (ЛЕБЕДЕВ А.П.), 1963: Генетические типы титаноносных магматических комплексов (Genetic types of titanium-bearing magmatic complexes *in* "Проблемы магматизма и генезиса изверженных горных пород" *('Problems of Magmatism and the Origin of Igneous Rocks')*, pp.111-118. Akad. Nauk SSSR Press Moscow.

LEBEDEV A.P. (ЛЕБЕЛЕВ А.П.), 1965: Расслоенные текстуры и титановая минерализация в Ангашанском габброидном массиве (Забайкалье) (Layered fabrics and titanium mineralization in the Agashan gabbroid massif (Transbaikalia)) *in* "Особенности формирования базитов и связанной с ними минерализации *('Features in the Formation of Basic Rocks and Their Associated Mineralization')*, pp. 5-113. Nauka Press, Moscow.

MALYSHEV I.I. (МАЛЫШЕВ И.И.), 1957: Закономерности образования и размещения месторождений титановых руд *(Patterns of Formation and Distribution of Deposits of Titanium Ores)*. Gosgeoltekhizdat, Moscow.

MYASNIKOV V.S. (МЯСНИКОВ В.С.), 1959: Некоторые особенности месторождений титаномагнетитовых руд Южного Урала и проявления в них метаморфизма (Some features of the deposits of titanomagnetite ores in the Southern Urals and manifestations of metamorphism in them). *Geologiya rudn. Mestorozh.,* No.2, pp.49-62.

NIYAZOV A.R. (НИЯЗОВ А.Р.), 1970: Рутил — основной рудный минерал титаноносных кор выветривания Джетыгаринского района (Rutile, the principal ore mineral of titanium-bearing weathering crusts in the Dzhetygara region). *Izv. Akad. Nauk kaz. SSR, ser. geol.,* No.4, pp.89-90.

OVCHINNIKOV L.N., PANOVA M.V., PODLESOVA R.G. *et al.* (ОВЧИННИКОВ Л.Н., ПАНОВА М.В., ПОДЛЕСОВА Р.Г. и др.), 1967: О калии-аргоновом возрасте некоторых роговых обманок Урала (The potassium-argon age of some hornblendes from the Urals). *Trudy Kom. Opred. absol. Vozr. geol. Form., 14,* pp.74-77.

SHTEINBERG D.S., KRAVTSOVA L.I. & VARLAMOV A.S. (ШТЕЙНБЕРГ Д.С., КРАВЦОВА Л.И., ВАРЛАМОВ А.С.), 1959: Основные черты геологического строения Кусинской габбровой интрузии и залегающих в ней рудных месторождений (The principal features of the geological structure of the Kusinsk gabbro intrusion and the ore deposits occurring in it). *Trudy gorno-geol. Inst. ural'. Fil., 40.*

SUSHON A.R. (СУШОН А.Р.), 1962: Условия формирования титановых россыпий в среднедевонских отложениях Тимана (Conditions of formation of titanium placers in the Middle Devonian deposits of Timan). *Izv. vyssh. uchebn. Zaved., Geol. Razv.,* No.6, pp.87-98.

SYSOLYATIN S.A. (СЫСОЛЯТИН С.А.), 1963: Технология получения высококачественных титановых концентратов титаномагнетитовых руд методом флотации (The technology of obtaining high-quality titanium concentrates from titanomagnetite ores by the flotation method). *Tsvet. Metallurg.*, No.5, pp.19-21.

VEKLICH M.F., DYADCHENKO M.G., KONDARCHUK V.Yu., KHOTUNTSEVA A.Ya. & TSIMBAL S.N. (ВЕКЛИЧ В.Ф., ДЯДЧЕНКО М.Г., КОНДАРЧУК В.Ю., ХОТУНЦЕВА А.Я., ЦИМБАЛ С.Н.), 1965: Этапы образования и вещественный состав россыпий Украины (Phases of formation and the material composition of the placers of the Ukraine), *in* "Геология россыпий" (*'The Geology of Placers'*), pp.219--227. Nauka Press, Moscow.

YUDIN B.A. & ZAK S.I. (ЮДИН Б.А., ЗАК С.И.), 1970: Титановые месторождения Северо-Запада СССР (восточная часть Балтийского щита) (The titanium deposits of the Northwest of the USSR (eastern portion of the Baltic Shield)). *Sov. Geol.*, No.9, pp.138-147.

ZHERDEVA A.N. & ABULEVICH K.N. (ЖЕРДЕВА А.Н., АБУЛЕВИЧ К.Н.), 1960: Минералогия Самотканского цирконо-рутило-ильменитового месторждения (The mineralogy of the Samotkan zircon-rutile-ilmenite deposit), *in* "Минеральное сырье" (*'Mineral Raw-Materials'*), vyp. 1, pp.26-36.

DEPOSITS OF VANADIUM

CLASSIFICATION OF VANADIUM DEPOSITS

Most of the deposits, from which vanadium is extracted at present, are complex. Vanadium is obtained along with the principal ore components, Fe, Ti, U, Pb, Zn, Cu, and P; in addition, vanadium is extracted during petroleum treatment. The greatest importance in reserves and quantity of output concerns the magmatic (titanomagnetite and ilmenite-magnetite) and sedimentary (carnotite and roscoelite in the varicoloured rocks) vanadium-bearing ores. Descloizite, cuprodescloizite, and vanadinite ores of the oxidation zone of polymetallic deposits play a substantial role in the production of vanadium. A few unusual deposits of vanadium are also known: patronite ores in asphaltites (Minas Ragra, Peru), and vanadium-bearing clays of the zone of argillization (Wilson Springs, USA).

In the USSR, the only raw-material sources of vanadium are the titanomagnetite and ilmenite-magnetite ores of the magmatic deposits (Yefimov *et al.*, 1969). In the capitalist countries about 50% of the vanadium is obtained from titanomagnetite and ilmenite-magnetite ores, 22.6% from carnotite and roscoelite ores in varicoloured rocks and phosphorites, 17.9% from clays of the zone of argillization, 8.7% from the oxidized polymetallic ores, and 0.8% from petroleum. In all in 1969, the capitalist countries produced 14,000 tonnes of vanadium (or 25,000 tonnes of V_2O_5).

The vanadium-bearing ores of the commercial endogenic and exogenic deposits (Table 31) belong to various genetic types (magmatic, sedimentary, and metamorphogenic deposits and a zone of oxidation of endogenic deposits). The concentration of vanadium in most types of ores of endogenic deposits does not exceed 1%, and usually comprises more than 0.1 - 0.2%. In the individual types of ores of the exogenic deposits, the amount of vanadium reaches a few percent. These ores contain, as a rule, specific vanadium minerals: carnotite, vanadinite, patronite, etc. However, in most of the sedimentary deposits, the concentrations of vanadium do not exceed a few tens of percent.

In addition to the genetic types of commercial deposits, shown in Table 31, there are others that deserve attention in respect to vanadium only as prospective types. To extract vanadium from them is unsatisfactory for several reasons: owing to small or unstable amounts of vanadium in the ore; the small scale of certain deposits; and the absence of a profitable technology. Among such deposits of vanadium, we must note:

1) the magmatic titanomagnetite deposits in gabbro-diabases;

2) the magmatic titanomagnetite deposits in olivinites;

3) the magmatic ilmenite-hematite, magnetite-ilmenite, rutile-apatite, and ilmenite-apatite (nelsonites) in anorthosites and gabbros;

4) contact-metasomatic deposits of magnetite (calc-skarn and albite-scapolite--skarns);

Table 31. *Genetic Types of Commercial Deposits of Vanadium*

Genetic type of deposit	Vanadium-bearing ores	Principal commercial component	Amount of vanadium in ore, %	Principal mineral-concentrators of vanadium (amount of vanadium, wt %)	Examples of deposits
Magmatic	Titanomagnetite in pyroxenites, peridotites, olivinites, hornblendites, and gabbros	Fe	0.05 - 0.17	Titanomagnetite (0.13 - 0.84)	Gusevogorsk, Pervoural'sk, and Volkovo (Urals)
	Ilmenite-magnetite in ilmenite-titanomagnetite types in gabbros	Fe, Ti	0.1 - 0.4	Magnetite, and titanomagnetite (0.31 - 0.62)	Kusinsk and Medvedevsk (Urals)
	Titanomagnetite in anorthosites, pyroxenites, and norites	Fe	0.9 - 1	Titanomagnetite (less than 1.5)	Bushveld Complex (South Africa)
Zone of oxidation of endogenic deposits	Descloizite, cuprodescloizite, and vanadinite in zone of oxidation of polymetallic deposits	Pb, Zn, Cu	1.1 - 1.7	Descloizite (12.3), cuprodescloizite (9.5 - 12.3), and vanadinite (10.6)	Berg-Aukas and Tsumeb (SW Africa)
Sedimentary	Carnotite and roscoelite types in mottled sediments ('Colorado Plateau' type)	U	0.5 - 2.8	Carnotite (11.2) and roscoelite (11.8 - 16.2)	Regions of Ambrosia Lake, etc. (USA)
	Phosphorites	P	0.05 - 0.5	-	Rocky Mountains (USA)
	Littoral-marine placers (black sands)	Fe, Ti	0.1 - 0.34	Titanomagnetite (0.25 - 1.7)	Kuriles (USSR); New Zealand
	Petroleum (high-sulphur types)	C	$\leq$ 0.05	Petroleum ash (0.6 - 58.4)	Urals - Volga Province (USSR); Venezuela, and Iran
	Patronite in asphalts	V	$\sim$ 6.2	Patronite (10.6 - 13.9)	Minas Ragra (Peru)
Metamorphogenic	Clays of argillization zone in crystalline schists	V	$\sim$ 1	Clay minerals ?	Wilson Springs (USA)
	Ilmenite-magnetite in amphibolites	Fe, Ti	0.24	Magnetite (0.8 - 1.3)	Otanmäki (Finland)

5)   hydrothermal deposits of magnomagnetite (the Angara-Ilim type);

6)   hydrothermal deposits of nolanite and uraninite;

7)   hydrothermal alunite deposits;

8)   ilmenite placers;

9)   deposits of sedimentary iron ores (iron oxide ores and marine sedimentary oolitic ores);

10) bauxite deposits;

11) deposits of carbonaceous-chert-clay and bituminous shales;

12) coal deposits (especially brown low-quality types);

The above endogenic and exogenic deposits of vanadium (commercial and prospective) were formed during various epochs.   Along with ancient (Precambrian) types there are vanadium deposits, the formation of which took place during the Tertiary and Quaternary periods.   The general trend in the evolution of ore--formation of the vanadium deposits appears to be as follows.   All the largest magmatic deposits of vanadium were formed during the Proterozoic (the Bushveld Complex, South Africa; the Adirondacks, USA, etc) or the Palaeozoic (Gusevogorsk, Kachkanar, Urals, etc.).   During the Palaeozoic, certain contact-metasomatic deposits of vanadium-bearing magnetite were also formed (Osokino-Aleksandrovsk, Urals, etc.).   The small-scale hydrothermal deposits of nolanite (near Beaver Lodge Lake, Canada) are restricted to ancient metamorphic complexes.   During the Mesozoic and Cainozoic, a few magmatic and hydrothermal deposits were formed, characterized by an increased vanadium content (Kamyshevsky Baikitik, Krasnoyarsk district).   The formation of a significant portion of the exogenic deposits of vanadium belongs to the Mesozoic and Cainozoic eras, although certain vanadium--bearing deposits, for example, carbonaceous-siliceous shales (Balasauskandyk, Kazakhstan), phosphorites (Karatau, Kazakhstan), and sedimentary iron ores (Wabana, Newfoundland), were formed during the Palaeozoic and Late Proterozoic.

MAGMATIC DEPOSITS

An increased vanadium content, as a rule, characterizes the titanomagnetite and ilmenite-magnetite deposits, spatially and genetically associated with ultramafic and basic rocks of the normal series.   The massifs, which consist of pyroxenites, peridotites, olivinites, gabbros, and anorthosites, occur on the ancient platforms, in the marginal portions of the Precambrian shields, and in the folded regions on the site of former linear geosynclines, where they are restricted to zones of deep-seated faults.   These massifs often form belts extending for hundreds of kilometres.

In the USSR, the most significant magmatic deposits of vanadium occur in the linear folded regions, amongst which the Urals are outstanding.   Abroad, all the principal magmatic deposits of vanadium have been concentrated in the crystalline shields and ancient platforms or in their immediate surroundings (the Canadian, Baltic, and Indian crystalline shields and the southern portion of the African crystalline shield).

Most of the intrusions of basic-ultramafic composition have been formed during the period of the early phases of development of the platforms or during the time of the specifically geosynclinal phases of development of the mobile zones (Smirnov, 1962; Kuznetsov, 1964).   In the Urals, for example, the formation of massifs of ultramafic and basic rocks of the platinum-bearing belt and of the titanomagnetite deposits restricted to them, has been associated with basaltoid magmatism, which was widely manifested in the eugeosynclinal region. Early activation of the platforms has been associated with the formation of the huge Bushveld lopolith and the unique deposits of vanadium-bearing titanomagnetite accompanying it.

Activation of the platforms of the later phases during the Palaeozoic, however, also contributed to the formation of quite large ultramafic massifs, containing titanomagnetite (the alkaline-ultramafic association of the Kola Peninsula), although the concentration of vanadium in them is small (on average 0.08%).

The following associations have been recognized, which are clearly associated with magmatic vanadium-bearing deposits (associations according to Kuznetsov, 1964).

1. Geosynclinal and orogenic regions:

1) gabbro-pyroxenite-dunite association; the Gusevogorsk, Kachkanar, Visim, Volkov, and Pervoural'sk deposits (Urals);

2) gabbro-diorite-diabase association; Kusinsk, Kopansk, Medvedevsk, and Matkal deposits (Urals);

3) gabbro-monzonite-syenite association; the Svarants and Kamakar deposits (Armenia);

4) gabbro-syenite association; the Yelet'-Ozero deposit (Karelia).

2. The crystalline shields and ancient platforms (activated sectors):

1) association of differentiated gabbroic and noritic intrusions; the Bushveld (South Africa);

2) gabbro-diabase associations of migmatites of the amphibolite facies and their associated anatectites; the Pudozhgorsk and Koikar deposits (Karelia);

3) anorthosite association; the Kruchinina and Chinei deposits (Transbaikalia), the Tsaginsk (Kola Peninsula), and Sanford Hill (USA).

The largest ultramafic and basic massifs, the area of which reaches several hundreds and even thousands of square kilometres, are restricted to the anorthosite association (Canadian Shield) and the associations of differentiated gabbroic and noritic intrusions (Bushveld Complex). The area of ultramafic rocks, which are associated with the deposits of vanadium-bearing titanomagnetite of the gabbro-pyroxenite-dunite association, reaches several tens of square kilometres (Urals). The reserves of vanadium in the individual deposits of these associations comprise several million tonnes.

Since the ultramafic and basic rocks which carry a vanadium-bearing titanomagnetite mineralization, were formed through differentiation of a basaltic magma, many of them have a layered construction. For example, the Bushveld lopolith is a complex layered body, in the lower portion of which chromite ores have been concentrated, and in the upper portion there are layered segregations of vanadium-bearing titanomagnetite, restricted to the zone of contact between the pyroxenites and the anorthosites. According to Reshit'ko (1961), the bodies of ultramafic rocks of the platinum-bearing belt

of the Urals have the form of layered brachy-folds, in which the ore pyroxenites
are arranged above non-ore olivine pyroxenites and dunites.   The layering of
the ultramafic rocks of this belt has also been pointed out by Vorob'ëva, Samoilova & Sveshnikova (1962).   Such layering has apparently resulted from crystallization differentiation within the intrusive chamber.   The intracameral formation of the intrusions also substantially influenced the formation of the large
segregations of titanomagnetite ores.

The ore bodies of the magmatic deposits of vanadium consist of layers
of segregated or schlieren-segregated ores, having the form of lenses, and also
vein-, layer-, and pipe-like shape.   Layer-like segregations of massive magnetite have been found, which usually lie conformably with the general 'stratification' of the basic-ultramafic massifs.   Discordant vein-like segregations of
massive magnetite are also known.   On the basis of method of formation, segregation and fusive deposits have been recognized (Zavaritsky, 1963).   Most of
the deposits of vanadium-bearing titanomagnetite belong to the fusive type.
Mineralization, as a rule, does not extend beyond the limits of the massifs of
basic and ultramafic rocks and is essentially their domain.   The magmatic origin
of these rocks and ores has been recorded by Sokolov (1957), Malyshev (1957),
Bogachev *et al.* (1963), Kuznetsov (1964), Godlevsky (1968), Borisenko *et al.*
(1968), Bogatikov *et al.* (1970), and other geologists.

The principal mineral-concentrators of vanadium in the magmatic deposits
are titanomagnetite and magnetite (usually less than 1% V).   The largest concentrations of vanadium have been observed in the ore minerals from rocks of basic
composition.   Even within one magmatic association, its amounts increase from
the pyroxenites to the gabbros (Table 32).   Very rarely, the amount of vanadium
in titanomagnetite reaches several percent (the deposits in the vicinity of
Maiurbhanj and Singhbhum in India).   In such cases, small inclusions of culsonite,
FeV_2O_4, are present in the titanomagnetite (decomposition texture of solid solution).   However, most commonly the inclusions in titanomagnetite consist of
ilmenite, hercynite, or other spinels.

The ferromagnesian minerals and feldspars contain a little vanadium
(less than 0.01 - 0.1%).   Low concentrations of vanadium have also been identified in ore minerals, associated with titanomagnetite (in the ilmenite, less than
0.16%;   and in chalcopyrite, bornite, and pyrite, less than 0.03% V).

The most important magmatic deposits of vanadium in the USSR are:
Gusevogorsk, Pervoural'sk, Kusinsk, Kopansk, Medvedevsk, and Matkal, and also
Visim, Kruchinina, Yelet'-Ozero, Pudozhgorsk, Koikar, Velikhovsk, Masal, Lyson,
and Kedran.   Abroad, they include the deposits of the Bushveld Complex (South
Africa);   Lac-Tio, Pyujelon, Mills, etc. (Canada);   Tahawus and others (USA);
Rodsend and Selvog (Norway);   Taberg (Sweden);   the vicinity of Maiurbhanj and
Singhbhum (India);   Burrumbi (Australia);   and the region of Tete (Mozambique).

The Gusevogorsk Deposit

The Gusevogorsk deposit of titanomagnetite is restricted to the pyroxenite massif of the same name, located in the eastern part of the larger Kachkanar
gabbro-pyroxenite massif.   The vanadium content of the ores was first described
by Malyshev *et al.* in 1934.

Table 32. *Content of Vanadium and Titanium in Titanomagnetite from Ultramafic and Basic Ores of the Gabbro-Pyroxenite-Dunite Association of the Urals (in wt %)*

Rocks containing titanomagnetite (deposit)	Vanadium			Titanium		
	Number of Samples	Amount		Number of Samples	from-to	average
		from-to	average			
Pyroxenite (Visim)	9	0.16-0.49	0.31	10	1.08-2.40	1.69
Pyroxenite (Gusevogorsk)	23	0.20-0.48	0.35	22	1.32-2.52	2.01
Hornblendite (Pervoural'sk	16	0.28-0.53	0.42	16	1.29-4.32	2.99
Gabbro (Volkovo)	11	0.36-0.84	0.72	8	0.90-4.20	2.94

The Kachkanar gabbro-pyroxenite massif is located approximately in the middle portion of the platinum-bearing belt of the Urals, which extends along the boundary between the Central Urals anticlinorium in the west, consisting of Upper Proterozoic - Cambrian slates, and the Ordovician and Silurian volcanogenic-sedimentary sequence of the Tagil megasynclinorium in the east. The massif occurs among metamorphosed volcanogenic and volcanogenic-sedimentary rocks of the Upper Ordovician and Silurian in the western limb of the Tagil megasynclinorium. According to N. Vysotsky, V. Kuznetsov, and V. Reshit'ko, the contacts between the massif and the country rocks are tectonic. In the zone of tectonic disturbances, the amount of amphiboles increased markedly in the gabbros and pyroxenites. The total area of the massif is about 110 km^2. The massif has a concentrically-zoned structure with a vaguely-defined stratification and a brachysynclinal form. In the central portion of the Kachkanar Massif, there are two large bodies of pyroxenites, surrounded by rocks of gabbroic composition. The axis of the brachysyncline extends from southeast to northwest; it plunges towards the centre of the massif at 30-35° in the northwest and 70-80° in the southeast. Such a structure is typical of the Kytlym, Svetloborsk, Nizhne-Tagil and other massifs of the platinum belt (Reshit'ko, 1961).

The Gusevogorsk pyroxenite massif, which is spatially and genetically associated with the vanadium-bearing titanomagnetite ores, is in plan a meridionally-oriented elongate body. Its length is about 8.5 km, width up to 4.6 km, area about 22 km^2, and its dip eastwards at 75-80°. The massif is located in the northeastern limb of a brachysyncline. On the west, the massif is bounded by a large tectonic fracture of submeridional strike. In this part of the massif, there are widespread hornblendites (Fig. 97).

In composition, the pyroxenites of the Gusevogorsk Massif consist of diallage, olivine, hornblende, and plagioclase types. Diallage pyroxenites in the main form the central portion of the massif, whereas the olivine pyroxenites predominate in its northeastern and southwestern parts. The hornblende and plagioclase pyroxenites usually occur in the zone of transition from the pyroxenites into the gabbros. Amongst the pyroxenites (mainly olivine types) there are segregations of wehrlites in the form of lensoid bodies, and occasionally olivinites. The structure of the ultramafic rocks is massive medium-grained, coarse-grained, and sometimes gigantic-grained. Although not everywhere clearly, there is a banding seen in the rocks. In individual sectors, the wehrlites and olivine pyroxenites have been intensely serpentinized.

Vein formations are widely distributed in the pyroxenites, and they consist mainly of plagioclasites, and less frequently, gabbros and fine-grained pyroxenites (gusevites).

The titanomagnetite mineralization has been associated mainly with the diallage and hornblende pyroxenites and to a lesser degree with other varieties of ultramafic rocks.   The vein rocks are usually barren.

Several ore segregations have been recognized in the deposit on the basis of exploration data: the Main, Western, Northern, Intermediate I, II, and III, Eastern, Southern, and Vyisk.   The shape of the ore segregation is complex.   The transition from the ore pyroxenites into the non-ore olivine pyroxenites is usually gradual.   The principal reserves of titanomagnetite ores have been concentrated in the Main, Northern, Western, and Intermediate I segregations.   The greatest vanadium content is found in the Western segregation (0.1% V).   On average, the ores of the deposit contain (in %): Fe, 16.7; V, 0.08; and Ti, 0.73 (Fominykh *et al.*, 1967).

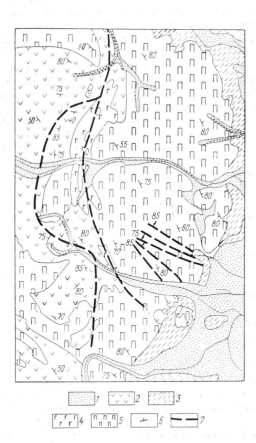

Fig. 97.
Diagrammatic geological map of the Gusevogorsk deposit
(After V. Fominykh *et al.*, 1967).

1) alluvial and deluvial deposits;
2) gabbros; 3) amphibolites;
4) hornblendites; 5) pyroxenites;
6) direction of banding in pyroxenites and gabbros; 7) main tectonic disturbances and zones

Segregated ores have an overwhelming distribution, and less frequently there are finely-segregated and schlieren types.   The bulk of the titanomagnetite in the ore ultramafic rocks fills the space between the ferromagnesian silicates (Fig. 98).

On the basis of coarseness of the principal ore mineral, the segregated ores have been divided into five types (Table 33): dispersed-segregated (less than 0.074 mm), very finely segregated (0.074 - 0.2 mm), finely segregated (0.2 - 1 mm), medium-segregated (1 - 3 mm), and coarsely segregated (more than 3mm). The concentration of the vanadium in the ores increases with increase in the dimensions of the titanomagnetite segregations.

The principal ore mineral of the Gusevogorsk deposit (titanomagnetite) contains up to 0.48% vanadium. In the rock-forming minerals, the vanadium concentration is lower: in the hornblende, less than 0.09%; in the diopside, less than 0.03%; and in the olivine, less than 0.003%. In the ilmenite (less than 0.1%), pyrite, bornite, and chalcopyrite, which are present in the ores as minor minerals, there is less than 0.03% vanadium.

Our investigation of the products from the enrichment plant of the Kachkanar Mining & Concentration Combine has shown that the titanomagnetite concentrate contains 0.35%, the agglomerate 0.4%, and the silicate tailings 0.037% vanadium.

Fig. 98.
Ore diallagite, sideronitic texture.
Gusevogorsk deposit, section 1094.
Magnification X 7

Table 33.   *Amount of Iron, Vanadium, and Titanium in Various Structural Types of Ores from the Gusevogorsk Quarry (in wt %)*
After V. Fominykh *et al.* (1967)

Type of ore by fabric	Iron	Vanadium	Titanium	Proportion of total ores, %
Dispersed-segregated	16.8	0.071	0.68	5.0
Very finely segregated	17.2	0.073	0.79	23.7
Finely segregated	17.8	0.084	0.82	44.3
Medium segregated	18.1	0.088	0.86	17.0
Coarsely segregated	17.4	0.089	0.87	10.0

The Pervoural'sk Deposit of Titanomagnetite in Hornblendites

This deposit occurs in the Pervoural'sk region, 44 km west of Sverd-lovsk. The deposit is restricted to the Revda Massif, which consists of an inter-associational intrusion, bordered on the west by the metamorphic rocks of the central Uralian anticlinorium, and on the east, by the Silurian rocks of the greenstone sequence. The massif extends in a meridional direction, with a width of 2000 - 3000 m, and is characterized by a zoned construction: the central portion consists of hornblendites, and the peripheral sectors, of rocks of gabbroic composition (Fig. 99). The hornblendites, to which the titanomagnetite mineralization is restricted, lie conformably with the gabbros. In the sector of the deposit, the belt of hornblendites has a width of from 1 km in the north down to 100 - 150 m in the south. The ore varieties of these rocks are concentrated in the zone of the eastern contact with the gabbros; their dip is close to vertical. Here, between the hornblendites and the normal gabbros there are sectors of a taxitic gabbro, and the hornblendites themselves are often coarse-grained or gigantic-grained.

The hornblendites consist mainly of hornblende ($\leqslant$95%). Titanomagnetite is less common, and is unevenly distributed in the body of hornblendites (from a few up to 30%, and rarely more). Feldspar, usually intensely altered, is present, and the other gangue minerals are epidote, chlorite, axinite, zoisite, garnet, and apatite, with the ore minerals, ilmenite, and sometimes chalcopyrite, bornite, and pyrite.

The hornblendites of the deposit are characterized by uneven grainsize. According to Malyshev (1957), we may recognize fine-grained (less than 1 cm), medium-grained (1 - 2 cm), coarse-grained (2 - 5 cm), and especially coarse--grained hornblendites with grainsizes in excess of 5 cm. The transitions between the individual varieties of hornblendites are also clear, and gradual.

The amount of titanomagnetite in the segregated ores is 20 - 30%, but in the massive ores, it reaches 60 - 80%. The most widespread are the lean--segregated ores. Less frequently, the hornblendites contain small schlieren, 0.2 - 0.3 m in length (sometimes up to 2.5 m), and also vein-like segregations of a massive titanomagnetite. The titanomagnetite usually contains a large quantity of small spinel inclusions; less frequently, there is ilmenite (break-down texture of solid solution).

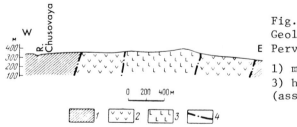

Fig. 99.
Geological section through the Pervoural'sk deposit.

1) metamorphic slates; 2) gabbros;
3) hornblendites; 4) rock contacts
(assumed)

The average amount of vanadium in the titanomagnetite from the schlieren is a little greater (0.49%) than in the segregated titanomagnetite (0.43%) from the hornblendites. The ilmenite contains 0.14% vanadium, the hornblende, 0.07%, the chalcopyrite, 0.004%, the feldspar, 0.003%, the axinite, 0.033%, and the epidote, 0.05%.

Table 34. *Amounts of Vanadium and Titanium in Titanomagnetite from Magmatic Deposits (in wt %)*

Magmatic association	Type of ore	Examples of deposit	Vanadium			Titanium		
			Number of samples	Amount from-to	average	Number of samples	Amount from-to	average
Gabbro-diorite-diabase	Massive and segregated in gabbros and gabbro-amphibolites	Kusinsk, Medvedevsk, Kopansk (Urals)	41	0.31-0.62	0.42	121	3.36-12.78	5.90
Gabbro-pyroxenite-dunite	Schlieren-segregated and massive in pyroxenites, wehrlites, gabbros, and hornblendites	Gusevogorsk, Kachkanar, Visim, Volkovo, Pervoural'sk (Urals)	136	0.05-0.84	0.35	257	0.61-4.32	1.81
Gabbro-monzonite-syenite	Segregated in olivinites, pyroxenites, and gabbros	Svarants, Kamakara (Armenia)	19	0.16-0.34	0.22	19	0.71-2.22	1.46
Gabbro-syenite	Schlieren-segregated and massive in olivinites, peridotites, pyroxenites, gabbros, and amphibolites	Yelet'-Ozero (Northern Karelia), Gremyakha-Vyrmes (Kola Peninsula)	17	0.11-0.45	.31	20	2.86-11.64	7.14
Differentiated gabbroic and noritic intrusions	Massive in norites, anorthosites, gabbros, and pyroxenites	Deposits of the Bushveld Complex (South Africa)	10	0.20-0.98	0.56	10	7.24-9.48	8.27
Gabbro-diabase of association of migmatites of amphibolite facies and associated anatectites	Segregated in gabbro-diabases	Pudozhgorsk and Koikar (Karelia)	68	0.40-0.73	0.55	29	6.0-13.0	9.20
Anorthosite	Segregated in massive types in gabbros, anorthosites, pyroxenites, and olivinites	Kruchinina (Transbaikalia), Chinei (Eastern Siberia), Tsgaina (Kola Peninsula)	20	0.12-0.55	0.35	19	6.2-13.6	7.45
Alkaline-ultramafic	Schlieren-segregated in olivinites and pyroxenites	Kugda, Bor-Uryakh, Odikhincha (Polar Siberia), Kovdor and Salma (Kola Peninsula)	53	0.010-0.274	0.081	51	1.00-12.15	5.46

The ores of the deposit are characterized by small amounts of iron (14 - 16%). However, their enrichment is easy, and the increased vanadium content of the titanomagnetite (up to 0.53%), with relatively low amounts of titanium (about 3%), and the possibility of using the hornblendite rubble for the production of concrete, make the exploitation of the Pervoural'sk deposit worthwhile.

In addition to the above magmatic deposits of vanadium, belonging to the gabbro-pyroxenite-dunite association of the Urals, others are known in the USSR. The fundamental data on the vanadium content of the magmatic deposits belonging to various associations, have been generalized in Table 34 (Borisenko & Lapin, 1972). Information on the ilmenite-magnetite deposits of the Kusinsk Group (Urals) has been presented in the chapter on 'Titanium'.

Abroad, the leading place on the basis of reserves and production of vanadium is occupied by the deposits of the Bushveld Complex, from the ores of which there is an annual production of about 35 - 40% of the vanadium from the capitalist countries.

DEPOSITS OF THE OXIDATION ZONE OF THE POLYMETALLIC DEPOSITS

Deposits of this type are quite widely distributed. They are known in Africa, Argentina, Mexico, USA, and Australia. In most cases, these are relatively small-scale deposits of vanadium, which has been concentrated only in the oxidized ores (up to 5.6%). The raw ores of the polymetallic deposits contain vanadium in very small amounts. In galena, sphalerite, chalcocite, tennantite, enargite, chalcopyrite, bornite, sulphides of iron, and other minerals of the primary ores, less than $0.00n - 0.0n$% of vanadium has been identified. Very rarely, sulphides with an increased vanadium content have been recorded, for example, germanite contains up to 2.9% vanadium (Geier & Ottemann, 1970).

The most significant deposits of this type occur in Africa: Berg--Aukas, Tsumeb, Abenab in Southwest Africa; Broken Hill in Zambia. The ore bodies have a pipe-like form; their upper portion consists of oxidized ores, containing cerussite, smithsonite, malachite, azurite, anglesite, and vanadium minerals (descloizite, cuprodescloizite, and vanadinite). The depth, to which the zone of oxidized ores extends, usually does not exceed a few hundred metres from the Earth's surface. However, sectors of oxidized ores also occur in the lower horizons, a feature associated with the circulation of subsurface waters in the horizons characterized by increased porosity.

The descloizite, cuprodescloizite, and vanadinite ores of the African deposits contain on average 1.1 - 1.7% vanadium. Almost the entire concentrate, now being extracted in Southwest Africa and containing about 1.2 thousand tonnes of vanadium (annual output), is sent for treatment in West Germany.

SEDIMENTARY DEPOSITS

Vanadium has been recorded in the ores of many sedimentary deposits, but increased concentrations are relatively rarely found. All the commercial deposits of vanadium of this type occur abroad.

The sedimentary deposits of vanadium, according to method of formation, have been divided into two groups (Kholodov, 1968): 1) syngenetic (placers,
iron-ore deposits, bauxites, coals, siliceous-carbonaceous shales, and layered
phosphorites); and 2) epigenetic (uranium-vanadium deposits in varicoloured
rocks, petroleum, solid asphaltites, etc.). The syngenetic deposits in section
and in area are controlled mainly by one facies, which is associated with the
common form of the process of ore- and rock-formation. This is most clearly
seen in the placers. The epigenetic deposits, on the other hand, usually cut
different facies, which is associated with processes of ore-formation superposed
later with respect to the surrounding rocks. The shape of the ore bodies for
deposits of this type is often extremely complex, and they may be restricted to
sequences of rocks of different age.

Littoral-marine placers (beach black sands) sometimes contain up to
0.87% of vanadium. The principal concentrator of vanadium is titanomagnetite.
In the USSR, present-day placer deposits of vanadium-bearing titanomagnetite
are known in the Black Sea coast of Georgia (Urek), on the shore of the Caspian
Sea (Lenkoransk and Astara regions), and in the Kurile Islands (Iturup, Kunashir,
and Paramushir islands). However, they are all characterized by relatively
small reserves. Abroad, large beach deposits are known in New Zealand (with
reserves of about 800,000 tonnes of V_2O_5). The accumulation of vanadium-
-bearing titanomagnetite in placers is, as a rule, associated with the breakdown of rocks of basic composition.

Alluvial placers of titanomagnetite occupy a subordinate position,
for example, in the valleys of the Rivers Kusa, Satka, Kuvasha, Chërnaya,
Kopanka, and Aya in the Urals. They were formed through the weathering of
the gabbroid rocks of the Kusinsk intrusion. Reserves of vanadium in deposits
of this type are limited.

The phosphorites of the USSR, which belong to various kinds (layered,
granular, concretionary, etc.), contain a little vanadium (in %): the layered
phosphorites of the Karatau, 0.005 - 0.018; the nodular phosphorites of the
Vyatka-Kama deposit, 0.0022 - 0.011; and the deposits in the vicinity of Bryansk
and Kursk, 0.0056 - 0.01 (Kholodov, 1968). However, in the USA, higher concentrations of vanadium are known, especially in the layered phosphorites of the
Rocky Mountains (the Permian phosphorite basin, up to 0.22%). In spite of the
relatively low concentrations of vanadium, experiments in the USA have shown
that it is possible to obtain V_2O_5 from plant waste from retreated phosphorites.

Carnotite and roscoelite deposits in the varicoloured rocks (the
'Colorado Plateau' type) are distributed in many countries, especially in the
USA (Utah, Arizona, New Mexico, and Colorado). On the Colorado Plateau, with
an area of about 609 X 720 km^2, the deposits are restricted to a sequence of
clastic rocks (siltstones, sandstones, gravelites, and conglomerates) of the
Palaeozoic-Mesozoic cover. The exploited deposits are, as a rule, complex
uranium-vanadium types. High concentrations of vanadium (up to 1.7%) are
characteristic of the sandstones, which contain in addition 0.18 - 0.34% U_3O_8,
and often copper (about 0.5%).

Carnotite, pitchblende, tuyamunite, and other ore minerals are
present in the matrix of the sandstones. The mineralization is restricted
to alluvial and alluvial-lacustrine deposits, often containing organic remains.
The shape of the ore bodies is varied: lenses, nests, and rolls. They usually
lie conformably with the bedding of the surrounding rocks, but in a number of

cases, discordant bodies are also found.   The dimensions of individual bodies are small (length up to 200 m, and thickness up to 10 m).   The most significant uranium-vanadium deposits are restricted to the Upper Jurassic Morrison Formation (Ambrosia Lake, New Mexico), the Triassic Shinarump (Monument Valley, Arizona), and Chinle formations (Big Indian Wash, Utah).   Up till now, ores of this type have been one of the principal sources of vanadium in the USA.

Petroleum.   The high-sulphur types of petroleum are characterized by an increased vanadium content, for example, those of the Urals-Volga province. However, even in the heavy types of oil, the concentration of vanadium does not exceed 0.03%.   Only in the petroleum ash, may the content of vanadium reach several tens percent.   For example, in the ash from the Ishimbaevo, Buguruslan, Tuimazy, Krasnokamsk and other deposits, 0.6 to 48% of vanadium has been identified (Gulyaeva, 1954).   Within the Volga-Urals province, several sectors have been recognized, where petroleum is distributed, containing an increased quantity of vanadium: the Zhigulevo-Pugachevo uplift, the Tatar arch, and the southeastern slope of the Russian Platform (Demenkova *et al.*, 1958).

Regional and stratigraphic factors influence the distribution of vanadium in petroleum (Katchenkov, 1963).   In the USSR, the most vanadium-rich ash from petroleum is that from the Carboniferous deposits of the Volga-Urals district (12.9%), and from the Devonian oils of Ukhta (16.4%).   In other regions, the concentrations of vanadium are lower : in the Cambrian deposits of Eastern Siberia 0.014%, in the Carboniferous deposits of the Ukraine 0.0013%, in the Jurassic of Mangyshlak 0.02%, in the Lower Cretaceous of the Northeastern Caucasus and Emba 0.06%, and in the Tertiary deposits of Turkmenia 0.2%, Sakhalin 0.16 - - 4.6%, and Fergana 0.17% (Flegontova, 1969).

In the USSR, vanadium is not being extracted from petroleum.   However, experiments in Canada, where there is a plant for the treatment of Venezuelan oil, indicate the possibility of obtaining vanadium on a commercial scale.   In 1966, this plant produced 160 tonnes of V_2O_5.   In Canada in recent years, it has been proposed to obtain vanadium from the oil sands of Alberta, in which there is 0.024% of vanadium (in the residual oil-products, 4%).   In 1974, Australia will construct a plant for the extraction of V_2O_5 from the bituminous shales in Queensland, near Julia Creek (about 5000 tonnes per annum).

Patronite in asphaltites.   In 1907, a natural deposit of this type at Minas Ragra, Peru, began to yield vanadium in significant amounts.   It consists of a vein-like segregation, containing solid bitumens, enriched in patronite (VS_4).   The segregation extends for 90 - 100 m, with a thickness of 8 - 12 m.   The surrounding rocks consist of shales, sandstones, and limestones of Cretaceous age.   The amount of vanadium in the ore is about 6.2%, and in the ash, up to 36.4%.   At present, the deposit is worked out.

In the USSR, increased amounts of vanadium have been identified in the asphaltites of the Satka, Velikhovsk, and Kaprovsk deposits of the Southern Urals (up to 37.4% in the ash).   However, the reserves of vanadium in them are small, for example, in the Satka deposit, it is not more than a few thousand tonnes (Kholodov, 1968).

METAMORPHOGENIC DEPOSITS

A few deposits of this type are known, from the ores of which vanadium has been extracted.   The best-studied and exploited at present is the Otanmäki

Table 35. *Types of Promising Vanadium-Bearing Deposits of the USSR*

Genetic type of deposit	Vanadium-bearing ores	Principal commercial component	Amount of vanadium in ore, %	Principal mineral-concentrators of vanadium (amount, wt %)	Examples of deposit
Magmatic	Titanomagnetite in gabbro-diabases	Fe	0.18 - 0.26	Titanomagnetite (0.40 - 0.73)	Pudozhgorsk (Karelia)
Hydrothermal	Magnomagnetite in dolerites (Angara-Ilim type)	Fe	0.2 - 0.5	Magnomagnetite (0.4 - 0.6)	Kamyshevsky Baikitik (Yakutia)
Sedimentary	Iron ores (oxide and marine sedimentary oolite ores)	Fe	0.011-0.11	---	Lisakovo, Kerchen, Ayat, and Bokchar
	Carbonaceous-siliceous-clay shales	V, Mo, Re, Pb, Ag, Cr	0.015-0.9	Sulvanite (13.0 - 14.2), Roscoelite (11.8 - 16.2)	Balasauskandyk (Kazakhstan)
	Bauxites	Al	0.03 - 0.095	---	Northern Urals and Tikhvinsk regions
	Ilmenite placers	Ti	---	Ilmenite (0.015 - 0.16)	Ukraine

ilmenite-magnetite deposit in the amphibolites of Finland.   The region of the deposit consists of ancient granite-gneisses with seams of amphibolites.   Two ore zones are restricted to the amphibolites;   lens-like bodies, consisting of magnetite and ilmenite, have been concentrated in these zones.   The ores are predominantly massive.   In the course of recent years, a magnetite concentrate has been obtained from the Otanmäki deposit which annually yields about 1300 tonnes of vanadium.   In the USSR, there is an increased vanadium content in the ilmenite-magnetite ores confined to the amphibolites of the Byelorussian crystalline massif, and the Novoselkovsk ore-show;   the magnetite of this ore--show contains 0.3% of vanadium.

Another metamorphogenic vanadium deposit is that of Wilson Springs in the USA.   This deposit is restricted to a zone of argillization, which is located on the contact between nepheline syenites and schists of the ancient Wichita Mountains metamorphic sequences.   The principal concentrators of vanadium (about 1%) are the clay minerals.   It is possible that weathering has effected the enrichment of these minerals in vanadium, as it influenced the metamorphosed rocks.   At present, about 4.5 thousand tonnes of V_2O_5 are obtained per annum from the ores of the deposit.

The principal reserves of commercial vanadium ores have been concentrated in the endogenic titanomagnetite, ilmenite-magnetite, and hematite--ilmenite deposits (Borisenko, 1970).   Petroleum (high-sulphur types) and phosphorites must be assigned to the number of other, most important raw--material sources of vanadium in the USSR, which have been subjected to experiments abroad for the extraction of vanadium.   In addition, the sedimentary iron ores, in which the amounts of vanadium are relatively low, must be regarded as promising with respect to this metal, with large reserves, and also the ilmenite placers (Table 35).

A practical interest for local commercial development may be directed to the titanomagnetite ores in gabbro-diabases (Pudozhgorsk), the magnomagnetite ores of individual deposits of the Angara-Ilim type (Kamyshevsky Baikitik), and certain other iron ores of endogenic deposits.   However, the majority of the hydrothermal and contact-metasomatic deposits are characterized by low concentrations of vanadium ($0.0n$%).   The group of promising vanadium-bearing ores include the carbonaceous-siliceous-clay shales of Kazakhstan and Middle Asia and the bauxites (see Table 35).   Possibilities of widely using the promising vanadium-bearing ores are mainly associated with the development of an appropriate technology, allowing the economically profitable extraction of vanadium and its accompanying useful components.

REFERENCES

BOGACHEV A.I., ZAK S.I., SAFRONOVA G.P. & ININA K.A.   (БОГАЧЕВ А.И., ЗАК С.И., САФРОНОВА Г.П., ИНИНА К.А.), 1963: Геогогия и петрология Елетьозерского массива габброидных пород Карелии *(The Geology and Petrology of the Yelet'ozero Gabbro Massif in Karelia).*   Akad. Nauk SSSR Press, Moscow.

BOGATIKOV O.A., GODLEVSKY M.N. & PETROV V.P. (БОГАТИКОВ О.А., ГОДЛЕВСКИЙ М.Н., ПЕТРОВ В.П.), 1970: Современные проблемы изучения базитового магматизма (Present-day problems in the study of basic magmatism). *Izv. Akad. Nauk SSSR, ser. geol.*, No.1, pp.3-15.

BORISENKO L.F. (БОРИСЕНКО Л.Ф.), 1970: Сырьевая база ванадия в зарубежных странах и основные тенденции в её освоении (The raw--material basis for vanadium in foreign countries and the main trends in exploiting it), *in* "Редкие элементы. Сырье и экономика" (*'The Rare Elements. Raw Materials and Economics'*), Vol.4, pp.22-31.

BORISENKO L.F. (БОРИСЕНКО Л.Ф.), 1972: О перспективных типах ванадийсодержащих руд (Promising types of vanadium-bearing ores), *in* "Редкие элементы. Сырье и экономика" (*'The Rare Elements. Raw Materials and Economics'*), Vol.6, pp.57-63. IMGRÉ Press, Moscow.

BORISENKO L.F. & LAPIN A.V. (БОРИСЕНКО Л.Ф., ЛАПИН А.В.), 1972: О концентрациях элементов-примесей в титаномагнетите и магнетите эндогенных месторождений различных типов (Concentrations of trace--elements in titanomagnetite and magnetite from endogenic deposits of different types). *Dokl. Akad. Nauk SSSR, 196*, pp.1441-1444.

BORISENKO L.F., LEBEDEVA S.I. & SERDOBOVA L.I. (БОРИСЕНКО Л.Ф., ЛЕБЕДЕВА С.И., СЕРДОБОВА Л.И.), 1968: О титаномагнетите и магнетите из железорудных месторождений различного генезиса (The titanomagnetite and magnetite from iron-ore deposits of different origin). *Geologiya rudn. Mestorozh., 10*, No.4, pp.40-53.

FLEGONTOVA, E.l. (ФЛЕГОНТОВА Е.И.), 1969: Микроэлементы в нефтях (Microelements in petroleum). *Trudy vses. nauchno-issled. geol.-razv. Inst., 279*, geokhim. Sb. No.10).

DEMENKOVA P.Ya., ZAKHARENKOVA L.N. & KURBATSKAYA A.P. (ДЕМЕНКОВА П.Я., ЗАХАРЕНКОВА Л.Н., КУРБАТСКАЯ А.П.), 1958: О связи ванадия и никеля с компонентами нефтей Волго-Уральской области (The association of vanadium and nickel in the components of petroleum from the Volga-Urals district). *Trudy vses. neft. nauchno-issled. geol.-razv. Inst., 117*, pp.186--213.

FOMINYKH V.G., SAMOILOV P.I., MAKSIMOV G.S. & MAKAROV V.A. (ФОМИНЫХ В.Г., САМОЙЛОВ П.И., МАКСИМОВ Г.С., МАКАРОВ В.А.), 1967: Пироксениты Качканара (*The Pyroxenites of Kachkanar*). Ural'sk Fil. Akad. Nauk Press, Sverdlovsk.

GEIER, B.N. & OTTEMAN, I., 1970: New primary vanadium-, germanium-, gallium-, and tin-minerals from the Pb-Zn-Cu deposit Tsumeb, South West Africa. *Miner. Deposita, 5* (1), pp.29-40.

GODLEVSKY M.N. (ГОДЛЕВСКИЙ М.Н.), 1968: Магматические месторождения (Magmatic deposits), *in* "Генезис эндогенных рудных месторождений" (*'The Origin of Endogenic Ore Deposits'*), pp.7-83. Nedra Press, Moscow.

GULYAEVA L.A. (ГУЛЯЕВА Л.А.), 1954: Микроэлементы нефтей и битумов перми и карбона Урало-Поволжья (The micro-elements of oils and bitumens from the Permian and Carboniferous of the Urals-Volga region). *Trudy Inst. Nefti Akad. Nauk SSSR, 3*, pp.188-206.

KATCHENKOV S.M. (КАТЧЕНКОВ С.М.), 1963: К геохимии элементов семейства железа в нефтях (The geochemistry of elements of the iron family in petroleum). *Trudy vses. neft. nauchno-issled. geol.-razv. Inst., 212*, geokhim. Sb., No.8, pp.23-26.

KHOLODOV V.N. (ХОЛОДОВ В.Н.), 1968: Ванадий *(Vanadium)*. Nauka Press, Moscow.

KUZNETSOV Yu.A. (КУЗНЕЦОВ Ю.А.), 1964: Главные типы магматических формаций *(The Principal Types of Magmatic Associations)*. Nedra Press, Moscow.

MALYSHEV I.I., PANTELEEV P.G. & PÉK A.V. (МАЛЫШЕВ И.И., ПАНТЕЛЕЕВ П.Г., ПЭК А.В.), 1934: Титаномагнетитовые месторождения Урала (The titano-magnetite deposits of the Urals). *Trudy ural'sk. Fil. Akad. Nauk SSSR*, 1.

RESHIT'KO V.A. (РЕШИТЬКО В.А.), 1961: Закономерности распределения пород и оруденения в брахисинклиналях габбро-перидотитовых массивов платиноносного пояса Урала (Distribution patterns of rocks and mineralization in brachysynclinal gabbro-peridotite massifs in the platinum-bearing belt of the Urals). *Razv. Okhr. Nedr*, No.9, pp.7-9.

SMIRNOV V.I. (СМИРНОВ В.И.), 1962: Металлогения геосинклиналей (The metallogenesis of the geosynclines), *in* "Закономерности размещения полезных ископаемых" *('Distribution Patterns of Mineral Deposits')* Vol.5, pp.17-81. Akad Nauk SSSR Press, Moscow.

SOKOLOV G.A. (СОКОЛОВ Г.А.), 1957: Типы и условия образования магматогенных железных руд СССР (Types and conditions of formation of magmatogenic iron ores of the USSR), *in* "Железорудная база черной металлургии СССР" *('The Iron-Ore Basis for Ferrous Metallurgy in the USSR')*, pp.34-48. Akad. Nauk SSSR Press, Moscow.

VOROB'ËVA O.A., SAMOILOVA N.V. & SVESHNIKOVA E.V. (ВОРОБЬЕВА О.А., САМОЙЛОВА Н.В., СВЕШНИКОВА Е.В.), 1962: Габбро-пироксенит-дунитовый пояс Среднего Урала (The gabbro-pyroxenite-dunite belt of the Middle Urals). *Trudy Inst. Geol. rudn. Mestorozh.*, *65*, pp.1-319.

YEFIMOV Yu.V., BARON V.V. & SAVITSKY E.M. (ЕФИМОВ Ю.В., БАРОН В.В., САВИЦКИЙ Е.М.), 1969: Ванадий и его сплавы *(Vanadium and Its Alloys)*. Nauka Press, Moscow.

ZAVARITSKY A.N. (ЗАВАРИЦКИЙ А.Н.), 1963: О классификации магматических рудных месторождений (The classification of magmatic ore deposits), *in* "Избранные труды" *('Collected Works')*, Vol.4, pp.380-390. Akad. Nauk SSSR Press, Moscow.

DEPOSITS OF ALUMINIUM

GENERAL INFORMATION

The principal raw-material source for the aluminium industry both in the USSR and in other countries is bauxite; about 90% of the alumina in our country is produced from it. In recent years, industry has also developed methods for producing alumina from nepheline and alunite ores. There is also promise of obtaining aluminium and its alloys from high-alumina rocks (kyanites, sillimanites, kaolins, and others) by electrometric means, by-passing the phase of alumina production.

Pre-revolutionary Russia did not possess an aluminium industry, and the small demand for aluminium was satisfied by imports, which in 1913 comprised 1600 tonnes. The problem of creating a local aluminium industry was tackled in the first years of the Soviet system. In 1917-1935, exploration was carried out on the Tikhvinsk bauxite deposit, on the basis of which the Volkhovsk aluminium plant was constructed in 1932, and the Dneprovsk in 1933, and in 1938, the Tikhvinsk alumina works.

In 1929-1931, exploration work in the Urals led to the discovery of the North Urals bauxite-bearing region, the largest in the Soviet Union. During World War II, an alumina industry was set up in the Urals. Here, the bauxite deposits of the Northern and Southern Urals regions were investigated, on the basis of which, the Uralian and Bogoslovsk aluminium plants were constructed. In post-war time, further prospecting for bauxites was carried out throughout the entire country. Especially significant discoveries were made in the 'fifties and 'sixties (Kirpal', 1971a).

In Northern Kazakhstan, bauxite deposits were discovered and examined in three bauxite-bearing regions: the Amangel'da, West Turgai, and Central Turgai. On the basis of these deposits, the Pavlodarsk aluminium plant was constructed and later expanded. In the Arkhangel'sk district, deposits of bauxites were discovered and examined in the North Onega bauxite-bearing region, on the basis of which bauxite mines have been constructed.

As a result of exploratory-research work, the new South Timan and Middle Timan bauxite-bearing regions have been discovered in the Komi ASSR. In the area of the Kursk Magnetic Anomaly (KMA), the new Belgorod bauxite-ore region has been discovered, with the Vislovsk, Belgorod, Belenikha, and other deposits. In the Krasnoyarsk region, the bauxite deposits in the Chadobets and Near-Angara bauxite-bearing regions, have been found.

In 1949, in the Soviet Union and for the first time in world practice on a commercial scale, alumina has been produced from the Kola nepheline concentrates. In the Krasnoyarsk region, proven reserves of high-quality nepheline ores (urtites) of the Kiya-Shaltyr deposit have resulted in the Acha alumina plant's being built.   In recent years, the commercial exploitation of the Goryachegorsk deposit of nepheline ores has been achieved.   On the basis of the Tezhsar deposit of nepheline ores in Armenia, the Rozdan heavy-chemical combine has been constructed.   The alunite ores of the Zaglik deposit have been used for the production of alumina in the Azerbaidzhan SSR.   However, bauxites will still remain the principal source of the aluminium industry for a long time.

Scientific-technical progress is impossible without the widespread use of aluminium in the national economy.

The world requirements of aluminium steadily increase, and on the basis of volume of production, aluminium has occupied a second place amongst the metals after iron.   The requirements of aluminium on a population basis in 1966 were (in kg):

USA	15.6	Great Britain	6.7
Belgium	11.9	France	6.1
Switzerland	10.8	Austria	6.3
Norway	10.5	Sweden	5.8
Canada	9.1	Japan	4.3
West Germany	7.4		

The production of primary aluminium in the capitalist and developing countries in the last 30 years, on the basis of rate of growth, has several times exceeded the extraction of other non-ferrous metals.   In 1968, 6.8 million tonnes of aluminium were obtained, that is, almost nine times more than in 1940, when its production equalled 0.74 million tonnes.   Since that time, the production of copper has increased 1.8 times, lead, 1.6 times, and zinc, 2.2 times.

In 1970, the extraction of primary aluminium comprised (in thousands of tonnes):

USA	3607	India	161
Canada	964	Italy	147
Japan	733	Spain	117
Norway	527	Ghana	108
France	381	Austria	90
West Germany	309	Other countries	710
Australia	190		
		Total	8044

In the near future, increase in production of aluminium will be recorded.   In the USA, the requirements for aluminium in 1980 will be 5.7 million tonnes, and in 2000, 14.7 million tonnes.

In order to ensure the increasing production of aluminium, the mining of bauxites will have to be continuously increased.   In 1938, in capitalist countries, 3.9 million tonnes of bauxites were extracted, in 1958, 20.0 million tonnes, and in 1969, 42.6 million tonnes.   It is likely that in the near future, the level of extraction of bauxites in the industrially developed capitalist and developing countries will increase to up to 50 million tonnes, and towards 1980, it will reach 100 million tonnes.   In the commercially developed capitalist and developing countries, the total reserves of bauxites in 1968 were 10 780 million tonnes, and proven and probable reserves, 3410 million tonnes.   About 80% of the reserves of bauxites occur in the equatorial regions, where there is a widespread development of deposits of the lateritic type (in Australia, Africa and Central and South America).

CLASSIFICATION OF BAUXITE DEPOSITS

The first classification, which took account of the genetic features of formation of the bauxites and their geological features, was presented in 1937 by Arkhangel'sky (1954), who divided all the bauxite deposits into two large groups, formed under continental and littoral-marine conditions.

In 1946, Vikulova separated the bauxites into five main types on the basis of conditions and locality of formation: 1) laterites of watershed plateaux; 2) lagoonal; 3) lacustrine; 4) valley (the heads of depressions of the gully type); and 5) in depressions of ancient karst.

Goretsky (1960) has recognized three groups of bauxite deposits, based on their association with different tectonic districts and their restriction to various structural zones.   In each group of deposits, subgroups are recognized within the tectonic districts, allowing for the nature of the surrounding deposits.   The types of bauxite deposits have been recognized on the basis of features of the pre-ore relief and their conditions of occurrence.

In all Goretsky has recognized three groups and seven subgroups of deposits.   On the basis of the pre-ore relief, he has recognized amongst the groups and subgroups, the following principal types of bauxite deposits:

1.   On stable sectors of the platforms:

1) in districts of anteclises — structural-erosional, karst-basinal, and structural-erosional on lava flows;

2) in the marginal parts of syneclises — structural-erosional, structural-erosional on lava flows, and valley-erosional.

2.   On the mobile sectors of the platforms:

3) in the marginal parts of syneclises, filled with coal-bearing sequences — structural-erosional on lava flows, karst-basinal, and valley-erosional;

4) in the marginal parts of intermontane and pre-montane downwarps — — structural-erosional and karst-basinal.

3.   In the geosynclinal districts:

5) in regions of mutual transition between geosynclines and platforms — — reefogenic-karst and reefogenic-basinal;

6) in marginal geosynclinal downwarps — reefogenic-karst, and karst-basinal.

Goretsky's classification is a great step forward, although it is not devoid of faults.   It does not reflect the group of deposits of lateritic bauxites, which provides more than half of the bauxite reserves of the world and has a vast commercial importance.   This classification was supplemented by Spirin *et al.* (1967), who recognized a group of deposits of the weathering crust.

Recently, in connexion with the development of broad geological-explor-atory work, classifications of these deposits for individual regions have been proposed for the platform bauxites of Mesozoic-Cainozoic age: for Kazakhstan, by Tyurin (1958); the Turgai downwarp, by Kirpal' (1964), the Mugodzhary region, by Kiselev (1963), and the Siberian Platform, by Pel'tek (1971), etc.

In Kazakhstan, amongst the primary-sedimentary chemogenic bauxites, Tyuri (1958) has recognized the following morphogenetic types: 1) supra-contact, or linear; 2) karst (carbonate and non-carbonate); 3) mantle-like; 4) trough-like; and 5) cobbly-residual (secondary, mechanically redeposited).

The bauxite deposits of the Turgai bauxite-bearing province have been divided by Kirpal' (1964) on the basis of shape and construction of the ore seg-regations, into three principal types: 1) karst; 2) karst-basinal; and 3) layer-like.

Kiselev (1970) has recognized two groups amongst the bauxite deposits of Kazakhstan: 1) deposits of stable sectors of the platforms; and 2) deposits associated with the mobile sectors of the platforms. Amongst the indicated groups, he has recognized two types: lateritic (pseudomorphous) and lateritic--sedimentary (clastic-pisolitic). The latter are divided into two subtypes: located in the weathering crust of silicate rocks (slope, basinal, and near-slope), and in the karst depressions in carbonate rocks (karst-supracontact, karst-polje, and karst-basinal).

On the Siberian Platform, Pel'tek (1971) has divided the bauxite deposits into primary (eluvial) and secondary (sedimentary) types, and amongst the latter, he has recognized primary-sedimentary and redeposited types. On the basis of conditions of formation, the primary-sedimentary deposits have been subdivided into two morphological types: 1) littoral open basins and lagoons; 2) basinal types of internal water-bodies. The redeposited deposits have also been sub-divided into two types: 1) karst-basinal; and 2) contact-karst with alumino-silicate rocks in the footwall (hanging wall).

Recently, Sapozhnikov (1971a, b) has proposed a new classification of the platform bauxite deposits. On the basis of conditions of formation, they are divided into two groups: sedimentary and residual; in addition, he has provisionally recognized a third type (superimposed deposits). Amongst the sedimentary deposits, bauxites have been recognized on the basis of position relatively to the elements of the ancient relief: 1) slope; 2) valley (deluvial and proluvial); 3) basinal (lacustrine and paludal); and 4) karst. The res-idual (laterite) deposits consist of one residual type, restricted to the ancient surfaces of planation.

Bushinsky (1971) has proposed a classification of bauxite deposits. He has divided them into four groups: laterite, complex, sedimentary clasto-genic, and sedimentary chemogenic. The sedimentary clastogenic deposits have been subdivided into two subgroups (deposits on silicate rocks and those on carbonate rocks). Three types have been recognized in the subgroup of sedi-mentary clastogenic deposits: slope, valley, and near-slope. The subgroup of deposits on carbonate rocks has been subdivided into two types: close-to--karst and distant-from-karst.

When recognizing various groups, subgroups, and types of bauxite dep-osits, it is necessary from our point of view to take account, first, of the conditions of formation (origin) of the bauxites, second, the restriction of the bauxite deposits to definite structural elements of the crust, and third, the morphology of the bauxite segregations, occurring in direct dependence on the pre-ore relief, and also the conditions of occurrence and the construction of the ore segregations.

We have divided the bauxite deposits into two groups on the basis of conditions of formation: weathering crusts (lateritic) and sedimentary (re-deposited). Each group is divided into subgroups according to the restriction of the bauxite deposits to the principal structures of the Earth's crust. Amongst the sedimentary bauxite deposits we have recognized two subgroups: deposits restricted to geosynclinal districts, and those of the platform dist-ricts.

Thus, amongst the bauxite deposits of the USSR, we have recognized and described the following groups:

I.   Weathering crusts (lateritic).

Deposits - Vislovsk (Voronezh Anteclise), Vysokopol (Ukrainian Shield), and Vezhayu-Vorykva (Timan Ridge).

II.   Sedimentary (redeposited).

1. Deposits of geosynclinal districts - Northern Urals, Southern Urals, the Salair group, and the Bokson deposit.

2. Deposits of ancient platforms - North Onega, Tikhvinsk, and South Timan groups (Russian Platform), the Chadobets and Near-Angara groups and the Tatar deposit (Siberian Platform and the Yenisei Ridge).

3. Deposits of the young platforms - Krasnooktyabr'sk, Belinsk, Naurzumsk, and Amangel'da group (Turgai Downwarp).

In the USSR, in addition to bauxites, other kinds of aluminous raw material are used for obtaining alumina. Amongst them are:

I.   Nepheline ores (Kiya-Shaltyr, Goryachegorsk, Tezhsar, and other deposits).

II.   Alunite ores (Zaglik and other deposits).

III. Kyanite ores (deposits of the Keiv group).

IV.   Kaolins and high-alumina clays (Angren deposit).

THE CONDITIONS AND REQUIREMENTS OF INDUSTRY IN RESPECT TO THE QUALITY OF BAUXITES

In all countries of the world, the principal mineral raw-material for the production of alumina by electrolysis, is bauxite. It consists mainly of free alumina, and also oxides of iron, kaolinite, titanium oxide, and other compounds. The principal industrial value of the bauxites is determined by the hydrates of aluminium in the form of monohydrates (diaspores and boehmite) and the trihydrates (gibbsite).

According to the amount of alumina and the proportion of the silica molecule ($Al_2O_3:SiO_2$), districts of utilization of the various bauxites are determined (*Instructions for Applying the Classification of Reserves to Bauxite Deposits*, 1962).   In the USSR, the technical requirements, applied to the bauxites, have been administered, with rare exceptions, by GOST 972-50* (Table 36).

Table 36.   *Technical Requirements for Bauxites*

Bauxite grade	Amount of Al_2O_3 calculated on dry material, wt %	$Al_2O_3:SiO_2$ (silica module)	Usage
B-C	52.0	12.0	Production of electrocorundum
B-0	52.0	10.0	
B-1	40.0	9.0	Production alumina, electrocorundum, and aluminous cement
B-2	46.0	7.0	Production fused refractories and aluminous cement
B-3	48.0	5.0	
B-4	42.0	3.5	Production alumina and refractories
B-5	40.0	2.6	
B-6	37.0	2.1	Production refractories (open-hearth)
B-7	30.0	5.6	Production alumina and aluminous cement
B-8	28.0	4.0	Production alumina

Bauxite, destined for the production of alumina, must not contain sulphur in excess of 0.7% in grades B-1, B-2, B-7, and B-8, and more than 1% in grades B-3, B-4, and B-5.   Bauxite of grades B-1, B-2, B-7, and B-8 is manufactured in two kinds according to the quantity of carbon dioxide: the first with up to 1.3% of carbon dioxide, and the second with more than 1.3% per dry weight of bauxite.

In the bauxite, destined for the production of alumina by the sintering method (grades B-3, B-4, and B-5), a lowered amount of aluminium oxide is achieved as a result of increasing the amount of calcium carbonate.

In the bauxite which goes into the production of electro-corundum, the following amounts of calcium oxide have been established: for bauxites of grades B-C and B-0, not more than 0.5%;   for grade B-1, not more than 0.8%; the amount of sulphur is not more than 0.3%.

---

* At present, a new GOST (All-Union State Standard) for bauxites is being developed.

In the bauxite for open-hearth production, the amount of sulphur must not exceed 0.2%, and the content of phosphorus, 0.6% (calculated as P_2O_5).

For the production of aluminous cement, the sulphur content must not exceed 0.5%. For the production of fused refractories, the amount of calcium oxide must not exceed 1.5%, and sulphur, 0.5%.

Specification GOST 972-50 sets out the technical requirements for bauxites, treated by the three principal methods employed at present in the USSR (hydrochemical (Bayer), sintering, and the combined method (Bayer-sintering)). The bauxites of grades B-C, B-O, B-1, and B-2 may be treated by the Bayer hydrochemical method at a temperature of 225°C; those of grades B-7 and B-8 are also treated by this method, but at a temperature of 105°C; the bauxites of grades B-3, B-4, and B-5 may be treated by the sintering method. Grade B-6 bauxites are not used for the production of alumina. Grades B-7 and B-8 are given to bauxites in which alumina occurs in the form of hydrargillite. The remaining grades include the monohydrate and trihydrate bauxites (Yershov, 1962).

In foreign countries, commonly accepted technical requirements for bauxites do not exist. There are only treatment grades for bauxites for individual regions and deposits. The aluminium industry in foreign countries has been based on the use of deposits of high-grade bauxites of the geosynclinal type in the Caribbean and Mediterranean basins and the lateritic type of Northern Australian, West Africa, and the northeastern portion of South America.

In the Soviet Union at present, mainly high-grade bauxites of geosynclinal type of the Northern Urals basin are employed, and to a lesser degree bauxites of the platform deposits of the Amangel'da and Tikhvinsk regions. In future, increase in output of bauxites will be achieved by exploitation of the bauxite deposits of the platform type in the West Turgai and North Onega regions.

EPOCHS OF BAUXITE-FORMATION IN THE HISTORY OF DEVELOPMENT OF THE EARTH'S CRUST

Bauxite deposits are unevenly distributed in the Soviet Union (Fig. 100). The formation of bauxites took place throughout almost the entire history of the Earth's crust, but it proceeded unevenly. The maxima of bauxite--forming events are replaced by periods in which bauxites did not form. The stratigraphical division of the bauxite reserves in the USSR demonstrates that the most intense bauxite formation took place in the Middle Devonian, the Early Carboniferous, and the Cretaceous, and in foreign countries, the maximum bauxite-formation apparently occurred during Neogene-Quaternary time. The formation of bauxites during the course of geological history, increased in pulses and progressively from the older to the younger epochs. This is explained by the increase on the surface of the Earth of areas with favourable palaeogeographical and palaeoclimatic conditions for the development of bauxites.

The formation of the various types of bauxite deposits also took place unevenly: during some epochs, mainly geosynclinal bauxites accumulated, and in others, mainly lateritic and sedimentary deposits of the platform districts.

The epochs with favourable climatic, palaeogeographical, and palaeotectonic environments for the formation and preservation of bauxites repeatedly

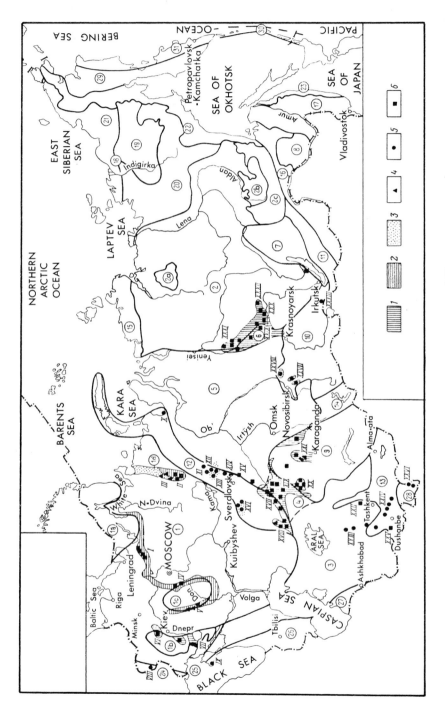

Fig. 100.   Map of distribution of bauxite deposits in the USSR.

1-3) districts of development of bauxite-bearing sediments: 1) Mesozoic and Cainozoic,
2) Carboniferous, 3) Devonian; 4-6) bauxite deposits: 4) residual (lateritic) weathering crusts,
5) sedimentary (redeposited) of the geosynclinal districts, 6) sedimentary (redeposited) of the
platform districts.

Groups or deposits of bauxites: I) Tikhvinsk, II) South Timan, III) Middle Timan, IV) Tula, V) Belgorod, VI) Smelyansk, VII) Vysokopol', VIII) Rudarnensk, IX) Basman-Kermen, X) Shchucha, XI) Ivdel', XII) Severoural'sk, XIII) Karpinsk, XIV) Alapaevo, XV) Kamensk-Ural'sk, XVI) Yuzhno--Ural'sk, XVII) West Turgai, XVIII) Mugodzhar, XIX) Central Turgai, XX) Amangel'da, XXI) Tselino-grad, XXII) Near-Chimkent, XXIII) Aktau, XXIV) South Fergana, XXV) Kairak, XXVI) Pamir, XXVII) Salair, XXVIII) Barzas, XXIX) Tatar, XXX) Near-Angara, XXXI) Bakhta, XXXII) Chadobets, XXXIII) Bokson

Principal structures. Ancient platforms (figures in circles): 1) Russian Platform (1a – Baltic Shield, 1b – Ukrainian crystalline shield, 1c – Voronezh Anteclise, 1d – Timan Ridge), 2) Siberian Platform (2a – Anabar crystalline massif, 2b – Aldan Shield, 2c – Stanovoi fold system). Young (epi-Hercynian) platforms: 3) Scythian-Turanian plate, 4) Turgai downwarp, 5) West Siberian plain. Folded districts and systems: 6) Yenisei Ridge, 7) Baikalian fold system, 8) Bureya median massif, 9) Kazakhstan fold district, 10) Altai-Sayan fold district, 11) Transbaikalian fold system, 12) Uralian fold system, 13) Tyan'-Shan' fold district, 14) Zaisan fold system, 15) Taimyr fold system, 16) Mongol-Okhotsk fold district, 17) Sikhoté-Alin fold system, 18) Verkhoyansk-Chukotka fold district, 19) Kolyma-Omolon median massif, 20) Yana-Kolyma fold system, 21) Chukotka fold system, 22) Okhotsk-Chukotka district, 23) East Sikhoté-Alin district, 24) Carpathian fold system, 25) Crimea, 26) Caucasian fold system, 27) Kopet-Dag, 28) Pamir, 29) Koryak-Kamchatka fold district, 30) Kuriles arc, 31) East Kamchatka district.

occurred during the history of the planet (Fig. 101). Strakhov (1947) first posed the question of the stratigraphical restriction of the bauxites to definite geological epochs. For the USSR, he recognized six epochs of bauxite-formation: 1) Late Silurian or Middle Devonian; 2) Early Carboniferous; 3) early Mesozoic (Jurassic); 4) Cretaceous (from Aptian to Senonian inclusive); 5) Palaeogene (especially Eocene); and 6) Late Neogene-Quaternary.

Discoveries in recent years have made it possible for Sapozhnikov (1971*a*) to be somewhat more precise about the earlier recognized epochs of bauxite-formation. He has recognized the following epochs: 1) Early Cambrian; 2) Middle-Late Devonian; 3) Early--Middle Carboniferous; 4) early Mesozoic; 5) Middle-Late Jurassic; 6) Albian-Cenomanian; 7) Late Cretaceous; 8) Palaeocene-Eocene.

An analysis of recent data demonstrates that for all the continents, it is convenient to recognize the following epochs of bauxite-formation: 1) Late Proterozoic – Cambrian; 2) Middle Devonian; 3) Late Devonian; 4) Early Carboniferous; 5) Middle--Late Carboniferous; 6) Permian; 7) Triassic; 8) Jurassic; 9) Cretaceous; 10) Palaeogene; 11) Neogene-Quaternary. In almost every epoch there was formation of both geosynclinal and platform bauxites. In some of them, the presence of lateritic bauxites has been established.

Late Proterozoic-Cambrian epoch. During this epoch, formation of bauxites took place mainly under geosynclinal conditions. The early Riphean (possibly, Middle Proterozoic) phase contains the oldest bauxite manifestations both in the USSR and in other countries. They occur amongst the ancient

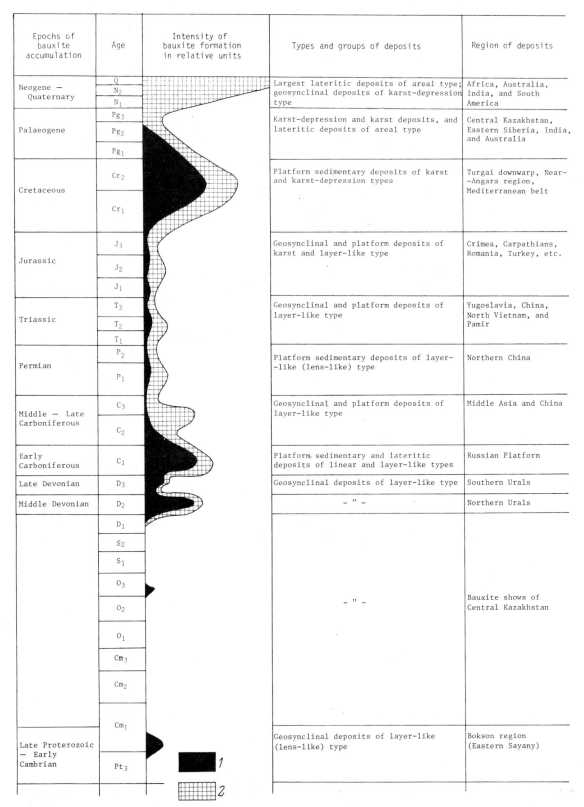

Epochs of bauxite accumulation	Age	Intensity of bauxite formation in relative units	Types and groups of deposits	Region of deposits
Neogene — Quaternary	Q / N_2 / N_1		Largest lateritic deposits of areal type; geosynclinal deposits of karst-depression type	Africa, Australia, India, and South America
Palaeogene	Pg_3 / Pg_2 / Pg_1		Karst-depression and karst deposits, and lateritic deposits of areal type	Central Kazakhstan, Eastern Siberia, India, and Australia
Cretaceous	Cr_2 / Cr_1		Platform sedimentary deposits of karst and karst-depression types	Turgai downwarp, Near--Angara region, Mediterranean belt
Jurassic	J_3 / J_2 / J_1		Geosynclinal and platform deposits of karst and layer-like type	Crimea, Carpathians, Romania, Turkey, etc.
Triassic	T_3 / T_2 / T_1		Geosynclinal and platform deposits of layer-like type	Yugoslavia, China, North Vietnam, and Pamir
Permian	P_2 / P_1		Platform sedimentary deposits of layer--like (lens-like) type	Northern China
Middle — Late Carboniferous	C_3 / C_2		Geosynclinal and platform deposits of layer-like type	Middle Asia and China
Early Carboniferous	C_1		Platform sedimentary and lateritic deposits of linear and layer-like types	Russian Platform
Late Devonian	D_3		Geosynclinal deposits of layer-like type	Southern Urals
Middle Devonian	D_2		– " –	Northern Urals
	D_1 / S_2 / S_1 / O_3 / O_2 / O_1 / Cm_3 / Cm_2 / Cm_1		– " –	Bauxite shows of Central Kazakhstan
Late Proterozoic — Early Cambrian	Pt_3		Geosynclinal deposits of layer-like (lens-like) type	Bokson region (Eastern Sayany)

1 ▮
2 ▦

Fig. 101. Stratigraphical distribution of intensity of bauxite-formation based on epochs of bauxite accumulation.
1) intensity of bauxite-formation in the USSR; 2) intensity of bauxite-formation in the world

platform formations of the Purpol' Group of the Teptorga Series, developed
along the eastern margin of the Patom plateau (Golovenok & Pushkin, 1964). Here,
amongst the argillaceous deposits there are concretions (up to 1 m across), thin
lenses, and seams of diaspore bauxites with an alumina content of up to 52%, and
an average silica content of 17%.

Recently, in the Yudoma-Mama region of Siberia at the base of the
Lakhanda Group of the Middle Riphean, good-quality bauxites with 2 - 5.3 silica
module have been discovered in six sectors.

During the Early Cambrian phase, bauxites were formed in the Bokson
deposit, where the segregations occur in a thick sequence of dolomites and limestones of the Lower Cambrian, being restricted to a large uplift of Lower Proterozoic and Archaean rocks.   In the Kuznets Alatau, Gornaya Shoriya, and the
Altai, bauxite manifestations and metamorphosed high-alumina rocks have been
found in Cambrian-Proterozoic sedimentary and volcanogenic-sedimentary rock
complexes.   In the Khankai region, fragments of bauxites of diaspore-boehmite
composition have been found in a deluvial breccia of limestones of the Markush-
evsk Group of Middle-Late Cambrian age.

The geosynclinal type of this epoch also includes the bauxite deposits
of China, restricted to thick sequences of Cambrian limestones (possibly also
Ordovician).

Middle Devonian epoch.   In the deposits of this epoch are concentrated
the highest quality and largest deposits of the geosynclinal type in the Northern
Urals, and also the Salair Ridge.   Recently, bauxites of Middle Devonian age have
been identified in the Shchucha synclinorium of the Polar Urals.

During this epoch, the formation took place of bauxite deposits exclusively in the geosynclinal districts of the Northern Urals and Salair bauxite-
-bearing province.   The Middle Devonian reefal carbonate facies, with which the
formations of the huge layered bauxite deposits of the Northern Urals are associated, are as a rule, restricted to the marginal parts and in part to the inner
rises in the Hercynian geosyncline.   Their formation occurred under conditions
of a warm and humid climate.   Commercial deposits of the Middle Devonian epoch
are known at present only in the Soviet Union.   The formation of bauxites during
the Devonian began, apparently, even at the end of Early Devonian time.   This
level has been provisionally recognized in the Upper Berdsk region of Central
Salair.

In the Urals, during Middle Devonian time, mainly three (in some cases
up to six), and in the Salair region, four levels of bauxite-formation have been
recognized.

The most productive, both in the Urals and in the Salair region, is the
lower Eifelian stratigraphical level (the Subrovsk horizon in the Urals).   The
largest and highest quality deposits of the Northern Urals basin are restricted
to this level, and also some deposits of the Ivdel' region.   In the Salair
region, this level includes the great majority of bauxite reserves of the Berdsk-
-Maya and Obukhovo deposits.

The second level in the Middle Devonian phase is the upper Eifelian.
In the Urals at this time, the bauxites of the Bogoslovsk horizon were formed,
to which have been assigned the deposits of the Karpinsky, Nizhne-Tura, Alapaevo,
Ivdel', and Nizhne-Serginsk regions.   In the Salair region, two levels of bauxite-formation have been recognized during late Eifelian time.   They have been

identified in the Obukhovo deposit.   Some investigators in the Urals have also recognized two levels (horizons) in the upper Eifelian sequence.

The third level of the Middle Devonian epoch of bauxite-formation belongs to early Givetian time.   At this time, certain deposits of the Ivdel' region and the northern flank of the Northern Urals basin were formed.

Late Devonian epoch.   This epoch of bauxite-formation is also manifested in the USSR only, within the western slopes of the Urals.   It was characterized by the accumulation of redbed sequences, consisting of arkosic and polymict sandstones and marls, and also carbonate deposits, mainly of dolomite composition.   These sediments have the features of a warm arid climate.

During the Late Devonian epoch, bauxites were formed in the geosynclinal districts.   Typical examples are the deposits of the Southern Urals bauxite-bearing region and other bauxite-shows on the western slopes of the Urals, although some geologists believe that the bauxite deposits of the Southern Urals region were formed in a transitional zone, including the eastern margin of the Russian Platform and the western portion of the Uralian geosyncline.

On the western slopes of the Urals, two stratigraphical levels of bauxite formation have been recognized: lower Frasnian (Pashiisk horizon) and upper Frasnian (Orlovsk horizon).   During early Frasnian time, formation of bauxites took place in the Chusovsk and Krasnovishersk regions.   The overwhelming majority of deposits in the Southern Urals bauxite-bearing basin, and also the Lower Serga region, are restricted to the Orlovsk horizon.

In 1970, Upper Devonian bauxite deposits were discovered in the southeastern portion of the Chetlas anticlinal structure in Central Timan.   These are the Vezhayu-Vorykva and Verkhne-Vorykva deposits, which on the basis of certain features may be assigned to the lateritic type.

Early Carboniferous epoch.   During this epoch intense bauxite-formation took place in the USSR.   It embraced the marginal parts of the ancient Russian Platform, where the bauxite-bearing formations surround the continental uplands, located within the Baltic Shield, and the Voronezh and Timan uplifts. The bauxite deposits in the depths of the districts of sedimentation are, as a rule, replaced by sand-clay carbonaceous sediments of the initial phases of a Viséan marine transgression.

This epoch is characterized by the formation of large sedimentary and lateritic deposits on the ancient platforms.   Bauxite formation at this time took place in many sectors of the Russian and North American platforms.   Especially intense bauxite formation has been manifested within the Russian Platform.   Here, large sedimentary deposits in the North Onega, South Timan, and Tikhvinsk regions were formed, and also the lateritic (residual) deposits of the Belgorod bauxite-bearing region of the KMA.

Four stratigraphical levels (horizons) have been recognized in the Early Carboniferous epoch: the Bobrikovsk, Tula, Aleksinsk, and Mikhailovsk.

Middle-Late Carboniferous epoch.   During this epoch bauxites were formed mainly in the USSR under geosynclinal conditions within the Southern Tyan'-Shan'.   Here, up to seven levels of bauxite formation have been recognized, assigned to the Namurian, Bashkirian, and Moscovian phases.   The bauxites have been restricted to reefogenic sequences, formed in the littoral zones of the geosynclinal districts.   However, deposits with commercial reserves of bauxite ores, formed during this epoch under geosynclinal conditions, have not so far been found in the USSR.

During this same epoch, large bauxite deposits were developed in the area of the China Platform in Liaoning, Shantung, Yunnan, and other provinces.

Permian epoch.   Deposits of bauxites of Permian age are not known in the USSR, with the exception of a small bauxite-show, situated in the Nakhich-evansk ASSR.   The formation of platform deposits, associated with coal-bearing sequences, continued during Permian time in Northern China and Korea in part.

Triassic epoch.   During the Triassic epoch, several phases and strat-igraphical levels of bauxite-formation were established, although large bauxite deposits of this age are not known in our country.   There are individual bauxite shows, the formation of which took place both under geosynclinal (Central Pamir) and platform conditions (in the Northern (Volchansk and Bogoslovsk areas) and Southern Urals (Orsk depression)), and also in Middle Asia (Konnov, 1972).

During the Triassic, bauxites were formed under platform conditions in Northeastern China and under geosynclinal conditions in Vietnam and Yugo-slavia.

Jurassic epoch.   During this epoch, the formation of bauxites took place during several phases both in the platform and the geosynclinal districts. The Early Jurassic phase includes the platform bauxites of the Mailisu deposit and manifestations of low-grade bauxites in other places in Kirgizia.   Bauxite--shows of the Middle Jurassic phase are known in the Eastern Carpathians (the Rudarnensk deposit of geosynclinal type) and on the Russian Platform (the Myach-kovsk deposit on the Moskva River of platform type).   During the Late Jurassic (Tithonian) phase, the formation of lean bauxites took place under geosynclinal conditions in the High Crimea (Basman-Karmen), where the bauxites are restricted to karst bands.

Abroad, the most intense bauxite accumulation during the Jurassic Period occurred in Romania, Turkey, and possibly Greece and Yugoslavia, where the formation of geosynclinal bauxite deposits took place.

Cretaceous epoch.   During the course of the Cretaceous period and the early Palaeogene epoch, bauxite formation occurred almost continuously over vast areas of the USSR and foreign countries.   During this time, laterite deposits were formed in the southern part of the Ukrainian Shield and sedimen-tary deposits in the Turgai, South Siberian, and Central Siberian bauxite--bearing provinces, and also in the Trans-Urals, the Near-Chimkent region of Middle Asia, and in other parts of our country (Kirpal', 1971b).   Bauxite deposits were formed mainly in the marginal parts of the young platforms. The bauxite-bearing sediments are restricted to the lower portions of the plat-form cover and occur in most cases in karst and erosional-karst depressions, located on the carbonate floor of the folded basement of the platforms.

Several phases of bauxite formation may be recognized in the Cretaceous System.

During the Aptian-Albian phase, bauxites were formed on the western and eastern slopes of the Mugodzhary, on the eastern slopes of the Urals, in the Trans-Urals, the Near-Salair region, the Chulym-Yenisei basin, Middle Asia, and on the Ukrainian Shield (the Smelyansk and Volodar-Volyn regions, and possibly the Vysokopol and South Nikol'sk deposits also).

The Albian-Cenomanian phase of bauxite formation was most intensely manifested in the central portion of the Turgai Downwarp (the Naurzum, Kushmurun, West Ubagan, and other deposits), and partially in the West Turgai region, in the Middle Urals, in the Yenisei Ridge, and the southwestern part of the Siberian Platform (the Chadobets region).

During the Cenomanian-Turonian phase, maximum bauxite accumulation occurred on the western margin of the Turgai Downwarp within the Krasnooktyabr'sk, Belinsk, Ayat, and other deposits.

During the Cretaceous period, intense formation of geosynclinal bauxites also took place in Hungary, Italy, France, and partly in Greece, Turkey and a number of other countries.

Palaeogene epoch.   The bauxite-bearing sediments of the Palaeogene epoch are clearly associated with the Cretaceous bauxite deposits.   As a rule, they are located in the same areas as the bauxites of Cretaceous (especially Late Cretaceous) age.   In the USSR, the Palaeogene epoch was especially intensely manifested on the eastern surroundings of the Turgai Downwarp, where commercial deposits of bauxites of the karst type were formed in the Amangel'da bauxite-bearing region of Central Kazakhstan.   In the Near-Angara region of the Central Siberian province, deposits mainly of karst type were also formed.

Abroad, lateritic bauxite deposits were formed during the Palaeogene epoch on the ancient platforms of Eastern Australia and Tasmania, and possibly India and the northeastern portion of South America, and also the Arkansas region of the USA.   At the same time, geosynclinal bauxite deposits were formed within the Mediterranean bauxite belt (Hungary, Yugoslavia, Greece, and other countries).

Neogene-Quaternary epoch.   During this epoch, the largest lateritic bauxite deposits in the world were formed, unfortunately not identified so far in our country.

In the deposits of this epoch, about 80% of the reserves of bauxites of the capitalist and developing countries are concentrated.   During Neogene--Quaternary time, large deposits were formed under both geosynclinal and platform conditions.   Geosynclinal deposits were formed in Jamaica, in the Dominican Republic, and in Haiti.   Huge platform deposits of this epoch are located in Australia, India, and the countries of Equatorial Africa (Tenyakov & Akaemov, 1972).

Thus, the most productive deposits for the USSR are the Upper Proterozoic — Cambrian, Middle Devonian, Lower Carboniferous, Cretaceous, and Palaeogene types.   Abroad, the most intense bauxite formation took place during the Triassic, Cretaceous, Palaeogene, and especially the Neogene-Quaternary epochs.

DEPOSITS OF THE WEATHERING CRUST (LATERITIC)

Lateritic bauxite deposits in foreign countries are of tremendous importance, their reserves comprise more than half of all the bauxite deposits, and from them more than half of the output of bauxites of the world is achieved. In the USSR, lateritic bauxite deposits have been discovered within the Belgorod bauxite-bearing region of the KMA and in Central Timan.   Recently, the Vysoko-pol deposit, early considered as sedimentary, has been assigned to the laterite type.

The laterites are the products of intense chemical weathering of aluminosilicate rocks of basic and intermediate composition, formed under conditions of a tropical and subtropical climate, as a result of intense and lengthy leaching of the rocks by warm rainwater and the removal from them of the alkali elements and silica, with the accumulation of free oxides of aluminium, iron, and titanium (Tenyakov, 1971).

Lateritic bauxites have been formed as a result of a combination of factors, principal amongst which are: 1) continental interruptions during the geological development of the region; 2) tropical, subtropical, or warm moderately humid climate; 3) peneplanation of the land (the laterites are restricted mainly to residuals of both young and old peneplains); 4) a quiescent tectonic regime; and 5) original aluminosilicate rocks, favourable for bauxite-formation.

The Vislovsk Deposit

The Vislovsk deposit of bauxites lies within the Belgorod bauxite--bearing region of the KMA, located in the southwestern part of the Voronezh anteclise.   It is the largest in the region and is restricted to the southern part of the Yakovlevsk iron-ore deposit.   In addition to the Vislovsk deposit in the Belgorod region, the Belenikha deposit has been discovered, and also a large number of bauxite-shows and promising sectors.   The overwhelming majority of deposits and bauxite-shows are concentrated around the deposits of rich iron ores (Klekl', 1969).

The Precambrian crystalline basement of the region of the Vislovsk deposit consists of rocks of two structural stages (Fig. 102).   The lower stage consists of intensely metamorphosed rocks of the Oboyan Series, amongst which there are biotite-plagioclase and garnet-biotite gneisses, plagioclase and microcline granites, and also plagiogranites and their migmatites of Early Archaean age.   The upper structural stage consists of rocks of Early Proterozoic age, amongst which there are formations of the Mikhailovsk and Kursk Series, plagiogranites and their migmatites, and also the Stoilo-Nikolaevo gabbro-diorite complex.

The Mikhailovsk Series in the lower part consists of albite-hornblende amphibolites and albite-epidote schists, and in the upper part, keratophyres, keratospilites, and their tuffs, and quartz-chlorite and other schists.

The rocks of the Kursk Series have been divided into three groups: the lower (sand-slate), middle (iron-ore), and upper (limestone-shale).   The lower group consists of quartzose gravelites, sandstones, quartzites, and also

phyllitic garnet-biotite and bimica schists.   The middle group consists of
ferruginous quartzites, and phyllitic, carbonaceous-chloritic, and chlorite-
-sericitic schists.   The upper group has been formed of quartz-sericitic,
carbonaceous-argillaceous, and phyllitic schists, and silty phyllites with
seams of marmorized limestones and carbonaceous dolomites.

Amongst the formations of the Lower Proterozoic sequences, there is
a considerable development of plagiogranites and their migmatites, and also
rocks of the Stoilo — Nikolaevo gabbro-diorite complex, which consist of serpentinites, talc and tremolite schists, biotite-hornblende syenites, microcline
granites, and their migmatites.

In the relief of the roof of the Precambrian crystalline basement,
there are linear-elongate ridges and plateaux with northwesterly strike, which
rise 30 - 100 m above the peneplained surface and are often dissected by gorges
and ravines 40 - 60 m deep.

An ancient weathering crust of linear type, from 5 to 170 m thick, is
ubiquitously developed on the rocks of the Precambrian crystalline basement.
In the tectonically weakened zones, a linear weathering crust, reaching 700 m
in thickness is developed.   The bauxite-bearing weathering crust of lateritic
type is most widely developed on the phyllitic schists of the lower and upper
groups of the Kursk Series and the metabasites of the Mikhailovsk Series.   The
highest-quality bauxites are restricted to the upper parts of the weathering
crust.

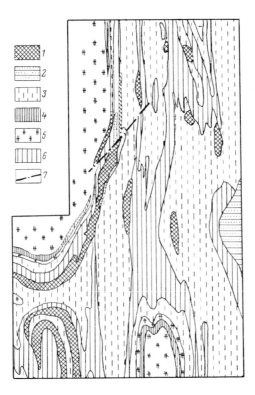

Fig. 102.   Diagrammatic geological
map of the Vislovsk baux-
ite deposit.

1) Bauxite-bearing weath-
ering crusts; 2) sedi-
mentary bauxites; 3)
phyllitic, quartz-sericitic,
carbonaceous-argillaceous
slates, and silty phyllites
of the upper Kursk Group
(the rocks are extremely
favourable for the forma-
tion of bauxites); 4)
keratophyres, keratospil-
ites, and their tuffs,
quartz-chlorite and other
schists of the Mikhailovsk
Series (rocks favourable
for the formation of
bauxites); 5) plagiogranites, their migmatites and
polymigmatites (rocks little
favourable and unfavourable
for the formation of bauxites); 6) ferruginous-
-quartzites and iron ores
(zones of greatest development of a linear weathering
crust); 7) tectonic faults.

On the Precambrian crystalline basement there are horizontal covering
formations, consisting of Lower Carboniferous, Middle and Upper Jurassic, Cretaceous, Palaeogene, and Quaternary rocks.   Their thickness varies from 350 to
400 m in the northern and up to 700 – 1000 m in the southern parts of the region
(Fig. 103).   The Lower Carboniferous deposits of the Viséan Stage in the lower
part consist of compact limestones with seams of argillites and jaspers, and
in the upper part, of silicified, often karsted limestones.   Their thickness
varies from a few metres up to 180 m, but is usually 30 – 70 m.

The Middle Jurassic sediments in the lower portion consist of grey
marine clays, and in the upper part, of an interlayering of siltstones and
sandstones.   The Upper Jurassic consists mainly of marine clays with intercalations of limestones and blue-grey sands, and also sandstones.   The widely
developed Cretaceous sediments (Neocomian-Campanian) consist of sandstones,
argillaceous quartz-glauconite sands, marls, and chalk.   The Palaeogene sediments are gaize-like clays and glauconitic sands, and those of the Neogene are
sands.   The Quaternary formations consist of sands, clays, and loams.

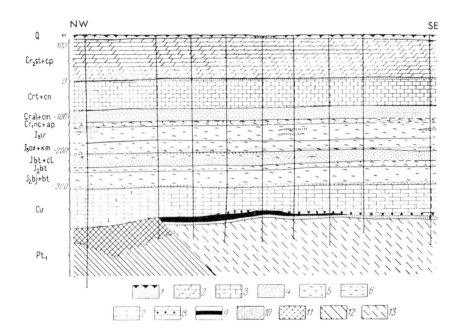

Fig. 103.   Geological section through the Vislovsk deposit.   (Compiled by
G. Kirpal' from data of the Belgorod Geological Research Expedition).

1) Quaternary deposits;   2) marls;   3) chalk;   4) sands;   5) clays;
6) argillaceous sands and sandy clay;   7) limestones;   8) sedimentary bauxites and ferruginous rocks;   9) residual bauxites;   10)
allites;   11) martite and martite-hydrohematite ferruginous ores;
12) magnetite quartzites;   13) phyllitic and chlorite-sericitic
schists.

In addition to the lateritic rocks, the presence of sedimentary bauxites has also been established in the region.   The former are developed everywhere, and the latter have been identified only in certain sectors.   The lateritic bauxites are restricted to the upper horizons of the weathering crust of phyllitic schists of the lower, middle, and upper groups of the Proterozoic Kursk Series, and also to the eruptive-sedimentary sequence of the Mikhailovsk Series.   The bauxite-bearing weathering crust in the main has a linear character of development, and in some sectors it occurs in the form of mantle-like segregations.   The bauxite segregations are usually located in the immediate vicinity of thick segregations of rich iron ores, forming with them rounded hilly shapes on the surface.   The bauxite segregations are wedge-shaped and inclined towards the slope of the linear-elongate hillocks.   The weathering crust occurs immediately below Carboniferous deposits of the Viséan Stage.   The bauxites are located at depths of 450 - 600 m.

Sedimentary bauxites fill elongated depressions or small wide hollows, located on the aluminosilicate rocks of the ancient basement or the terrigenous-carbonate rocks of Tournaisian age.

The Vislovsk bauxite deposit is the largest in the region, and there are four sectors within its boundaries: the Eastern, Western, Southern, and Belgorod.   The largest are the Eastern and the Belgorod sectors.   The segregations of the Eastern sector have been given a preliminary investigation.   The best quality bauxites are concentrated in this sector, containing SiO_2, 7.1 - - 11.2%, Al_2O_3, 49.4 - 50.6%; Fe_2O_3, 6.5 - 8.1%; FeO, 16.6 - 19.6%; and TiO_2, 2.6%;   the silica module varies from 2.6 up to 7.   The quality of the bauxite ores of the Southern and Belgorod sectors is somewhat lower.

The bauxites of the Vislovsk deposit usually have a boehmitic and boehmitic-gibbsitic composition.   In mineral composition, the following varieties may be recognized amongst the bauxites and bauxitic rocks of the Vislovsk and other deposits of the Belgorod region of the KMA (Nikitina *et al.*, 1971): boehmite (chamosite-boehmite, and kaolinite-boehmite),   gibbsite (chamosite-gibbsite, chamosite-boehmite-gibbsite, and kaolinite-gibbsite) and chamosite-hematite-diaspore-boehmite.   The first of these is widely distributed and is the principal type, the second is less widely distributed, and the third occurs rarely.

The bauxites usually have a banded, shaly, and pseudo-pisolitic structure, inherited from the parent rock.   Less frequently, there are uniform pseudo-pisolitic bauxites.

The Vysokopol' Deposit

The Vysokopol' deposit is located on the southern slopes of the Ukrainian crystalline shield and is restricted to the upper portion of a lateritic weathering crust.   The rocks of the Precambrian crystalline basement of the Vysokopol' deposit, after which the bauxite-bearing lateritic weathering crust is developed, consist of amphibolites, chloritic, hornblendic, micaceous, and other schists, ultramafics, diabases, granites and their migmatites.

The lateritic bauxites of the deposit form the upper zone of the weathering crust of crystalline rocks of basic composition.   The largest segregations of lateritic ores are restricted to amphibolites, and chloritic and hornblendic schists.   Features of bauxite occurrence have also been identified in the weathering crust of alkaline rocks (Fig. 104).

The Precambrian crystalline basement and the lateritic weathering crust are overlain by horizontally arranged sediments of the platform cover, which in the lower part consist of marine greenish-grey laminated clays of Middle Eocene age.   On the latter rest the sediments of the Kiev Group, consisting of grey marly clays, marls, silts, and siltstones.   The overlying Khar'kov Group is formed of glauconitic sands, green clays, and a manganese ore.   The Neogene deposits consist of limestones, marls, sands, and clays of Tortonian, Sarmatian, Maeotian, and Pontian age.   The formations of the Quaternary system consist of loess-like loams, and brown and reddish-brown clays. The total thickness of the covering sediments is 60 - 80 m.

Fig. 104.
Diagram of distribution of bauxite segregations of the Vysokopol' deposit (After Yu. Bass).

1-4) weathering crust: 1) of amphibolites and subordinate ultramafic rocks;   2) of gneissose micaceous rocks;   3) of plagioclase granites and their migmatites;   4) of plagioclase-microcline granites (Zhitomir) and their migmatites;   5) plagioclase granites and their migmatites;   6) bauxites.

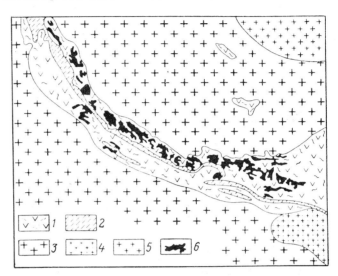

The Vysokopol' deposit consists of 10 small segregations of bauxite ores, which form an irregular band.   According to Bass *et al.* (1971), the bauxite segregations are restricted to the upper part of a lateritic weathering profile of amphibolites, schists (chloritic and hornblendic), and serpentinites. The segregations are located mainly along the northern part of the amphibolite belt.   The dimensions and shape of the segregations depend on the composition and relief of weathering crust of Precambrian rocks (Fig. 105).

In the buried relief, the area of distribution of the weathering crust has been recognized in the form of a ridge-like uplift about 30 km long and on average 2 km wide.   Along both sides of the uplift there are large depressions, from which a branching network of smaller tributaries pass out in various directions, separating the uplift mentioned into individual elevated sectors.   The bauxite segregations are restricted to the northern, most dissected portion of the ridge-like uplift.   They cover like a mantle the raised sectors of the ancient relief, being absent, as a rule, in the depressions. The segregations usually have an irregular shape, and their contours have numerous projections and embayments.   The bauxite segregations rest horizontally and only in individual cases do they slope at 3 - 5°, and rarely up to 10°.

The bauxites consist of stony, friable, and less frequently, argillaceous varieties.   Amongst them are bauxites with the relict fabric of the parent rocks (most commonly schists), a pseudobreccia-like fabric, transitional fabrics from breccia-like to pisolitic, a pisolitic fabric, and a pipe-like fabric.   Often, near the surface of the large bauxite blocks, a typical pisolitic texture is clearly seen, and towards the centre of the block, the quantity of pisolites decreases, so that in a number of cases, bauxites with fabrics of the parent textures are observed in the middle of the block. This suggests a laterite origin for the bauxites of the Vysokopol' deposit.

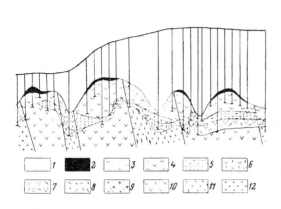

Fig. 105.
Geological section through the Vysokopol' deposit (After M. Slavutsky).

1) deposits of the Quaternary, Neogene, and Palaeogene systems; 2) bauxites; 3) mottled ferruginous non-quartz kaolins; 4) light-coloured kaolins with quartz; 5) montmorillonite rocks; 6) hydromica-kaolinite rocks; 7-9) decomposed, leached rocks: 7) amphibolites, 8) gneisses, 9) granites; 10) amphibolites and associated schists (chloritic, hornblendic, etc.); 11) biotite gneisses; 12) plagioclase granites.

Besides the described ores of the weathering crust, sedimentary bauxites are known in the deposit amongst the continental formations of the Buchak Group, but they have no practical value owing to the low quality of the ores and the insignificant reserves.

Three horizons have been recognized in the section of lateritic bauxite segregations. In the upper part there are brown ironstones with gibbsite, 0.5 m thick, containing Al_2O_3, 20%; SiO_2, 7%; and Fe_2O_3, 55%. In the middle part, there is a bauxite horizon of gibbsite composition, 1.5 m thick, containing Al_2O_3, 38.05%; SiO_2, 8.55%; and Fe_2O_3, 25.25%. The lower horizon consists of a gibbsite-kaolinite-iron rock, 1 - 1.5 m thick, containing about 35% Al_2O_3, Fe_2O_3, 30%, and having an increased amount of SiO_2 (15 - 17%). The total thickness of the segregation is usually 2.5 m.

In mineral composition, the bauxites belong to the gibbsite, gibbsite--hydrogoethite, and gibbsite-kaolinite types. The average amounts of individual minerals in the bauxites are as follows (in %): gibbsite 43; boehmite 4; kaolinite 18; hydrogoethite, hydrohematite, and goethite 31; ilmenite 2; and quartz 1. The gibbsite in the bauxite occurs both in the crystalline and the dispersed form.

The Vezhayu-Vorykva Deposit

This deposit has been discovered by the Ukhta territorial geological survey during exploratory work in Central Timan. The Central Timan bauxite--bearing region is characterized by the presence of provisionally lateritic bauxite deposits of early Frasnian age, previously unknown in this region and in our country, and also sedimentary bauxite-shows of Viséan age, analogous to the South Timan types. Up till now, two deposits, the Vezhayu-Vorykva and the Verkhne--Vorykva, have been discovered in this region.

The Vezhayu-Vorykva deposit occurs in the southeastern portion of the Chetlas horst-anticlinal structure. The bauxite segregations are located in upper Frasnian dolomites, dolomitized limestones, and shales of the Bystra Group.   These rocks are strongly decomposed in the upper part, and have been converted into blocks and clasts, enclosed in dark-brown, yellow, brown, and greyish-green clays.   In the carbonate and shaly rocks of the Bystra Group there is a wide development of products of a weathering crust, consisting of red, reddish-brown, yellow, and yellowish-green silty and sandy clays of hydro-mica and kaolinite-hydromica composition, up to 20 m thick.   Portion or all of this sequence is probably sedimentary (Fig. 106).

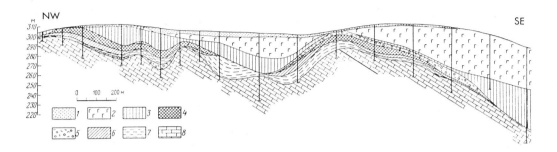

Fig. 106.   Geological section through the Vezhayu-Vorykva deposit.

1) unconsolidated Quaternary deposits;   2-6) lower Frasnian form-ations: 2) basalts,   3) sandstones, clays, and siltstones, 4) bauxites, 5) tuffs, 6) allites;   7) kaolinite-hydromica clays of the weathering crust;   8) dolomites and limestones of the Bystra Group of the Upper Riphean.

The rocks of the Bystra Group or their weathering crust are overlain by a bauxite-enclosing sequence, consisting of fresh or intensely weathered tuffs and tuffites, allites, bauxites, siltstones, and sandstones with seams of argillite-like, silty, and carbonaceous clays.   In places, the silts and clays have been intensely kaolinized.   The sequence is 30 m thick.

According to geologists of the Ukhta geological survey, the bauxites of the Vezhayu-Vorykva and Verkhne-Vorykva deposits comprise the lateritic weathering crust of tuffs and tuffites of early Frasnian age.   The bauxites are light and porous, and have a brownish-red colour.   Under the microscope, the lithoclastic texture of the tuff is seen, in the fragments of which there are plagioclase porphyroblasts, replaced by boehmite.

The bauxites or the terrigenous sediments are overlain by compact aphanitic basalts, which are also of early Frasnian age.   The thickness of the basalts, which rest on the bauxites, is 10 - 20 m, and beyond the limits of the ore segregations, 60 - 70 m.   On the basalts rest sedimentary rocks, also of early Frasnian age, consisting of sandstones, siltstones, and clays.

In the Vezhayu-Vorykva deposit, there is one large segregation, elon-gated in a sublatitudinal direction.   The depth of occurrence of the top of the bauxites is 0.5 - 25.8 m, and the absolute level of the base is 249 - 281 m. The content of alumina according to individual samples varies from 36.6 - 55.2%; Fe_2O_3, 20.2 - 35%; TiO_2, 2.1 - 4.0%;   CaO, 0.11 - 0.56%;   and S, up to 0.03%; the silica module is 1.5 - 11.5.

In mineral composition, the bauxites have been assigned to the hemat-
ite-chamosite-boehmite and diaspore-hematite-chamosite-boehmite types.   The
bauxites consist mainly of boehmite (45 - 50%), diaspore (up to 10%), and also
chamosite, goethite, hematite, and hydrohematite.

SEDIMENTARY (REDEPOSITED) DEPOSITS OF THE GEOSYNCLINAL DISTRICTS

Peive (1947) recognized for the first time in the Urals an independent
genetic type of bauxite deposit of the geosynclinal districts.   These deposits
occur on the karsted surface of carbonate rocks and have been formed during the
time of continental breaks in the deposition of marine sediments.   Geosynclinal
deposits are known both in the USSR and abroad, being restricted to Hercynian and
Alpine fold complexes, developed on the site of the Uralian, Altai-Sayan, Tyan'-
-Shan', Mediterranean, and other geosynclines (Grigor'ev, 1968).

The geosynclinal bauxites, as noted by Bushinsky (1967), occur amongst
marine carbonate deposits on the surface of carbonate rocks, devoid of terrigenous material.   The formation of bauxite deposits took place after the break in
sedimentation.   A cyclicity in the formation of the sediments has been observed
in this group of deposits.   Each cycle begins with a bauxite horizon, which is
replaced upwards by dark-grey bituminous and then grey reef-like limestones.   The
formation of the pre-ore karsted surface on the carbonate rocks in the various
geosynclinal districts took place during the various phases of their development.
The karst sometimes developed shortly after the emplacement of the geosyncline,
and its development continued through almost all the phases of its history.   The
formation of karst and the accumulation of bauxites occurred at the time of continental interruptions in limestones of shallow-water origin.   In the shallow-
-water regions of sufficiently small uplift of the Earth's crust, to cause retreat
of the sea, the land emerged and weathering and the formation of karst began.
Then a small depression occurred and the bauxite-bearing sediments formed sank
below sea level and were overlain by a new layer of rocks.

The bauxites of the geosynclinal group are located both in the marginal parts of the geosynclinal districts, which is typical of the deposits of
the Southern Urals bauxite-bearing region, the Crimea, and the Carpathians, and
also towards the inner parts of the geosynclinal districts (Sapozhnikov, 1971a,
b) — the deposits of the Northern Urals region, Southern Fergana, Berdsk-Maya
in Salair, and the bauxite deposits of Yugoslavia, Austria, etc.

In the geosynclinal districts, the bauxites, as a rule, occupy a definite stratigraphical position.   They are restricted to breaks in sedimentation
and are always located above a surface of unconformity.   These breaks are usually
temporary and have been traced in a sequence of rocks of a single stage of horizon.
In some cases the formation of bauxites has been associated with longer breaks,
when the bauxites occur at the boundary between the basement and the sedimentary
cover of the platform districts.

The most favourable regions for prospecting for bauxites are those
where ore-bearing carbonate rocks are located near uplifts, consisting of rocks
from which a weathering crust, serving as a source of the ore components, may be
formed, during the development of bauxites (eruptive rocks of basic and intermediate composition, various shales, etc.).

The formation of the deposits took place within narrow, linearly-elongated bauxite-bearing zones, which are sometimes repeated, changing during time from the geosyncline to the nearby platform.   Deposits of this type are distributed in the Lower Cambrian, Devonian, Jurassic, Cretaceous, and Palaeogene sequences (Tenyakov, 1971;   Tenyakov *et al.*, 1972).

The upper surface of the layer-like segregations of bauxites is usually smooth, and the lower surface, uneven, dependent on variations in the reefal limestones and pre--ore karst.   The bauxite-bearing sequences are usually crumpled into folds and broken by faults.

The bauxites of the geosynclinal districts are distinguished by high and persistent quality and constancy in thickness of the ore bodies.   There is a predominance of monohydrate diaspore-boehmite, diaspore, and boehmite varieties of bauxites, and the weakly deformed deposits of Cretaceous and Palaeogene age contain gibbsite.

Deposits of this type are developed in the Northern Urals, Southern Urals, Bokson, and other bauxite-bearing regions of the USSR, and also in Hungary (Bardossy, 1957), France, Yugoslavia, and other countries.   The largest province of geosynclinal bauxite deposits in the USSR is the Uralian bauxite province, within which there are several bauxite-bearing regions: the Northern Urals, Southern Urals, Ivdel', Karpinsk, Alapaevo, and Nizhne-Serginsk.   The largest regions are the Northern Urals and Southern Urals.

THE NORTHERN URALS GROUP OF DEPOSITS

The deposits of the Northern Urals group were discovered in 1931 by N. Korzhavin.

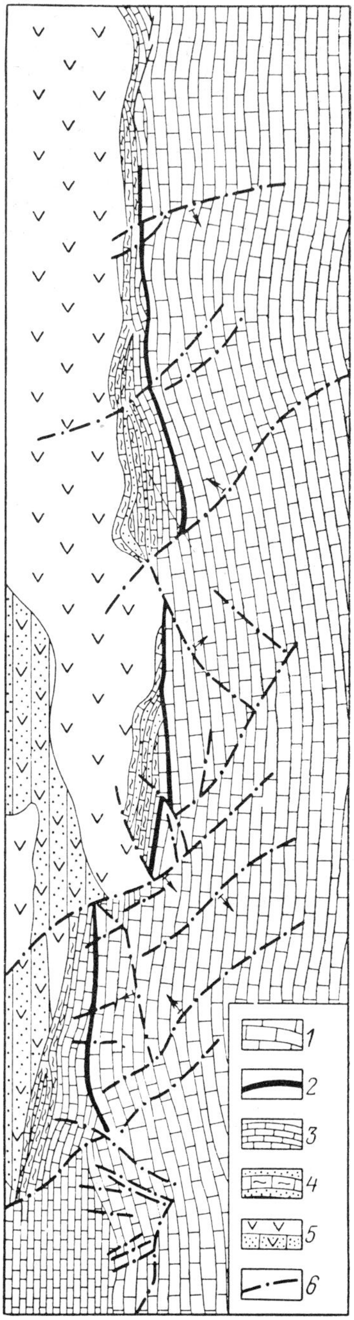

Fig. 107.   Diagrammatic geological map of the Northern Urals bauxite-bearing region. (After E. Gutkin and Yu. Rodchenko with additions by G. Kirpal'.).

1) carbonate deposits (reefal limestones, and bituminous, clay--calcareous shales, D_2); 2) bauxite-bearing deposits; 3) light--coloured, reefal limestones (Petropavlovsk Group, D_1^1); 4) limestones, shales, sandstones, and conglomerates (Sarainaya Group, $S_2^2 - D_1^1$); 5) porphyrites with seams of sandstones and limestones (Sos'va Group, S_2^2);   6) tectonic faults.

The deposits of this region are located on the eastern slopes of the Northern Urals within the western limb of the Shegul'tan syncline of the Tagil synclinorium. The bauxite-bearing layer has a meridional strike and dips eastwards at 20 - 30°. The geological boundaries of this group of deposits in the north and south are the places of gradual thinning-out of the ore horizon, and the western boundary is marked by the points of outcrop of the horizons on the surface, and in the eastern part of the deposit, the bauxites have been traced by exploratory drilling to depths of up to 2000 m.

The Northern Urals group of deposits has been subdivided by east--west faults and non-ore intervals into a series of deposits and bauxite-shows (Fig. 107). On the northern continuation of the Northern Urals group is the Ivdel' deposit, and on the southern continuation, the Karpinsk deposit. These three groups form a genetically unique bauxite-bearing province, elongated meridionally.

The Northern Urals group of bauxite deposits is restricted to a simple meridionally elongated depression, bounded on the west by the pre-Uralian ridge, which consists mainly of basic eruptives (andesite-basalt and diabase porphyrites), cut by intrusions of gabbros and granodiorites. In the eastern part of the region, there is a band of low ridges, consisting of andesite-basalt porphyrites (Bol'shun, 1971; Gladkovsky & Sharova, 1951). According to the scheme, accepted by most geologists (Rodchenko, 1964), the stratigraphical section of the deposits of the region displays the following form (Fig. 108).

Frasnian	D_3^1b	Light-grey, massive, reefal limestones
	D_3^1a	Slates, with seams of dark-grey and platy limestones
Givetian	D_2^2c	Grey and light-grey, massive, reefal limestones
	D_2^2b	Dark-grey, bituminous, bedded limestones
	D_2^1e	Light-grey, massive reefal limestones
Eifelian	$D_2^1d_{1+2}$	Dark-grey, platy, and finely-platy limestones with seams of clay and clay-siliceous shales
	D_2^1c	Light-grey, massive, reefal limestones
	D_2^1b	Dark-grey, bituminous, bedded limestones with *Amphipora*
	D_2^1a	Subrovsk ore horizon
Gedinnian and Ludlovian	D_1^1	Pink and light-grey, reefal limestones (Petropavlovsk Group)
	$S_2l_3 - D_1^1$	Dark-grey, bituminous limestones, shales, sandstones, and conglomerates (Sarainaya Group)

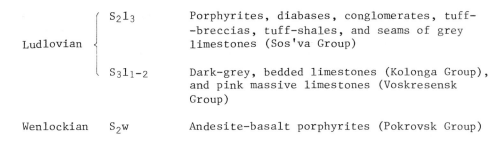

Ludlovian $\begin{cases} S_2l_3 & \text{Porphyrites, diabases, conglomerates, tuff-} \\ & \text{-breccias, tuff-shales, and seams of grey} \\ & \text{limestones (Sos'va Group)} \\ \\ S_3l_{1-2} & \text{Dark-grey, bedded limestones (Kolonga Group),} \\ & \text{and pink massive limestones (Voskresensk} \\ & \text{Group)} \end{cases}$

Wenlockian S_2w Andesite-basalt porphyrites (Pokrovsk Group)

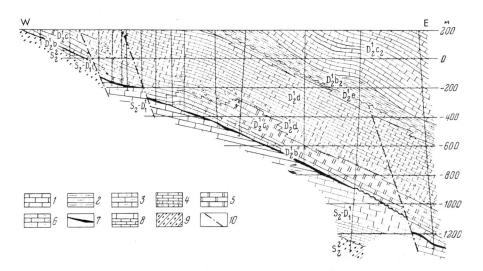

Fig. 108.   Geological section through Kal'insk bauxite deposit along traverse
line 21 (After G. Bol'shun)

1) Upper Eifelian, massive, often reefal limestones ($D_2^1c_2$);   2)
similar dark-grey, bituminous limestones with *Amphipora* ($D_2^1b_2$);
3) Lower Eifelian, light-grey, reefal limestones (D_2^1e);   4) similar
grey, platy limestones, silicified in the lower part (D_2^1d);   5)
similar light-grey, reefal limestones (D_2^1c);   6) similar dark-grey,
bituminous limestones with *Amphipora* (D_2^1b);   7) bauxite horizon;
8) light-grey, and pink, reefal limestones of the Petropavlovsk
Group, stratified in the lower part ($S_2 - D_1^1$);   9) tuff-breccias
of the Sos'va Group (S_2^2);   10) tectonic faults

        The age of the bauxite horizon is early Eifelian, but within the
Northern Urals group ore-shows of late Eifelian age are also known, which have
no commercial importance.

        Within the bauxite-bearing belt, there is a wide development of dis-
junctive fractures (mainly steeply-dipping faults and reversed faults) with
different amplitudes of displacement.   The largest of them have a sublatitudinal or diagonal (*i.e.* oblique-latitudinal) strike, and a vertical throw of
about 200 - 400 m.   The boundaries of the deposits and sectors are, as a rule,
large pre-ore tectonic fractures (Gutkin & Rodchenko, 1965).   At present, 11

pre-ore faults have been recognized, breaking up the area of the basin into a series of depressed and uplifted blocks.   The blocks depressed during pre-ore time were the most favourable structures for bauxite accumulation.   In addition to the faults noted, geological exploration has defined more than 240 post-ore tectonic fractures, which detracts from the quality of the bauxites in the sectors adjoining the fault planes.

The commercial mineralization has been associated with the Subrovsk Horizon, which occurs among Devonian limestones (Eifelian Stage).   The base of the ore horizon consists of reefal massive limestones of the Petropavlovsk Group (Gedinnian deposits) with a very uneven surface.   At the top there are bitumin-ous limestones, and Eifelian clay, in places, carbonaceous shales (Vagransk Group).   The ore horizon has been subdivided into two subhorizons: a lower horizon consisting of high-grade, low-grade, and jasper-like bauxites, and an upper, of mottled pyritized bauxites.   The red high-grade bauxites fill hollows in the limestones of the cover, and the red low-grade and jasper-like bauxites are concentrated on the slopes of the depressions, with the mottled bauxites distributed everywhere.   The red bauxites have the principal commercial value, their reserves in individual deposits comprising 85 - 97% of the total.

At the base of the ore bodies in places is a limestone-bauxite breccia (ore), 0.2 - 0.5 m thick, which is distinguished by the small content of alumina and has no commercial value.

The bauxite deposits of the Northern Urals region are restricted to a vast ancient karst province.   The bulk of the commercial ores occur in large depressions in the pre-ore karst relief.   The non-ore sectors have irregular outlines, and are confined to the pre-ore rises, on which bauxite was not pre-cipitated or was removed from them into nearby sectors.   During the formation of the bauxites, removal of material took place from the west.   The source of the principal bauxite-forming components was the weathering crust of massifs of Upper Silurian eruptive-clastic rocks.

The bauxite segregations of the individual deposits, being confined to the negative forms of the karst relief of the Palaeozoic limestones, are var-iable in thickness and have complex outlines in plan.   Their structure is quite complicated as a result of interlayering of bauxite varieties (red high-grade, low-grade jasperoid, mottled, and also porphyritic and pyritic (bauxite-pyrites)), occurring in the form of individual nests.

According to the number and shape of the projections of the roof, the ore fields have been divided into three types: the first is characterized by discontinuous mineralization, non-ore windows occur sporadically, and they have small dimensions, from a few square metres up to several hundred;   the second type is transitional, and here the non-ore windows occupy 20 - 30% of the area; and the third type is distinguished by discontinuous mineralization, the non--ore windows occupying 30% and more of the total area, being grouped in elong-ate belts or areas of different aspect of considerable dimensions.

In mineral composition, the red bauxites belong to the diaspore type; the jasperoid and mottled types are diaspore-boehmite;   and the bauxite-pyrites consist of pyrite-diaspore-boehmite.

The red bauxites are of high quality and are characterized by the following composition (in wt %): Al_2O_3, 53 - 55; SiO_2, 2 - 6; Fe_2O_3, 23 - 25;

CaO, 1.6 - 2.5; S, 0.12 - 0.40 (up to 1.1); CO_2, 1.9 - 3.6; TiO_2, 2.0 - 2.5.
The high-sulphur pyritic and mottled bauxites, containing from 1 to 15 wt % of
sulphur (on average 5.4 - 7.3%), comprise 5% of the total reserves.

THE SOUTHERN URALS GROUP OF DEPOSITS

This group of bauxite deposits, located on the western slopes of the
Southern Urals, is elongated in a northeasterly direction. About 50 deposits
and bauxite-shows are known, being confined to the Orlovsk Group of the Frasnian Stage of the Upper Devonian.

The Southern Urals deposits were discovered by A. Belousov in 1933,
in the vicinity of Kukshik and Novaya Pristan' villages.

Three subgroups have been recognized within the Southern Urals group
of bauxite deposits: the Novopristan', Vyazovsk, and Kukshik, and also
individual deposits and bauxite shows.   Each subgroup combines several deposits.

In tectonic respects, the Southern Urals bauxite-bearing region is
located in the pre-Urals foredeep (first-order structure) and lies within the
Karangau uplift (second-order structure), which consists of folds of lower
orders.   The bauxites of the Southern Urals group are confined to the Uluirsk
synecline and the Sim trough, separated by the Suleimanovsk brachysyncline.

The deposits of the Novopristan' and Kukshik subgroups lie on the
southeastern limb of the Uluirsk syncline, and on the limbs of the Suleimanovsk
anticline is the Vyazovsk subgroup.   In the extreme southwestern part of the
region the Serpeevsk bauxite-show has been discovered, confined to the southern limb of the Sim trough.

The geological structure of the Southern Urals group of bauxite
deposits involves the participation of two complexes of formations. The lower
consists of unfossiliferous Upper Proterozoic deposits, intensely deformed and
metamorphosed.   The rocks of this complex form uplifts, located within the
bauxite-bearing region and on its periphery.   The upper complex consists of
faunally defined sediments of the Devonian, Carboniferous, and Permian, confined exclusively to depression structures.

The Upper Proterozoic rocks consist of quartzose sandstones, and
various shales, limestones, and dolomites of the Avzyan, Zil'merdak, Katovsk,
Tsizersk, and Min'yarsk groups up to 4000 m in total thickness.   The lower
Palaeozoic consists of formations of the Ashinsk Group, being mainly quartzose
sandstones, siltstones, and shales, 40 - 60 m thick.

The bauxite-bearing rocks belong to the Devonian System.   In the
lower part they consist of uniform quartzose sandstones, siltstones, argillites,
dolomites, and limestones of the Takatinsk, Vanaizva, and Vdovesk beds of early
Eifelian age, resting transgressively on formations of the Upper Proterozoic
and lower Palaeozoic.   Higher up come the limestones of the Kal'tsedov and
Biya beds of the upper Eifelian substage, up to 100 m thick.   The Givetian
Stage unites the Gusovsk beds, which consist mainly of sandstones, and the
the Isoyavsk beds, consisting of carbonate rocks, up to 40 m in total thickness.

Deposits of Late Devonian age rest transgressively on Middle Devonian formations and consist in the lower part of quartzose hematitized limestones of the Pashiisk Group and limestones of the Kynovsk beds, with a total thickness of up to 20 m. The middle Frasnian deposits consist of limestones of the Serpeevsk, Domanik, and Samsanovsk beds, the last being strongly karsted. In the lower portion of the upper Frasnian deposits there are the Orlovsk beds, including bauxites, allites, bauxite rocks, sandstones, conglomerates, siltstones, argillites, and limestones. They are overlain by limestones and dolomites of the Famennian Stage, up to 800 m thick.

The Carboniferous deposits consist mainly of limestones with rare seams of clays, about 2000 m thick. The lower division of the Permian System consists of limestones, shales, sandstones, and conglomerates, up to 600 m thick, on which rest Quaternary formations. Magmatic rocks are absent within the region.

An analysis of the palaeogeography demonstrates that in the region during the extent of Palaeozoic time, a shallow-water marine environment predominated, the existence of which terminated in the Early Permian.

The bauxite-bearing layer extends from east to west and is separated by non-ore sectors and faults into segregations of varying size.

The ore bodies of layer- and lens-like shape are usually irreguar, extended in various directions according to the pre-ore relief. The lower boundary of the ore layer is uneven, depending on the degree of karstification of the underlying limestones, and the upper boundary is even and sharp. The thickness of the ore layer depends on the depth of the karst cavities. The ore layer has been complicated by faults and fold movements.

The ore bodies consist of red, grey, mottled, clastic, and oolitic (diaspore-chamosite) bauxites. The bauxite ores consist of diaspore, forming pisolites and clasts, boehmite, kaolinite, and chamosite, comprising the matrix, hematite, minerals of titanium oxide, pyrite, and in the cracks and veinlets, calcite and dolomite. The amount of pyrite in the grey bauxites reaches 40%.

The N o v o p r i s t a n ' s u b g r o u p is restricted to the southeastern limb of the Uluirsk syncline, which has been complicated by smaller folds. It combines a series of deposits: the Mezhevoi Log, Barsuchii Log, Sosnin Log (Pervomaisk), Blinovo-Kamensk, Kurgazak, Alekseevak, Ivanovo-Kuz'ma, and Ailinsk.

The bauxite segregations have a layer-like form and a gentle dip (5 - - 12, less frequently 20 - 25°). The bauxites consist mainly (60%) of a high--grade red variety; in the upper parts of the segregations there are grey and mottled bauxites, the reserves of which are regarded as non-commercial. In mineral composition, the red bauxites belong to the hydrohematite-diaspore-boehmite type.

The V y a z o v s k s u b g r o u p of deposits extend in the form of a narrow band. Within it, the Vyazovsk deposit is separated into five sectors: the Vuryuzan 1, 2, and 3, and the Chukur, and Permsk. In each of the sectors there are 1 - 3 ore bodies.

The principal rock-forming minerals are boehmite, diaspore, and kaolinite. The following principal components occur in the bauxites (in wt %): Al_2O_3, 50.9; SiO_2, 13.0; Fe_2O_3, 20.4; TiO_2, 2.3; and CaO, 0.7.

The deposits of the K u k s h i k s u b g r o u p are located in the Bashkirian ASSR;   they include the Krasno-Kama, Pokrovsk, Aisk, and Novoe deposits.

The bauxite rests on the karst surface of light-grey limestones. The bauxite-bearing sediments are overlain by interstratified light- and dark--grey bituminous limestones.   The bauxites consist of numerous varieties, distinguished in colour, density, and textural features.

The amount of basic components varies within wide limits (in wt %): Al_2O_3, 39 - 78; SiO_2, 1 - 12; Fe_2O_3, 1 - 30; TiO_2, 2.1 - 4.   In mineral composition, the bauxites belong to the boehmite-diaspore type.

THE SALAIR GROUP OF DEPOSITS

The Salair group of bauxite deposits is located in the central part of the Salair bauxite-bearing region, which embraces the region of Central Salair, and its steep northeastern slope and the gentle southwestern slope, gradually sinking below the sediments of the Biya-Barnaul basin, also known as the Southwestern Near-Salair region.   Here, there are the well-known Berdsk--Maisk, Obukhovo, Novogodnee, and Oktyabr'sk deposits and a number of bauxite shows.

The area of the Salair group of bauxite deposits consists of sand--shale deposits of the Cambrian and Ordovician, limestones of the Silurian and Lower and Middle Ordovician, sand-shale deposits of the Middle and Upper Devonian and Lower Carboniferous, and Quaternary formations.   The sedimentary sequences of the Palaeozoic basement have been cut by granitoid intrusions (Fig. 109).

The rocks of the region form a system of alternating anticlinal and synclinal folds, which have specific names.   The Berdsk-Maisk, Novogodnee, and Oktyabr'sk deposits are confined to the Verkhne-Berdsk synclinal structure. The limbs of this structure consist of Silurian and Devonian deposits.   At the base of the Middle Devonian, there is an ore horizon, consisting of baux-ites, and leptochlorite and clay shales, up to 20 m thick, which has been term-ed the Berdsk bauxite horizon.

The Obukhovo anticline, within which the deposit of the same name is located, consists of Lower and Middle Devonian limestones.   Two further baux-ite horizons have been identified here in addition to the Berdsk layer.

Recent work in the Salair region in the Devonian deposits has estab-lished the presence of four levels of bauxite accumulation: Gedinnian, lower Emsian, and lower and upper Eifelian (Ageenko, 1970).

The Gedinnian level of bauxite formation has been identified in the Verkhne-Berdsk, Talits-Yelban, Vetokha, and Matrenka structures.   It is con-fined to the base of the Lower Devonian, rests on the eroded surface of Sil-urian limestones, and consists of allite-sialite shales with free alumina.

The second, lower Emsian level of bauxite accumulation is defined by grey and dark-grey mica-corundum bauxites and allites, containing from 17 to 25% SiO_2 and from 39 to 53% Al_2O_3.   It occurs in the limestones of the cover of the Berdsk ore horizon and has been identified in the Obukhovo bauxite deposit.

The third, lower Eifelian level is the most productive.   It has been established in all the deposits of the Salair group.   It comprises a layer of diaspore and corundum bauxites, resting on light-grey and white limestones of the Lower Devonian and covered by lower Eifelian dark-grey and black limestones. Its thickness is 15 m.   The average amount of silica in the bauxites varies from 12 to 18%, and alumina, from 46 to 52%.

The fourth, upper Eifelian level has been established in the Obukhovo deposit, where it is defined by an horizon of corundum bauxite and sheared corundum-mica rocks up to 3.3 m thick.   The amount of silica in the bauxite is 20.46% and alumina, 46.5%.

All the levels of bauxite formation are characterized by the occurrence of the bauxite layers on light-grey and white, often banded limestones, and their covering by black or dark-grey limestones.

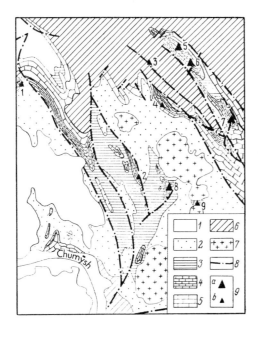

Fig. 109.
Diagrammatic geological map of the Salair group of bauxite deposits (After N. Ageenko)

1) Quaternary deposits;   2) Middle and Upper Devonian and Lower Carboniferous sand-shale formations; 3) Givetian sand-shale deposits; 4) Lower and Middle Devonian limestones;   5) Silurian limestones; 6) Cambrian and Ordovician sand-shale deposits;   7) granitoids; 8) faults;   9) bauxite deposits (a) and shows (b): 1 - Chupa, 2 - Gunikha, 3 - Matrenka, 4 - Chudinovsk, 5 - Oktyabr'sk, 6 - Novogodnee, 7 - Berdsk-Maisk, 8 - Obukhovo, 9 - Bobrovo

The morphology, structure and attitude of the ore bodies in all the deposits of the Salair group have many features in common. The ore bodies have a layer-like form. Their lower boundary is very uneven with numerous depressions, sink-holes, and pockets; the upper boundary is relatively even, and conformable with the stratification of the limestones. Frequently, interlayering of the bauxites with the limestones of the roof has been observed. In the

lower portion of the ore layer, there is a 'bauxite-breccia', formed as a result of fillings in the uneven base of the layer, the cracks, and cavities. The
dimensions of the layer-like ore bodies of bauxite vary within wide limits.
The bauxite layers, along with the surrounding rocks, have been crumpled into
folds of various kinds with dips from a few degrees up to vertical.

The morphology of the ore bodies has been determined by the nature
of the pre-ore relief of the limestones that underlie the ore horizons, post-
-ore erosion, tectonic deformations, and sag phenomena, associated with karst
formation.   In the pre-ore relief, highs in the reef construction and inter-
-reefal depressions have been observed.   The rises amount to 5 - 15 m and
rarely up to 50 m above the top of the ore horizon and they divide it into
individual portions.   The normal dimensions of the rises are 1000 X 200 -
- 100 m.   The ore layer is absent in the axial portion of the rises, and on
the slopes its thickness falls to less than 1 m.   The inter-reefal depressions
were the most favourable areas for the formation of the highest-grade bauxites
of maximum thickness.   All the good-quality ore segregations are located in
them.

Post-ore erosion of the bauxites was quite widespread.   Proof of
this is the discovery in the supra-ore limestones of weakly rounded bauxite
pebbles.   Karst sag phenomena usually disrupt the continuity of the bauxite
and lead to increase in its thickness.

On the basis of composition and types of ores, the Berdsk-Maisk deposit is markedly distinguished from the Obukhovo deposit, which is the result
of metamorphic effects of the granite intrusion on the latter, with the conversion of the diaspore bauxites into corundites.

The Berdsk-Maisk deposit consists of two types of ores: leptochlorite-diaspore and chlorite-diaspore.   The leptochlorite-diaspore bauxites have
a finely and cryptocrystalline texture, a dark-grey and black colour, and consists of diaspore, dispersed leptochlorite, and chloritoid.   The average chemical composition of these bauxites is as follows (in wt %): SiO_2, 8.3 - 11.2;
Al_2O_3, 61.2 - 63.4; TiO_2, 2.2 - 2.4; Fe_2O_3, 9.0;   FeO, 4.0 - 16.6;   MgO, 0.8 -
- 1.5;   calcination loss, 11.7 - 12.9.

The chlorite-diaspore bauxites have a grey, dark-grey, and black tint,
and a micro- and crypto-oolitic texture.   They consist of diaspore, chloritoid, leptochlorites, and clay-micaceous matter.   The average amounts of the
principal components are as follows (in wt %): SiO_2, 15.8 - 19.0; Al_2O_3,
45.3 - 51.0; TiO_2, 0.7 - 1.8; Fe_2O_3, 2.7 - 17.0;   MgO, 0.1 - 1.15;   calcination loss, 8.4 - 10.9.

The bauxites of the Obukhovo deposit consist of corundum, mica-
-corundum, and corundum-mica varieties.   The corundum bauxites have a dark-
-grey and black tint, and a compact fabric.   The average chemical composition
is as follows (in wt %): SiO_2, 0.16; Al_2O_3, 64.5; Fe_2O_3, 1.6;   FeO, 14.0;
TiO_2, 2.87;   CaO, 0.73; Fe_2O_3, 1.61;   S, 0.63;   calcination loss, 4.41.

The mica-corundum bauxites are sheared ores of grey colour, consisting of corundum, and a small amount of diaspore, leptochlorites, and micas.
Their chemical composition (in wt %) is: SiO_2, 20.5; Al_2O_3, 53.4; Fe_2O_3,
4.3;   FeO, 8.4; TiO_2, 2.5;   CaO, 4.1;   MgO, 2.3;   S, 2.3;   calcination loss,
4.7.

The corundum-mica varieties of bauxites have a grey, and rarely brown colour, and are intensely sheared.   The average amounts of the principal components are as follows (in wt %): SiO_2, 16.4; Al_2O_3, 28.4; Fe_2O_3, 9.2; TiO_2, 1.5;   CaO, 6.4; calcination loss, 5.4.

The distribution of the various varieties of bauxite ores in all the deposits of Salair is identical.   In the central portion of the bauxite segregations, the highest-grade ores normally occur, and on the periphery, they are replaced by bauxites with a lower content of alumina and allites.   Mutual transitions have been observed between the recognized types of ores both along strike and dip, and they are sometimes mutually interstratified with a predominance of one type.

The Berdsk-Maisk Deposit

The Berdsk-Maisk deposit is located in the central part of the Salair Ridge and is confined to the southeastern portion of the Verkhne-Berdsk syncline, complicated on the limbs by later folding.   The peripheral portion of the syncline is composed of Upper Silurian limestones, the limbs consist of Lower Devonian limestones, and the core, of Middle Devonian limestones.   The bauxite--bearing horizon occurs on the boundary between the Lower and Middle Devonian limestones.   All the rocks have been cut by dykes, associated with intrusive processes of the Hercynian cycle.   In the northeastern limb of the syncline, a conformable reverse fault brings to the surface a substantial block of the underlying rocks and causes duplication of the ore horizon over an extent of about 600 m along strike with a throw of the beds along the fault of 400 m and more. The ore horizon has been traced along both limbs of the synclinal structure.

The Obukhovo Deposit

The Obukhovo deposit is located in the central portion of the Zalesovsk basin, which has been filled with upper Givetian sand-shale and carbonate deposits, cut by adamellite intrusions.   All the rocks have been intensely metamorphosed; the limestones have been recrystallized, the sand-micaceous rocks have been hornfelsed, and the bauxites converted into corundum-mica rocks.

The deposit is a complicated brachyanticlinal structure, in the core of which Gedinnian rocks crop out, and along the limbs and in the core of ancillary synclines, Coblenzian, Eifelian, and lower Givetian limestones.   In the southeastern portion, the brachyanticlinal structures of the deposit have been disrupted by tectonic fractures (Fig. 110).

Two ore horizons have been revealed in the deposit: the lower, Obukhovo, and the upper, Berdsk.   The horizons are separated by a sequence of limestones, from 160 to 250 m thick.   The Obukhovo horizon is of commercial importance.   It rests on the uneven surface of Lower Devonian marmorized limestones. In the limestones of the surface, pockets are often observed, which have been filled with bauxite.   The ore horizon dips at from 0 up to 90°, and on average 30°.   At the top, the ore horizon is interstratified with black bituminous limestones.   The Berdsk horizon has no commercial importance, it is located in the limestones of the Khvoshchevsk Group, and is up to 8 m thick.

Besides solid ores, clastic types are known in the deposit.   The latter form an eluvial segregation, consisting of fragments and blocks of bauxite amongst the mottled clays.

Fig. 110.
Diagrammatic geological map of
the Obukhovo bauxite deposit
(After N. Ageenko).

1) unconsolidated Mesozoic-
-Cainozoic deposits; 2) Givetian
sand-shale sequence; 3-5) bauxite-
-bearing complex of reefal limestones of the upper (3), and lower
(4) Eifelian, and Lower Devonian
(5); 6) dark-grey limestone horizon; 7-9) bauxite horizons: 7)
upper Eifelian, 8) lower Eifelian
(Berdsk and Obukhovo), 9) lower
Emsian (in section); 10) granitoids; 11) disjunctive faults;
12) segregation of clastic ores.

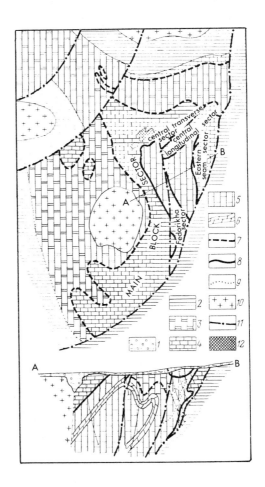

        The principal ore-forming mineral in the bauxites is corundum. Dep-
ending on its amount, we may recognize corundum, mica-corundum, and corundum-
-mica varieties of ores. The first varieties predominate (comprising about
90%). All the varieties are mutually associated by gradual transitions.

The Bokson Deposit

        The Bokson deposit is located in the eastern portion of East Sayan,
in the basin of the River Bokson. Bauxites in this region were first discovered in 1931 by F. Golovachev, who had discovered bauxite pebbles in the eluvium of the river. The Bokson deposit is situated in the eastern periclinal
closure of the Bokson-Serkhoi synclinorium and is confined to the area of two
synclines, separated by an anticlinal uplift. It has a fold-block structure.

        The nature of the sedimentary rocks, which form the region of the
Bokson deposit, their great thickness, and also the development of tuffogenic
and eruptive formations indicates that at the time of bauxite formation here,
there existed a geosynclinal or similar environment.

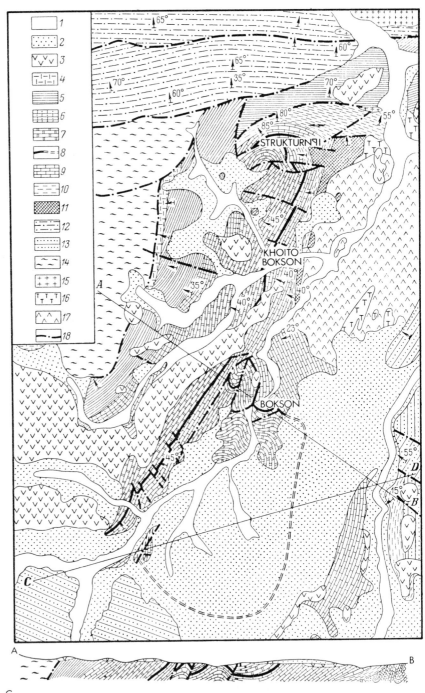

Fig. 111.

Diagrammatic geological map of the Bokson bauxite deposit (After N.Il'ina).

1) fluvial deposits; 2) glacial deposits; 3) basalts; 4) Yamata Group; 5) Mangatgol' Group; 6-9) Bokson Group: 6) crystalline limestones and dolomites, 7) upper dolomites, 8) dolomites, surrounding bauxite (continuous line – definite; dashed line – assumed), 9) lower dolomites; 10) Sarkhoi Group; 11) Diba Group; 12) Gargan Group; 13) Oka Group; 14) Archaean gneisses; 15) granites; 16) serpentinites; 17) diabases, diorites, and gabbros; 18) tectonic breaks.

The area of the Bokson deposit consists (Il'ina, 1958) of Archaean gneisses, Upper Proterozoic (Gargan Group) carbonate rocks (mainly dolomites), and clay-sericite and chlorite-graphite schists, and also Upper Proterozoic (Oka Group) green schists.   The grey shales, and sandstones with seams of conglomerates and dolomites of the Diba Group belong to the Upper Proterozoic- -Cambrian sequence.   This group has been cut by diabases and has a limited distribution.   Sinian formations (the Sarkhoi Group) consist of acid and basic eruptives and their tuffs, shales, tuffogenic and polymict sandstones, conglomerates, breccias, and rare seams of dolomites (Fig. 111).

The Bokson bauxite-bearing group (Sinian - Lower Cambrian) rests on the Sarkhoi Group without any definite angular unconformity.   In its lower part, there are conglomerates and breccias, consisting of material from the underlying rocks.   They are overlain by dolomites with seams of algal reefo- genic dolomites and marls, with an overall thickness of about 700 m.   Higher up in the sequence, there are bauxites and overlying clay graphite shales. It is possible that the bauxites and shales may belong to the Lower Cambrian, that is, the bauxites are confined to the stratigraphical break between the Upper Proterozoic and the Cambrian.   The bauxites are overlain by dolomites of the adjacent roof, with an overall thickness of 30 - 140 m, with seams of ferruginous sandstones, marls, and carbonate breccias.   The latter are over- lain by dolomites with flints, marmorized dolomites, and dolomitized limestones, about 200 m thick.   The total thickness of the Bokson Group is 2600 - 3000 m.

The Middle Cambrian deposits (Mangatgol' Group) consist of clay- -carbonate shales with seams of limestones and sandstones.   The Yamata Group of Middle-Late Cambrian age consists of redbed rocks: breccias, conglomerates, sandstones, shales, tuffs, and eruptives with rare seams of dolomites.

The intrusive formations, underlying the Bokson Group, consist of basic and ultramafic rocks.   Granites cut the Bokson Group in the western portion of the deposit.   Palaeogene basalts form a cover along the watersheds of the plateau.   The Mesozoic-Cainozoic formations consist of pebbles and basalts, and in places, they have been found on the Palaeozoic rocks below the basalts.   The Quaternary deposits consist of glacial, alluvial, eluvial, and deluvial formations.

The deposit consists of a layered segregation, occurring within reefogenic, algal, streaky, bedded, and banded dolomites.   The base of the ore body is uneven.   Pockets, nests, and cracks have been observed in the underlying dolomites, which are filled with red or green bauxite.   The top of the bauxite layer is smooth.   The bauxites, markedly or gradually chang- ing in composition, pass into black graphitized or clay slates, fine-grained red dolomitic breccias, and fine-grained ferruginous sandstones.

The bauxites consist of red and green varieties.   The former occur in the lower part of the layer, and the latter in the upper part, and in some cases, a repeated alternation of these varieties has been observed.

The deposit consists of 11 sectors.   In the central and eastern sectors, there is a widespread distribution of pockets, nests, and cracks, filled with bauxites; the western sectors are characterized by a simpler con- struction of the bauxite layer, which is almost not karsted at all at the top. The bauxite ores of the Bokson deposit are low-grade.   The amount of principal components in them is as follows (in wt %): Al_2O_3, 41.2; SiO_2, 20.4; Fe_2O_3, 25.4; silica module, 2.19.   In mineral composition, the bauxites belong to the chlorite (kaolinite)-diaspore-boehmite type.

SEDIMENTARY (REDEPOSITED) DEPOSITS OF THE ANCIENT PLATFORMS

On the ancient platforms, the sedimentary bauxite deposits are often confined to the areas where the marginal portions of the syneclises adjoin the large platform uplifts and anteclises and, as a rule, they are located near the sloping portions of the uplifts.   In some cases, the deposits are limited to the basins or fields of development of ancient (Mesozoic) karst within basement highs.

The bauxite deposits of the platform districts are often associated with coal-bearing sediments, although bauxite-formation and coal-accumulation, as a rule, are somewhat separated in time and space.   The coal-bearing sediments in the sequence are usually located above the bauxite-bearing sediments or are removed at a distance from the basement highs over a greater distance than the bauxites.

Typical of this group are the deposits of the northern part of the Russian Platform (North Onega, Tikhvinsk, and South Timan groups) and the southwestern portion of the Siberian Platform and the Yenisei Ridge (the Chadobets, Tatar, and Near-Angara groups).

The bauxite deposits of the Russian Platform have been concentrated in the western and northern peripheral zone of the Moscow syneclise.   Their formation has been associated with initial phases of a Viséan transgression of the Early Carboniferous.   During Viséan time, a warm and humid climate prevailed in this area, and in this respect, almost complete coincidence in time of bauxite formation and coal-accumulation has been observed, although the bauxite-bearing sediments, in relation to the coal-bearing types, occupy the most marginal portions of the Moscow syneclise.

On the Russian Platform (Kirpal', 1972), the bauxite segregations usually have a layer-like form and in a number of cases, a substantial area and thickness (the North Onega and South Timan groups).   In the Tikhvinsk region, the bauxite segregations have an elongate valley-like form.   In the southwestern portion of the Siberian Platform and the Yenisei Ridge, the bauxite deposits are confined to areas of development of carbonate rocks of the folded basement. Their formation took place during Cretaceous and Palaeogene time in karst funnels and karst-basinal depressions.   The form of the bauxite segregations is usually lens-like, pocket-like, and nest-like.

Abroad, bauxite deposits of this type have been discovered on the China and North American platforms.   Here, the bauxites are also associated with coal-accumulation and occur at the base of the coal-bearing sediments.

THE NORTH ONEGA GROUP OF DEPOSITS

The deposits of the North Onega group, discovered in 1949, are located in zone of transgressive overlap of the Lower Carboniferous deposits of the Moscow syneclise on to the crystalline rocks of the eastern slopes of the Baltic Shield (Fig. 112).

The crystalline rocks of the basement have been divided into two associations.   There is an older Karelian association, consisting of basic igneous rocks and chlorite-amphibolite schists, and a post-Karelian association, consisting of diabase porphyrites and basalts, cut by peridotites and pyroxenites. An ancient weathering crust has been developed everywhere on the rocks of the crystalline basement.

In the depressional hollows in the crystalline basement, there are horizontally, almost unmetamorphosed Cambrian formations, consisting of greenish-grey argillitic siltstones, resembling varved clays.   Higher up on the Cambrian formations, Upper Devonian deposits are transgressively distributed, consisting in the lower part of greenish-grey clays and grey siltstones, and in the upper part, mottled clays and quartzose siltstones.   The Upper Devonian deposits are also transgressively overlain by a Lower Carboniferous sand-shale bauxite-bearing sequence.   The distribution of the bauxite-bearing deposits in the region is determined by the relief of the ancient crystalline basement.

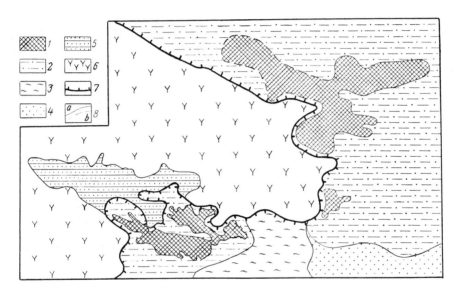

Fig. 112.   Geological-lithological diagram of the pre-ore basement of the
central part of the North Onega bauxite-bearing region.   Compiled
by G. Kirpal' from data of the Plisets ore administration.

1-4) sediments of Tula-Aleksa age: 1) bauxite-bearing sediments,
2) clays, siltstones, bauxite rocks, and argillaceous sands, 3)
kaolinite clays, 4) quartzose sands, silts, and alternating clays
and sands; 5) Upper Devonian sand-clay rocks; 6) Upper Proterozoic
volcanogenic rocks of basic and ultramafic composition: picritic
porphyrites, agglomeratic lavas, metadiabases, and shales; 7)
boundary of uplift of ridge of the Vetrenyi belt; 8) contours of
bauxites: *a*) definite, *b* assumed.

The region of the north Onega group of bauxite deposits towards the beginning of the Viséan stage was a plateau, the surface of which was dissected by old valleys, inherited from the relief of the basement, with a depth of 60 -- 70 m and a width of 1 - 3 km.   They opened into vast basin-like depressions, weakly inclined towards the east.   The length of the basin is 10 - 15 km, width 4 - 5 km, and depth up to 80 m.   These closed basins, and also the deltaic sectors of the old valleys are the reservoirs of the largest bauxite bodies.   The bauxite-bearing sediments are confined to the old valleys, being simple basins or trough-like hollows with a width of 1 - 3 km and a depth of 60 - 80 m.

The bauxite-bearing sediments are limited to the lower portion of the Lower Carboniferous sequence (Fig. 113).   Their thickness is 45 m in the depressions, and it decreases on the highs in the relief.   The deposits have been subdivided into four sequences (in upward succession): 1) sub-ore, 2) bauxite, 3) iron-pisolite, and 4) supra-ore.   The bauxite-bearing sediments are overlain by the carbonate rocks of the Devyatinsk Group of the Protvinsk horizon of the Namurian Stage.   The sub-ore sequence belongs to the Tula horizon of the Yasnopolyansk substage of the Viséan Stage.   It consists of sands, silts, argillaceous sandstones, siltstones, and less frequently clays.   On the slopes of the shield, basal conglomerates with seams of carbonaceous clays and thin coal seams have been found in it.

The bauxite sequence fills the planar wide valley-like hollows in the erosion relief and rests on the sub-ore sequence and on Devonian and Cambrian sediments, and on the slopes, sometimes on the weathering crust of metapicrites and chlorite-amphibolite schists.   It consists of kaolinite clays, siallites, allites, and bauxites.   The kaolin clays occur mainly at the top of the sequence, and towards the central part they are replaced symmetrically by siallites, and then separated by allites.   The sequence reaches 25 m in thickness.

The bauxite segregations are usually located near the slopes of the crystalline basement;   at distances of 4 - 5 km from the foot of the ridge of the Vetrenyi belt, commercial bodies of bauxites are not found.   As it sinks, the bauxite sequence splits and is replaced by argillaceous quartzose sands.

The bauxites are stony, porous, brittle, and jointed rocks with horizontal or cross-bedding.   In the upper and lower parts, the bauxite-bearing sequence is red or dark-pink in colour, and in the middle portion, light-pink, grey, yellow, and white colours predominate.   Throughout the entire section of the sequence, there are plant remains in the form of hollow ferruginized tubes, scraps and impressions of stigmarian roots and stems.

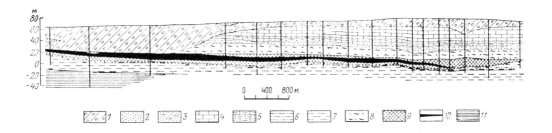

Fig. 113.   Geological section through the Iksa bauxite deposit (After É. Kal'berg)

1) cobbly loams;   2) sands;   3) argillaceous sands;   4) limestones; 5) argillaceous limestones;   6) clays;   7) silty clays;   8) clays with ferruginous pisolites;   9) bauxite rocks;   10) bauxites;   11) argillites.

The iron-pisolite sequence consists of clays of kaolinite-goethite composition with numerous seams of iron ores of pisolitic structure; its thickness varies from 3 to 33 mm.

The supra-ore sequence belongs to the uppermost part of the Viséan Stage and consists of siltstones and sandy clays, sands, and sandstones.

The Iksa Deposit

The Iksa deposit is limited to the basin of the same name (a second-order structure) in the large ancient buried North Onega depression.   The basin is elongated northwesterly, has an asymmetrical construction, and a maximum depth of 225 m.   The depression has been filled with Cambrian, Upper Devonian, and Lower Carboniferous rocks.

The depth of the bauxites varies from 39 to 137 m, and on average for the various segregations comprises 45 - 110 m.   The bauxites contain (in wt %): Al_2O_3, 49 - 53; SiO_2, 16.7 - 18.5; Fe_2O_3, 7.3 - 14.4; TiO_2, 2.4; silica module, 2.1 - 2.9.   An increased amount of Cr_2O_3 in the bauxites (on average 0.57%) indicates an association between the bauxites and the basic rocks of the Vetrenyi belt.   The sub-ore clays contain an increased amount of vanadium pentoxide.

The bauxites consist of argillaceous, stony, and saccharoidal varieties, have a clastic, finely-dispersed (pelitic), and oolite-pisolite texture. The coarse-clastic varieties occur in the deep-seated portions of the basins, and in the middle portion of the sequence, the bauxites consist of medium- and fine-pisolitic varieties, and in the upper portion they have a pelitic texture with a large quantity of plant remains.

In mineral composition, the bauxites belong to the kaolinite-boehmite, kaolinite-gibbsite, and kaolinite-gibbsite-boehmite types.

A regular distribution of the principal ore-forming components has been observed in the segregations.   The maximum amount of alumina is concentrated in the central portions of the segregations, and the minimum amount, in the marginal portions;   the distribution of silica has an inverse pattern. In addition to the bauxites, there are segregations of allites in the Iksa deposit, occurring in the soil and in the roof, and also segregations of lean iron ores, located above the bauxites.

THE TIKHVINSK GROUP OF DEPOSITS

The deposits of the Tikhvinsk group, discovered in 1916 by T. Timofeev, are confined to the belt of Lower Carboniferous sediments on the northwestern limb of the Moscow Syneclise.   The bauxite-bearing zone extends in a submeridional direction for 260 km (Fig. 114).   Within this zone there are more than 30 deposits: the Bat'kovsk, Sinensk, Podsosnensk, Malyavinsk, Yartsevsk, Zapol'sk, Maksimovsk, Gorsk, etc.

The pre-ore basement of the deposits of the Tikhvinsk group consist of a mottled group of the Famennian Stage and two sedimentary sequences of the Lower Carboniferous.   The Upper Devonian sediments consist of sandy, often micaceous, laminated clays of bluish-grey colour with lenses of sands and mottled sandstones.   In the southern part of the region, the upper portion of the sandy clay sequence contains marl horizons, limestones, and dolomites, which belong to the Tournaisian Stage.

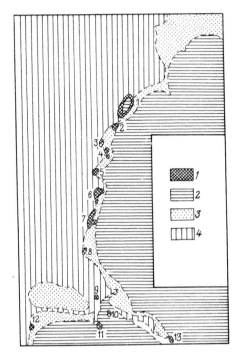

Fig. 114.
Diagrammatic geological map of the Tikhvinsk group of bauxite deposits.

1) bauxites; 2) Viséan limestones; 3) Lower Carboniferous sand-clay deposits; 4) Upper Devonian sand-clay deposits.
Bauxite deposits: 1 - Bat'kovsk; 2 - Segol'sk; 3 - Radynsk; 4 - Podsosnensk; 5 - Gubsk- -Pochaevo; 6 - Novo-Usadinsk; 7 - Krasnor- ucheisk; 8 - Nikomlya rivers; 9 - Pareevsk; 10 - Tabash; 11 - Zapol'sk; 12 - Terebezh; 13 - Rudnogorsk

The bauxite-bearing sediments are confined to the middle portion of the Tula horizon of the Oka substage of the Lower Carboniferous (Fig. 115). They rest with an erosional break on Devonian mottled clays.   The bauxite- -bearing sequence is also overlain by mottled clays with lenses of sands, and in some cases by cobbly clays and other glacial deposits or a soil-vegetation layer.   The rocks of the Tula horizon are overlain by sand-clay rocks with seams of limestone and rare dolomites of the Oka and Serpukhovo substages of the Lower Carboniferous.   Higher up are Quaternary clays and loams with cobbles of solid rock.

The pre-ore relief has been developed on the Devonian deposits and comprises a broad upland, elongated in a meridional direction and dissected by river valleys, and broad and deep gorges.   The length of these valleys is usually 1 - 4 km, rarely 7 km.   Their maximum width in the lower reaches is up to 500 - 1000 m, and the minimum in the upper reaches is 100 m.   The depth is normally 10 - 30, rarely 40 m.   The valley slopes have a gently smoothed surface and gradually merge with the watershed area.   The commercial bauxite deposits are confined to the ravine-beam valleys and depressions in the pre-ore relief.   The size of the deposits and the morphology of the segregations depends on the dimensions and shape of these depressions.

Two palaeogeographical zones have been recognized in the Tikhvinsk region: continental (watershed) and littoral-marine, elongated in a submerid- ional direction.   The bauxite-bearing sediments are confined to the valleys of the continental palaeogeographical zone, and in the valleys, having connexion with the sea, bauxites are absent.

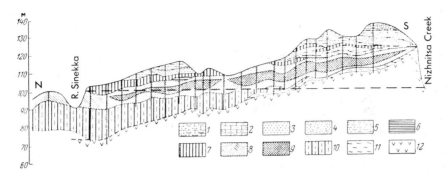

Fig. 115.   Longitudinal geological section of the Bat'kov deposit (After S.
Vishnyakov).

1) organic-clastic limestones;   2) marls;   3) sands and sandstones;
4) argillaceous sands;   5) sandy clays;   6) laminated clays;   7)
grey and carbonaceous clays;   8) siallites;   9) allites and bauxites;
10) carbonaceous, sub-bauxite clays;   11) mottled clays;   12) Devonian
deposits.

The ore segregations have a narrow elongate lens-like form with a
simple weakly concave or uneven top surface and a downwardly convex floor. The
segregations slope to the dip of the surrounding Devonian sequences at 10 - 20 m
per km.   The depth of occurrence of the bauxites is from 1 - 2 to 100 - 150 m,
but is usually not more than 40 m.

In cross-section, the bauxite segregations have a concentric structure.
In the central portion there are good-quality bauxites, which are replaced
towards the periphery by allites, and then siallites.   In this respect, the
amount of alumina and iron oxide gradually increases, and silica decreases from
the periphery towards the centre of the segregations.

The bauxites are characterized by a reddish-brown colour, weak sorting
of the clastic material, the presence of a substantial quantity of plant remains,
and a large amount of secondary calcite.   Amongst the bauxites there are strong,
friable, and argillaceous varieties.   In textural features, we can distinguish
amongst them, clastic, finely dispersed (pelitic), and oolite-pisolite types.

The original material during the formation of the bauxites was a
weathering crust of Devonian clays.   In mineral composition, the bauxites
belong to the gibbsite-boehmite-kaolinite type with varying ratios of the ore-
-forming minerals.   The bauxites, on the basis of individual deposits contain
(in wt %): Al_2O_3, 35.7 - 48.8; SiO_2, 11.0 - 17.9; Fe_2O_3, 10.1 - 19.5; TiO_2,
1.6 - 2.8;   CaO, 0.6 - 10.3;   average value of silica module 3.5 (2.8 - 4.1).

THE SOUTH TIMAN GROUP OF DEPOSITS

The bauxites in South Timan were discovered in 1951 at a depth of
177 m in Viséan deposits (Abramov, 1970).   In this region, the oldest formations are rocks of the Upper Proterozoic (Riphean) metamorphic series, on which
almost horizontal Devonian, Carboniferous, and Permian rocks rest with marked
angular unconfirmity, and these are almost everywhere overlain by Quaternary
sediments (Fig. 116).

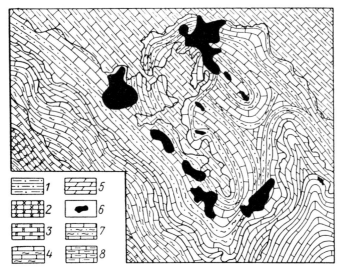

Fig. 116.
Geological-lithological map of the southern part of the South Timan bauxite-bearing region.

1) mottled clays, marls, and sandstones of the Ufimian Stage; 2) Lower Permian limestones and dolomites with flint concretions; 3) Upper Carboniferous limestones with flint concretions; 4) Muscovian limestones and clays; 5) Bashkirian limestones, dolomitized limestones, and clays; 6) bauxite-bearing deposits; 7) Viséan-Namurian dolomites, dolomitized limestones, and at the base clays, siltstones, argillites, bauxites and bauxite rocks; 8) Famennian limestones, argillaceous limestones with rare seams of clays, and siltstones.

The Riphean metamorphic series is known in the western part of the region and consists in the lower part of interstratified limestones and dolomites with seams of clay-chlorite-quartz and quartz-carbonaceous shales, and in the upper part, of a uniform sequence of interstratified grey, greenish-grey, and black chlorite-sericite and feldspar-quartz schists (the Ochparm Group).

Amongst the Devonian deposits there are rocks of the middle and upper divisions. The Eifelian, Givetian, and lower Frasnian deposits consist of clays, sandstones, and siltstones with rare seams of limestones in the upper part. The upper Frasnian deposits consist in the lower part of organogenic limestones with seams of clays, and the upper part is characterized by an alternation of beds of gypsum, anhydrite, de-gypsumized dolomites, limestones, clays, and gypsum-dolomite rocks (Ukhta Group). The lower Famennian deposits consist of limestones, dolomitized limestones, and dolomites with seams of sandstones and clays. The Devonian deposits are 300 - 350 m thick, including the lower Frasnian section which is 130 - 180 m thick.

The Carboniferous deposits consist of Viséan, Namurian, Bashkirian, and Moscovian rocks. The Viséan rocks are bauxite-bearing, and are widely developed, resting on the eroded surface of Famennian, and in the southeastern part of the region on Tournaisian deposits. They consist of terrigenous sediments of the Tula and lower Aleksa horizons, clay-carbonate deposits of the upper Aleksa horizon, and dolomites and dolomitized limestones of the Mikhailovsk, Venevsk, Tarus, and Steshevsk horizons and the Namurian Stage.

The bauxites are associated with the terrigenous deposits of the lower Aleksa sub-horizon of the Tula horizon. They rest on carbonate and carbonate-clay rocks of the Upper Devonian or on a deluvial bed, and are overlain by Viséan carbonate deposits up to 70 m thick.

The bauxite-bearing sediments crop out very rarely on the present surface, and mainly in the river valleys. About 15 outcrops are known, consisting of kaolinite clays and argillites.

The terrigenous sequence of Tula-early Aleksa age has been subdivided into four members (in upward succession): the sub-bauxite, the deluvial bed, the bauxite-bearing, and the carbonaceous-mottled.   The thickness of the terrigenous sequence within the uplifts varies from a few up to 30 metres, and in the basins in the pre-Viséan relief, its thickness reaches 100 m and more as a result of increase in thickness of the sub-bauxite member (up to 75 m).

The deluvial bed is known only within the bauxite segregation in small, but relatively deep-seated (10 - 15 m) erosional hollows in the relief of the Devonian limestones.   It consists of poorly rounded cobbles, pebbles, blocks, and rubble of a finely-stratified argillaceous limestone of Devonian age.   The matrix consists of clay and argillite of varying colour, often with an oolite-pisolite-clastic fabric.

The bauxite-bearing member is arranged in bands of varying width along the slopes of the uplifts and rests on Devonian limestones or the deluvial bed, and in the basins, on the deposits of the sub-bauxite member.   It consists of kaolinite clays, allites, and various lithological varieties of bauxites. The bauxite segregations are located in the middle portion of the member.   Upwards and also on the periphery, the bauxite horizon passes into allites, and then into a kaolinitic argillite.   The member varies in thickness from 0.8 to 12 m, and is usually 4 - 6 m (Fig. 117).   The bauxite segregations have an elongate form with wavy outlines.   They occur at depths of 10 - 180, and most commonly 40 - 100 m.   The average thickness of the exposed rocks varies from 57 to 77 m.

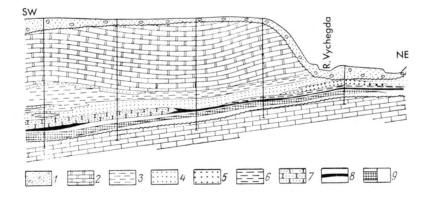

Fig. 117.   Geological section of the bauxite segregation in the South Timan region.   (After P. Abramov).

1) loams, sandy loams, sands, and gravel-pebble-sand mixtures of the Quaternary system;   2) dolomites;   3) mottled silty clays; 4) mottled argillaceous siltstones;   5) dark-grey carbonaceous siltstones;   6) silty dark-grey carbonaceous clays;   7) sandstones; 8) black carbonaceous argillites;   9) bauxite-bearing member: a) bauxites, b) allites;   10) Famennian argillaceous limestones.

The carbonaceous-mottled member is distributed more widely than the bauxite-bearing member.   In its lower portion there is a layer of carbonaceous argillites, and above come silty or argillite-like platy carbonaceous clays,

gradually passing upwards into dark-grey and grey siltstones.   Above the carbonaceous rocks, and in the north of the region above the bauxite-bearing member,
there are mottled silts, argillaceous siltstones and silts, and silty clays. The
total thickness of the member is 8 - 28 m, most commonly 15 - 20 m.

The bauxites of the South Timan region belong to the high-alumina,
high-silica, low-iron, and often high-sulphur type.   They contain (in wt %):
Al_2O_3, 40 - 70; SiO_2, 12 - 28; Fe_2O_3, 3.6 - 12.72;   S, 0.38 - 3.0, the sulphur
being mainly pyritic.   The silica module based on samples varies from 2 to 28,
and on average through the segregations, comprises 2.4 - 3.2.

In mineral composition, the bauxites belong to the kaolinite-gibbsite-
-boehmite and kaolinite-boehmite types.   The former is developed in the northwest, and the latter, in the southeastern part of the region.

THE CHADOBETS GROUP OF DEPOSITS

In the southwestern portion of the ancient Siberian Platform and the
Yenisei Ridge, there are large numbers of predominantly small deposits, mainly
of the karst type, which are of Cretaceous-Palaeogene age.   These deposits are
confined to areas of development of carbonate rocks of the pre-Cretaceous basement and occur in the karst funnels of the erosional-karst and suffusion-karst
depressions.

Recently, pebbles and fragments of high-quality lateritic bauxites
were discovered on the interfluve of the Vel'mo and Podkamennaya Tunguska on
residuals of the high ancient plateau, in the Chadobets region and the Altai-
-Sayan district.   These discoveries indicate that in addition to the karst
deposits on the Siberian Platform, lateritic deposits may be found, the bauxites of which, as a rule, are of high quality.

Within the Siberian Platform and the Yenisei Ridge, there are the
Chadobets, Tatar, and Near-Angara groups of bauxite deposits.   In addition,
there are a large number of small deposits and bauxite-shows, located outside
these groups (Krivtsov, 1968-1969).

The Chadobets group includes the Central, Punya, Ibdzhibdek, and also
the Verkhne-Tera, Chuktukon, Polpodsk, Medvedkovsk, Tsembas, Nokum, and other
bauxite shows.   The deposits of this group are confined to areas of development
of limestones of the Chadobets dome-like uplift, which represents a brachyanticlinal fold, elongated in a meridional direction over a distance of 45 km, with
a width of 30 km.   The core of the fold consists of Upper Proterozoic rocks,
and the limbs, of Cambrian formations.   The Upper Proterozoic rocks crop out
on the surface in two places, forming two uplifts, the northern and southern.
These cores were surrounded initially by an inner ring of terrigenous rocks of
the Taseev Series of Late Proterozoic - Early Cambrian age.   Then, along the
periphery of the uplift, there is an outer ring, consisting of carbonate rocks
of the Lena Stage of the Lower Cambrian and terrigenous rocks of the upper Lena
group of the Upper Cambrian (Fig. 118).

The bauxite-bearing deposits are best developed in the central portion
of the brachyanticline; in the areas of the outer ring, they are much less developed.   They occur in vast erosional-karst depressions and karst funnels in
the weathering crust of Upper Proterozoic and lower Palaeozoic carbonate sequences.

Fig. 118. Geological-lithological map of the Chadobets group of bauxite
deposits. (After E. Pel'tek).

1) Quaternary alluvial deposits; 2) Upper Tertiary (?) sand-clay
deposits; 3) Upper Cretaceous bauxite-bearing deposits; 4)
bauxites; 5) weathering crusts; 6-13) basement rocks (Proterozoic
and Palaeozoic): 6) sandstones, 7) silt-clay, phyllitized slates
and silts, shales, and argillites, 8) limestones, 9) dolomites,
10) tuffs and tuff-breccias, 11) traps, 12) vein formations of
ultramafic rocks (picritic porphyrites and micaceous pyroxenites),
13) eclogitic and kimberlitic rocks.

Bauxite deposits (Roman numerals): I) Central, II) Punya, III)
Ibdzhibdek, IV) Chuktukon.

The bauxite-bearing deposits have been subdivided into three horizons: 1) sub-ore, consisting of yellow clays, up to 50 m thick; 2) middle, the ore proper, formed of various types of bauxites, bauxitic rocks, and allites; 3) upper, the supra-ore, consisting of grey clays.   In plan and section, the bauxites usually pass into allites.   In the middle portion of the sequence in some cases, two and more bauxite horizons have been observed.   The sub-ore horizon is distributed considerably more widely than the ore horizon.

On compositional features, stony bauxites, usually forming the central portions of the large segregations, friable, which surround the stony variety, and argillaceous types, confined mainly to the flanks, and also to the soil and top of the ore bauxites, have been recognized.

The Central Deposit

The Central deposit is confined to the central portion of the Chadobets uplift and lies on the southwestern slopes of a vast erosional-tectonic basin, elongated meridionally.   The datum levels of the basinal margins are 220 - 300 m, and the minimum datum level of the floor of the basin in the northern part is 130 m, and in the southern part, about 30 m.   The basin has been filled with bauxite-bearing sediments, which rest on a weathering crust of rocks of the ancient basement and they have been subdivided into three horizons: the sub--ore, ore, and supra-ore.

The sub-ore horizon consists of deluvial formations, represented by ocherous-yellow sandy clays with nests and lenses of kaolinitic clays and inclusions of clasts of a brown ironstone and solid rocks; the horizon is 30 - 40 m thick.

The ore horizon rests with a break on the uneven surface of the sub--ore sedimentary horizon and has been divided into two subhorizons.   The lower ore subhorizon consists of silty clays, and high-alumina brown ironstones, with lenses of ferruginous bauxites.   The upper ore subhorizon consists of friable, stony, and argillaceous litholigical varieties of bauxites, and also bauxitic rocks with lenses of kaolinitic and lignitic clays.   Between the lower and upper ore subhorizons there is an inter-ore member of kaolinitic clays, 30 - 40 m thick.

The supra-ore horizon is distributed with a break on the bauxite-bearing horizon and consists of brown silty clays with clasts of cherty rocks, brown ironstone, and stony bauxite.

In the Central deposit, the ferruginous products of redeposition of the lateritic cap have been distributed in the lower portion of the sedimentary bauxite-bearing sequence.   Selected bauxites of high quality are located in the upper parts of the sequence, forming lens-like segregations, separated from the ferruginous horizon by seams of bauxitic kaolinite clays, and also coals.   Such structure of the bauxite-bearing sequence is seemingly a mirror reflexion of the sequence of the laterite weathering profile.

In the Central deposit, several ore segregations have been discovered, located at two topographical levels.   Most of them are associated with the upper ore subhorizon, and the others, with the lower subhorizons.   Each segregation consists of a series of ore bodies.   The large ore bodies are characterized by a complex layer-like form with numerous swells and pinches, and small lensoid and pocket-like form (Fig. 119).

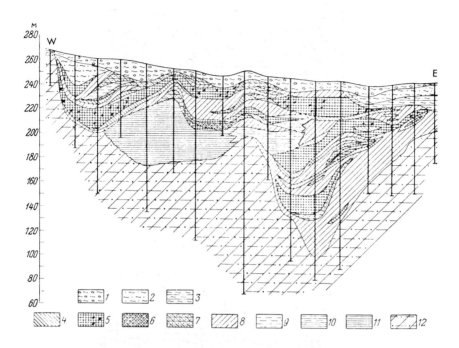

Fig. 119. Geological section through the Second ore body of the Central
deposit (After A. Krivtsov).

1-2) Quaternary deposits: 1) deluvial clays with clasts of solid
rock,   2) lacustrine-paludal clays, often muddy, frequently sandy,
sometimes with peat; 3-12) Cretaceous bauxite-bearing deposits:
3) reddish-brown, compact, viscous clays, 4) allites, frequently
with pisolitic texture, 5) friable bauxites, 6) stony, intensely
ferruginous bauxites, 7) aluminium-iron rocks, 8) light-grey or
white kaolinite clays, 9) mottled clays, 10) dark-grey, lignitic
or carbonaceous clays, sometimes with remains of carbonized wood,
11) yellow laminar clays, 12) Mesozoic-Cainozoic weathering crust
of sand-clay formations.

The bauxite ores consist of friable (47%), stony (33%), and argillaceous (22%) varieties. In mineral composition, they belong to the gibbsite-
-hematite and kaolinite-gibbsite types.

The bauxites contain (in wt %): Al_2O_3, 24 - 36; SiO_2, 5 - 16;
Fe_2O_3, 15 - 40; TiO_2, 5 - 18; silica module 5.3. The bauxite ores of the
deposit have a pisolitic texture (Fig. 120, 121), and are characterized by an
increased amount of titanium minerals and a comparatively low content of alumina.

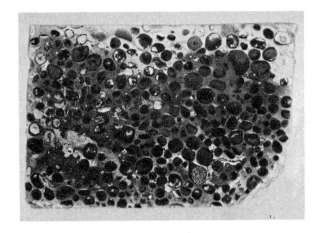

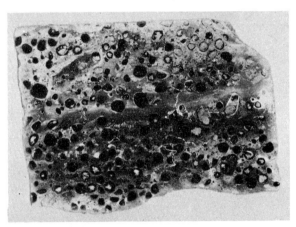

Fig. 120.   Stony bauxite.   Central deposit.   Drill-hole 795, depth 19 m.
Reduction X 1.5.

Fig. 121.   Stony bauxite, bleached.   Central deposit.   Drill-hole 808, depth
55 m.   Reduction X 1.5.

The Ibdzhibdek Deposit

The Ibdzhibdek deposit rests on the pre-ore basement, which consists
of limestones and dolomites of the Lena Stage of the Lower Cambrian.   Red argillites and siltstones of the Évenka Group of the Upper Cambrian are developed
in the southern part of the deposit.

Bauxite-bearing sediments fill karst funnels, located in the area of
development of carbonate rocks of the basement.   The thickness of the bauxite-
-bearing sediments is normally 25 - 35 m, sometimes 70 - 96 m.   The bauxites

belong mainly to the argillaceous variety (about 50%), and also to the stony and friable varieties.   The quality of the bauxites is low, with an alumina content of 37.7 wt %, and silica, 11.0 wt %.

The Punya Deposit

The pre-ore basement of this deposit consists of carbonate rocks of the Lena Stage of the Lower Cambrian, and in the northwestern part, a layered dolerite intrusion.   In the areas of carbonate rocks, karst fields have been developed and individual funnels with depths from 10 - 15 up to 80 - 100 m, have been filled with Cretaceous-Palaeogene bauxite-bearing sediments.

Twenty-one ore bodies have been found in the deposit, which have a nest-like, pocket-like, and lensoid shape.   The thickness of the bauxites in the karst funnels is 50 - 60 m.   The bauxites consist of friable (47%) and stony (46%) varieties.   The principal ore-forming minerals are gibbsite (40 - - 50%) and hydrohematite (24 - 37%).

THE TATAR GROUP OF DEPOSITS

The Tatar group includes the Tatar, Murozhna (Ivanovsk), Indygla, and Sulaksha deposits, consisting of ore sectors.   The bauxite deposits of this group are located within the Central anticlinorium of the Yenisei Ridge, which consists of a complex of strongly deformed, metamorphosed Proterozoic sedimentary rocks, cut by intrusions of acid and basic composition.   The ancient basement is overlain by bauxite-bearing mottled argillaceous deposits of the Murozhna Group of Palaeocene-Eocene age, and Neogene and Quaternary formations.

The bauxite-bearing sediments occur in the karst funnels and erosional-karst depressions, which are located in areas of distribution of marmorized carbonate rocks of the Pechenga Group close to amphibolites.   Individual deposits represent groups of segregations, located in the erosional-karst depressions and karst funnels (Fig. 122).

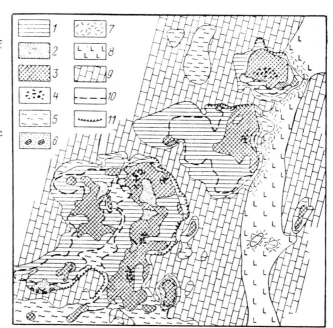

Fig. 122. Diagrammatic geological map of the central portion of the Tatar bauxite deposit (After E. Pel'tek).

1-4) productive subgroup (Palaeogene): 1) mottled, kaolinitic clays, 2) bauxitic clays, 3) friable bauxites, 4) stony bauxites; 5-7) sub--ore subgroup: 5) breccia--like clays and silt-clay deposits with clasts of basement rocks and bauxites, 6) calcareous quartzose sand, 7) weathering crust of amphibolites; 8-9) Proterozoic deposits: 8) amphibolites, 9) limestones; 10) contours of buried ore bodies; 11) benching in karst relief.

The Tatar Deposit

The Tatar deposit consists of a series of bauxite-bearing sectors. The largest of them are the Tatar, Sredne-Tatar, Sokhata, Murla, Berezovo, Podgolechnyi, and others, and in addition, small bauxite-shows are known.

The deposit belongs to the karst type, consisting of complex ore bodies of small dimensions, having a complex nest- and lens-like shape, which are confined to ancient karst funnels, located in the zones of contact between the amphibolites and amphibolite schists, and the carbonate rocks of the Pechenga Group of the Upper Proterozoic. The depth of the karst funnels reaches 230 m. In the Tatar sector there are 14 ore bodies, in the Sredne-Tatar sector, 11, and in the Sokhata sector, 6.

Amongst the bauxite-bearing sediments in the various sectors and segregations, there are from one to three productive bauxite-bearing horizons, consisting of bottled and bauxitic clays, and friable and argillaceous bauxites with an abundance of clasts and nodules of stony bauxite. The bauxite-bearing horizons are separated mutually by mottled clays or are superposed on one another. At the base of each of these horizons there is an increased amount of sandy or rubbly material. Carbonaceous clays and brown coals are distributed in the roof of the bauxites.

The bauxites consist mainly of friable, and also argillaceous varieties. The argillaceous bauxites and bauxitic clays form the lower portions of the ore segregations.

The average amount of the principal components in the bauxites (in wt %) is as follows: SiO_2, 6.5 - 9.2; Al_2O_3, 38.5 - 37.4; Fe_2O_3, 30.5 - 31.7; TiO_2, 4.2 - 4.7. The principal minerals are gibbsite, hematite, maghematite, goethite, magnetite, kaolinite, halloysite, hydromicas, ilmenite, rutile, quartz, etc.

The Indygla Deposit

The Indygla deposit is located to the east of the Tatar deposit and is confined to the basin of the same name. Within it are the Dolgozhdannyi, Eastern, Mitrofanovsk, Talovsk, and other ore sectors. The ancient basement of the basin consists of phyllitized argillaceous and sand-clay slates of the Uderei Group of the Lower Silurian, which have been cut by layered amphibolite intrusions. The ancient basement is overlain by bauxite-bearing sediments of Palaeogene age, consisting of carbonaceous and kaolinitic clays with lens-like segregations of bauxites, allites, and bauxitic rocks, which fill the Indygla basin. The bauxites occur at depths of from 3 to 35 m, and the thickness of their bodies varies from 1 to 8 m. They are distinguished by the low quality. They contain (in wt %); SiO_2, 10; Al_2O_3, 32; FeO, 38; TiO_2, 4.

The Murozhna (Ivanovsk) Deposit

The Murozhna (Ivanovsk) deposit includes the Murozhna, Chikal', Northern, Afinogenovo, and Malomurozhna bauxite-bearing sectors and a number of bauxite shows. It is located in the Malomurozhna basin, which is well defined morphologically in the present-day relief. The ancient basement of the basin consists mainly of crystalline limestones with seams of quartz-mica schists, and

in the peripheral parts of the basin, of chlorite-actinolite schists, amphibolites and phyllites.   The bauxite-bearing sediments consist of the Murozhna
Group, which is separated into two horizons: the lower, sub-ore, and the
upper, ore horizons.

The sub-ore horizon has been formed by sandy and silty mottled and
ocherous clays, containing a gravel of cherty rocks and clasts of a pisolitic
stony bauxite.

The ore horizon consists of mottled kaolinitic clays, allites, gibbsite-kaolinite clays with lensoid and nest-like bauxite segregations, and also
carbonaceous clays with intercalations of coals in the middle portion of the
sequence.   In the sequence there are two layers of bauxitic ores, separated
by clays, and containing in places carbonaceous material and in the marginal
parts overcrowded with deluvium.

In the Murozhna deposit, eight segregations of bauxites have been
discovered, which have a lens-like, and less frequently pocket-like form;
their thickness in the central portions is 20 - 50 m, and in the peripheral
parts, 1 - 15 m.   The bauxite ores consist mainly of friable and argillaceous
varieties, with blocks and nodules of a stony bauxite.   They contain (in wt
%): Al_2O_3, 35.3; SiO_2, 8.1; TiO_2, 2.8; and Fe_2O_3, 34.4.

The Sulaksha Deposit

The Sulaksha deposit is located to the north of the Tatar deposit,
and in the immediate vicinity of it are the Gurakhta, Malo-Kadra, Il'insk,
Balanda, and other bauxite-shows.

The ancient basement of the Sulaksha deposit consists of limestones
of Proterozoic age and amphibolite intrusions.   In the contact zone between
the carbonate rocks and the amphibolites, and also in the limestones, at some
distance from the aluminosilicate rocks, there are karst funnels, filled with
bauxite-bearing sediments.   In the lower part, they consist of coarsely-clastic
deluvial material, and in the upper part, mottled, finely-dispersed kaolinite
clays with lenses of brown ironstones, bauxites, and bauxite-bearing rocks.

Ten ore segregations of complex construction and from 1 to 35 m thick,
have been discovered in the deposit.   Three of them, having the largest dimensions, are confined to a single karst band, with a maximum depth of 142 m.   The
bauxites of the Sulaksha deposit are characterized by a large content of iron
oxides (up to 45 - 50 wt %), silica (8 - 25%), titanium oxide (4 - 12%) and a
relatively low quantity of alumina (24 - 48%).

THE NEAR-ANGARA GROUP OF DEPOSITS

The bauxite deposits of the Near-Angara group are confined to the
Angara-Pit synclinorium and the Irkineevsk rise in the Yenisei Ridge, and
consist of intensely deformed Proterozoic formations.   In the latter, in
karst funnels and erosional-karst depressions, elongated linearly along the
structures of the ancient basement, are concentrated mottled bauxite-bearing
deposits of Cretaceous-Palaeogene age.   The dimensions of the karst funnels
are from 25 X 50 up to 150 X 600 m, and the depressions from 50 X 200 up to

1200 X 12,000 m.   The karst funnels are mainly confined to areas of development of carbonate rocks, and the erosional-karst depressions are located in the contact zones between aluminosilicate and carbonate rocks.   The extent of the contact zone between the carbonate rocks of the Dzhursk sequence and the terrigenous rocks of the Krasnogorsk sequence of the Potosku Group is about 350 km. However, the bauxites in this extended zone are known only in individual sectors. The bauxite segregations in plan have the shape of complex elongate lenses, in which the length is normally several times that of the width, with a marked variation in thickness along the longitudinal axis and across the elongation.   Ubiquitously, with the elongate segregations in all deposits there are individual isolated segregations, equant in plan.

The thickness of the bauxite-bearing sediments that fill the depressions and funnels, reaches 120 m.   In most cases they have been divided into two sequences: a lower, sub-ore sequence up to 50 m thick, and an upper, bauxite-ore sequence with a thickness of up to 100 m.   The sub-ore sequence consists of hydromicaceous and kaolinite-hydromicaceous clays with clasts of weathered rock.   The upper bauxite-bearing sequence consists of kaolinite and kaolinite-hydromicaceous clays and various litholigical varieties of bauxites.

Within the Near-Angara group, several bauxite deposits are known, such as the Kirgitei, Verkhoturovsk, Porozha, and Yenda, and also individual segregated ore-shows, which occur between the Irkineevsk rise and the Chadobets uplift.   Each deposit usually includes several ore sectors.

The Kirgitei Deposit

The Kirgitei deposit is confined to the western limb of the Angara-Pit synclinorium and includes the Kirgitei and Verkhne-Kirgitei bauxite sectors, and also the Nizhne-Kirgitei, Sredne-Kirgitei, Illerkon, Rybinsk, Gremyacha, Kopchenga, and other bauxite-shows.   The ore sectors and bauxite-shows are located in the depression which extends in a meridional direction for more than 30 km.   Over 17 ore segregations have been counted in it.

The ancient basement of the deposit consists of sand-clay phyllitized schists and carbonate rocks of the Uderei Group of the Lower Sinian, adjacent along a tectonic contact.   Along the contact, there are karst funnels and erosional-karst depressions, filled with mottled sand-clay sediments of Cretaceous and Palaeogene age, containing several bauxite-bearing horizons, separated by clays.

In the Kirgitei sector there are 16 ore bodies.   The bauxites include stony, friable, and argillaceous varieties, with a predominance of the last. The standard bauxites contain (in wt %): Al_2O_3, 34 - 48; SiO_2, 2 - 16; Fe_2O_3, 15 - 28; TiO_2, up to 3.5.

The Verkhoturovsk Deposit

This deposit is located on the southern limb of the Angara-Pit synclinorium.   Eighteen ore bodies have been recognized in the deposit, and they have a lens-like form.   The bauxites are restricted to the upper portion of the bauxite-bearing deposits of Cretaceous-Palaeogene age which fill the depressions in the zone of contact between the terrigenous and carbonate rocks.

The ore bodies of the bauxites are located in a single bauxite-ore zone, about 12 km long.   The bauxites contain (in wt %): Al_2O_3, 44.5; SiO_2, 14.8; silica module, 2.4 - 4.2.   The bauxites have pisolitic and layered textures (Fig. 123).

Fig. 123. Stony bauxite.   Southern sector, Verkhoturovsk deposit.   Reduction X 15.

The Porozhna Deposit

The Porozhna deposit of bauxites is confined to the Irkineevsk rise. Within it are the Porozhna and Ul'da bauxite sectors, and also the Artyuga, Lunchsk, Nizhne-Tera, Akinsk, and other bauxite shows.

The Porozhna sector includes 19 ore bodies, occurring among mottled, brick-red kaolinitic and hydrargillite-kaolinitic clays.   Their shape is complex: pocket-like in the karst funnels, and lens-like in the erosional--karst depressions.   The ore bodies extend for 130 - 600 m.   The ore bodies with the highest-quality bauxites are confined immediately to the contact zones between the shales of the Krasnogorsk sequence and the carbonate rocks (Fig. 124).

The Yenda Deposit

The Yenda deposit is located on the eastern limb of the Angara-Pit synclinorium, in the zone where the latter merges with the Siberian Platform. It includes the Yenda, Verkhne-Yenda, and Sukhoi bauxite sectors, and also the Velinda, Yenda, Yel'chimsk, Katalanga, and other ore-shows.

The ore bodies of the Sukhoi sector are located in the contact zone of the Krasnogorsk shaly and the Dzhursk carbonate sequence, filling a large valley-like karst-basinal depression, elongated along the contact of the above sequences for a distance of 20 km.

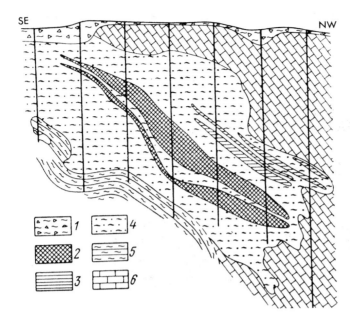

Fig. 124.
Geological section across the strike of the First ore body of the Porozhna deposit (After I. Kusov)

1) deluvial deposits; 2) bauxites; 3) bauxitic clays; 4) clays; 5) shales and argillites; 6) limestones and dolomites

The sub-ore sequence in the deposit is distributed on the karsted wavy surface of the carbonate and in part terrigenous rocks of the Potosku Group and consists of ferruginous hydromica-kaolin and sandy clays, 25 - 80 m thick.   The ore horizon consists of kaolinitic mottled clays, allites, and bauxites.

Twenty bauxite ore bodies of lens- and pocket-like form have been recognized in the deposit.   They contain (in wt %): Al_2O_3, 40.5; SiO_2, 13.8; Fe_2O_3, 24.0; TiO_2, 2.78;   silica module, 2.9.

SEDIMENTARY (REDEPOSITED) DEPOSITS OF THE YOUNG PLATFORMS

Bauxite deposits on the young (epi-Hercynian) platforms are confined to erosional-karst and suffosional-karst depressions and karst funnels, located in areas of development of carbonate rocks, and also to the valley-like erosional depressions, developed on aluminosilicate rocks.   The deposits have been concentrated, as a rule, on a folded basement, and at the base of the platform cover of horizontal unconsolidated sediments of Mesozoic-Cainozoic age (Kirpal' & Khatskevich, 1971).

The largest deposits of the young platforms are known within the Turgai downwarp;   small deposits have been found on the eastern slopes of the Middle and Northern Urals, in the Mugodzhary, and the Tselinograd, Ékibastuz-Pavlodar, and the Near-Chimkent bauxite-bearing regions.

In the Turgai downwarp, the bauxite deposits are grouped into three major bauxite-bearing regions: West Turgai, Central Turgai, and Amangel'da (Kirpal', 1966).   In all the regions of Turgai, a spatial and genetic association has been observed with the composition of the underlying rocks of the Palaeozoic (pre-Cretaceous) basement.   It is expressed in the fact that the bauxite deposits are confined to the karst belts of development of limestones or to the zones of contact between the carbonate and aluminosilicate rocks.   In this respect, they form bauxite-bearing zones, extending along the structures of the

Palaeozoic folded basement.   In the Amangel'da region, there are five bauxite-
-bearing zones: the Arkalyk, Nizhne-Ashut, Verkhne-Ashut, Ushtoba, and Aktas.
The first of them has a ring-like form, and the other four extend in a submeridional direction.

         In the West Turgai region, three bauxite-bearing zones (the Central,
Livanovsk, and Taunsor) extend in a submeridional direction.   In the areas
between the bauxite-bearing zones there are, in places, karst funnels, filled
with bauxite.

         The L i v a n o v s k   bauxite-bearing zone in the western portion
of the West Turgai region extends meridionally.   Within it are the Slavyansk,
Varvara, Batola, Ak-Aul'sk, Pokrovsk, North Livanovsk, South Livanovsk, Sakharovo,
and other deposits and bauxite-shows.   This zone is characterized by the fact
that the overwhelming majority of the deposits and bauxite-shows are confined
to the Livanovsk deep-seated fault, and belong to the karst type.   They are
located on Silurian, Middle and Upper Devonian, and Lower Carboniferous carbonate rocks.   The individual deposits consist of numerous small isolated, typical
karst segregations, located at datum levels of +180 and +240 m.

         The C e n t r a l bauxite-bearing zone occupies the central position
in the region.   It extends along the structures of the Palaeozoic folded basement.   Within it are the Belinsk, Ayat, Timir, Karabaital, Klochkovo, and other
bauxite deposits.   A characteristic feature of this zone is the fact that its
deposits are restricted to comparatively narrow belts of carbonate rocks, occurring amongst vast areas of volcanogenic-sedimentary rocks of intermediate composition of Viséan-Namurian age.   The deposits of this zone belong to the karst-
-basinal and karst types.   The bauxites occur at datum levels of from +150 to
+210 m.

         The T a u n s o r bauxite-bearing zone is located in the eastern
and southeastern part of the West Turgai region.   It extends in a submeridional
direction.   The Taunsor, Ozernoe, Kyzylkol', Batmankol', and North Urkach deposits are confined to the southern portion of this zone, and in the northern
portion there are the Mamyrkul', Kuzhukul', and other deposits and bauxite-shows.
The deposits of this zone, as a rule, consist of small segregations of the karst
type, in some cases of the contact-karst and layer-like type, restricted to the
areas of development of carbonate rocks of Tournaisian - early Viséan age.   The
bauxite-bearing sediments occur at datum levels of from +80 up to +190 m.

         The most characteristic deposits of the young (epi-Hercynian) plat-
forms are the Krasnooktyabr'sk, Belinsk, and Naurzum, and the Amangel'da group,
differing in features of geological construction, conditions of location of the
ore segregations, the morphology and dimensions of the ore bodies, the age of
the bauxite-bearing sediments, and other characters.

The Krasnooktyabr'sk Deposit

         Deformed rocks of the Palaeozoic basement and horizontally overlying
sediments of the Mesozoic and Cainozoic participate in the geological structure
of the deposit.   The rocks of the basement and the bauxite-bearing sediments do
not crop out on the present surface and they are overlain by sand-clay sediments
of Cainozoic age, from 27 to 60, and on average 40 m thick.

         The Palaeozoic basement consists mainly of middle Viséan and Namurian
sediments.   They consist of a sequence of carbonate and volcanogenic rocks

(andesitic porphyrites of intermediate composition and their tuffs, tuff-breccias, and tuff-sandstones).   The Palaeozoic sediments have been cut by small diorite and plagiogranite-porphyry intrusions.   The levels of the top of the Palaeozoic basement in the deposit vary from + 140 to + 200 m.

The surface of the basement has been complicated by fractures of sub-latitudinal strike, and also by karst and erosional-karst depressions and karst funnels, filled with bauxite-bearing sediments and bauxites (Fig. 125).

Fig. 125.
Geological-lithological map of the Palaeozoic basement and bauxite-bearing sediments of Krasnooktyabr'sk deposit (After D. Venkov & G. Kirpal')

1) bauxites and bauxitic rocks; 2) bauxite-bearing sediments: bauxitic, kaolinitic, and mottled clays; 3) weathering crust of eruptive-pyroclastic rocks; 4) weathering crust of intrusive rocks; 5-8) Viséan--Namurian formations: 5) limestones, 6) inter-bedded limestones and other sedimentary rocks with tuffs and porphyrites, 7) porphyrites and their tuffs, 8) diorites, dioritic porphyrites, and plagiogranites of Early Carboniferous age; 9) tectonic disturbances.

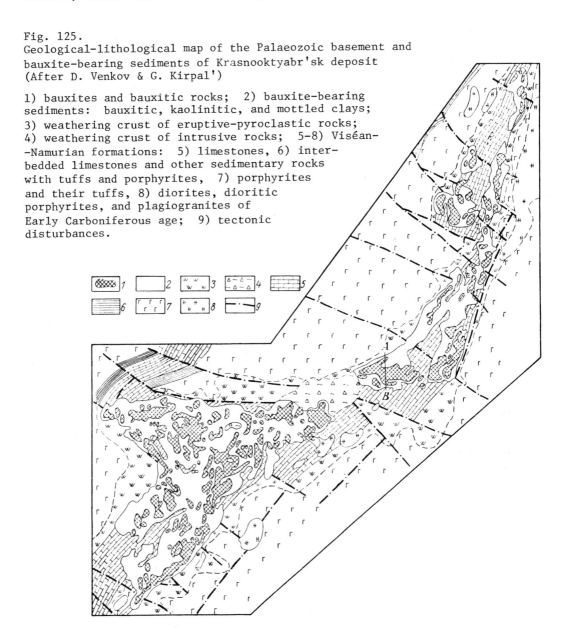

The Mesozoic-Cainozoic deposits in the lower part consist of formations of a weathering crust, and bauxite-bearing sediments, which are overlain by laminated marine clays of the Chegan Group, and also by continental sand-clay sediments of the Kutan-Bulak, Chilikta, Chargai, Naurzum, and Aral' groups, and Quaternary loams.

The bauxite-bearing sediments occur in erosional-karst and suffosion-karst enclosed valleys, karst basins and karst funnels, located in the area of development of carbonate rocks, extending northeastwards. Twenty-four ore segregations or sectors have been discovered in the deposit, and these are separated from each other by highs of carbonate rocks or mottled clays of the sub-ore sequence.

On their lithological features and age, the bauxite-bearing sediments of the deposit have been subdivided into two horizons, the lower, sub-ore, and the upper, ore horizons.

The sub-ore horizon consists of mottled clays, often with a clastic texture, reddish-brown, brown, yellow, and sometimes grey in colour. The clays have a hydromica-kaolinite composition. In the upper part of the horizon, there are sometimes small lenses of better-quality ores. The sub-ore horizon is Cenomanian in age.

The ore horizon in the mottled clays of the sub-ore horizon is associated with a break and consists mainly of stony, friable, saccharoidal, and argillaceous varieties of bauxites. Its thickness varies from a few metres up to 120 m. Kaolinite-lignite clays up to 20 m thick occur in the upper portion of the ore horizon in the largest segregations. They are kaolinitic in composition with seams and lenses of grey, dark-grey, and black lignitic clays. The sediments of the ore horizon, on the basis of spore-pollen data, belong to the Turonian-Santonian interval (Fig. 126).

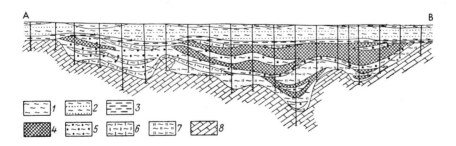

Fig. 126. Section through segregation III of the Krasnooktyabr'sk bauxite deposit (After G. Kirpal')

1) loams and sandy loams; 2) vari-grained, quartzose sands, and varicoloured clays (Chilikta Group); 3) laminated, dark-green clays (Chegan Group); 4-7) Cenomanian-Turonian bauxite-bearing sediments: 4) stony and friable bauxites, 5) argillaceous allites (bauxitic clay), 6) white kaolinitic clays, 7) mottled clays; 8) Viséan-Namurian limestones.

The Krasnooktyabr'sk deposit consists of two ore fields: the Northern and the Southern. Fifteen bauxite segregations are located in the Northern field, and 9 in the Southern. All the bauxite segregations are confined to erosional-karst, karst-basinal, and karst depressions in a basement with extremely complicated relief. The largest of them were formed in the zone of contact between limestones and andesitic porphyrites. On the basis of the form and structure of the pre-ore relief, the deposit belongs to the karst-basinal type, and some segregations in the deposit are typically karst in type. The morphology of the ore bodies is directly related to the pre-ore relief. The ore segregations are a combination of linear-elongated or equant shapes of the ore bodies, splitting on the flanks both transversely and longitudinally. They have a complicated structure, dependent on the interlayering and facies transitions of the better-quality bauxites, allites, and clays. The ore bodies and bauxite segregations recognized on the basis of sampling data and chemical analyses possess fantastic contours in plan and section, and also variable chemical and lithological composition of the ores.

Amongst the lithological varieties in the deposit, we may recognize stony (35%), friable (57%), and argillaceous (8%) bauxites. The principal ore-forming minerals are: gibbsite (hydrargillite), hydrohematite, hematite, and kaolinite. The most widely distributed minor minerals are siderite and calcite, and less frequently, chlorite (chamosite), pyrite, and marcasite. The bauxites of the upper portions of the ore bodies of the deposit have been sideritized, and the amount of siderite varies from 1.05 up to 10 - 12%, and sometimes reaches 21 - 25%. This significantly lowers the quality of the bauxite, since the carbon dioxide, present in the siderite, is a harmful additive. Its amount in the bauxites varies from 1.48 up to 3.8%, and is on average 2.65%.

The chemical composition of the bauxites is characterized by significant variance (in wt %): Al_2O_3, 41.78 - 49.86; SiO_2, 5.23 - 14.00; and Fe_2O_3, 7.21 - 24.22.

The following types or kinds of ores have been recognized according to the technological features and uses of the bauxite ores in industry (the production of alumina and heavy metallurgy): 1) hydrochemical (Baierovsk), 2) clinker, and 3) metallurgical.

The Belinsk Deposit

The Belinsk deposit is located in the northern part of the West Turgai bauxite-bearing region. The Palaeozoic folded basement of the deposit consists of limestones, andesitic porphyrites of intermediate composition, and their tuffs, tuff-siltstones, tuff-sandstones, tuff-conglomerates, argillites, and siltstones. The basement rocks have been cut by numerous intrusions of granodiorites, diorites, diorite porphyries, and also dykes of albitophyres and diabase porphyrites. They have been crumpled into folds with gentle limbs and have a submeridional strike.

The limestones, in the area of which the bauxite segregations occur, occupy the central part of the deposit (Fig. 127). The limestones have been broken up into blocks of varying size by submeridional and sublatitudinal tectonic disturbances. As a result, the limestone belts in the southern part of the deposit alternate with belts of eruptive-sedimentary rocks.

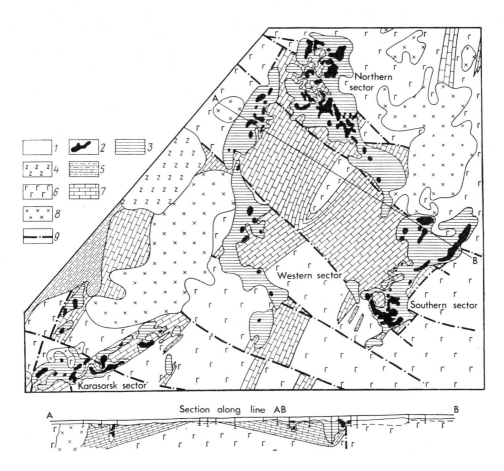

Fig. 127. Geological-lithological map of the Palaeozoic basement and the
bauxite-bearing sediments of the Belinsk deposit. Compiled by
G. Kirpal' from data by the Temir Geological Research Expedition.

1) clay-sand covering sediments of Palaeogene and Quaternary age;
2) bauxites and bauxitic rocks; 3) bauxite-bearing sediments, and
mottled bauxitic and proluvial-deluvial clays of Cenomanian-Turonian
age; 4) Triassic basalts and basaltic porphyrites; 5) Middle and
Upper Carboniferous redbed polymict conglomerates, sandstones,
siltstones, and argillites; 6-7) middle and upper Viséan rocks:
6) andesitic and andesite-basaltic porphyrites, tuffs, tuffites,
tuff-breccias, and tuff-sandstones, 7) limestones; 8) Lower
Carboniferous diorites, quartz diorites, and dioritic porphyrites;
9) assumed tectonic disturbances.

A weathering crust, consisting of kaolinite and hydromica-montmorillonite clays is everywhere developed on the rocks of the Palaeozoic basement.
The maximum thickness of the weathering crust (up to 60 m) has been recorded
in the zones of tectonic faulting and in the contacts between rocks of varying
composition. The depressional basins, filled with bauxite-bearing sediments,

were formed in the carbonate rocks in the zone of stratigraphical and tectonic contact between the latter and the pyroclastic rocks of basic composition.

The bauxite-bearing sediments are overlain by laminated marine clays of the Chegan Group of Late Eocene age, and where the bauxite-bearing rocks are not present, the clays rest directly on the folded Palaeozoic basement; the same applies to the continental sandy and argillaceous sediments of Oligocene age and Quaternary loams. The total thickness of the sediments overlying the bauxites is from 6 to 55 m.

The bauxite-bearing sediments consist of an interlayering of mottled kaolinitic and bauxitic clays, allites, carbonaceous clays, and stony, friable, and argillaceous bauxites. Argillaceous sediments predominate in the lower part. The bauxite bodies are confined to the upper part of the sequence. The total thickness of the bauxite-bearing sediments is 170 m. The age of the bauxite-bearing sediments is Cenomanian-Turonian, but the major portion of the bauxite ores was apparently formed in Turonian time.

Twenty-six ore-enclosing basins, varying in shape and size, have been discovered in the deposit. Their dimensions in plan vary from a few karst funnels 50 - 60 m in diameter up to relatively circular erosional-karst depressions up to 1500 m long and 300 - 700 m wide. In each isolated basin, there is a single segregation, consisting of two to three main and several subsidiary ore bodies. These bodies have a lensoid shape.

The bauxite segregations of the deposit have been grouped into four sectors: the Southern, Northern, Western, and Karasorsk, extending along the contacts between the tectonic limestone blocks and the volcanogenic rocks. In all, 230 ore bodies have been counted in the deposit, of which 128 are of commercial importance.

The bauxites consist of stony (30.2%), friable (30.4%), and argillaceous (36.5%) varieties, which in vertical section are interlayered with varicoloured clays and allites. Towards the edges of the depressions, the standard bauxites are replaced facies-wise by bauxitic and mottled clays. The stony bauxites, as a rule, form the upper portions of the bauxite-bearing sediments, and the argillaceous types are located in the lower, deepest portion of the ore-enclosing depressions.

In mineral composition, the bauxites belong to the trihydrate (gibbsite) type. From data supplied by the All-Union Institute of Aluminium and Magnesium (VAMI), the principal ore-forming minerals are gibbsite, the amount of which varies from 50 to 70%, kaolinite (2 - 20%), and hematite and hydrohematite (10 - 23%). There are also small amounts of: corundum (1 - 5%), quartz (up to 3%), calcite (up to 1.5%), siderite (up to 10%), and rutile (1.7 - 2.9%).

In chemical composition, the bauxites of the Belinsk deposit are the best in the region. They contain (in wt %): Al_2O_3, 40.4 - 46.8; SiO_2, 5.68 - - 9.64; Fe_2O_3, 12.83 - 23.86; TiO_2, 1.84 - 2.34; CaO, 0.30 - 1.64; silica module, 4.2 - 7.1. In the main, the ores are hydrochemical, and they may be treated by the Bayer method or by Bayer-sintering, and only a small portion of the ores belongs to the metallurgical type.

THE AMANGEL'DA GROUP OF DEPOSITS

The Amangel'da group of deposits of bauxites is located on the eastern margin of the Turgai downwarp in the upper reaches of the River Ashchitasty-turgai.   This group includes five large bauxite deposits, which form the Central ore field.   These are the Arkalyk, Northern, Verkhne-Ashut, Nizhne-Ashut, and Ustoba deposits (Fig. 128).

Rocks of Palaeozoic, Mesozoic, and Cainozoic age are involved in the geological structure of the deposit.   The Palaeozoic basement consists of deformed metamorphic, igneous and sedimentary rocks of Precambrian and Palaeozoic age.   A sequence of Mesozoic-Cainozoic sediments, consisting of continental sand-clay formations, lies horizontally on the uneven surface of the intensely-eroded folded basement.

The oldest Upper Proterozoic and Lower Cambrian formations, forming the western and southern parts of the region, consist of schists, gneisses, quartzites, basic eruptives, their tuffs, and cherty slates.   These rocks have been cut by Caledonian granitoids.

The central portion of the region of the deposits consists of Middle and Upper Devonian and Lower Carboniferous sand-clay and carbonate rocks.   In the lower part, the Devonian sediments consist of conglomerates, and arkosic and polymict sandstones, and above come Frasnian redbed sand-clay and hydro-mica slates and argillites.   The latter are replaced directly, without breaks, by Famennian and lower Tournaisian limestones, and dolomitized and argillaceous limestones.   Transgressively above come cherty limestones and marls of the upper Tournaisian, which are gradually replaced by an alternation of argillites, siltstones, limestones, and sandstones of the Viséan Stage.   The Palaeozoic sequence terminates with Manurian deposits, consisting of grey sandstones, siltstones and argillaceous limestones.

An ancient weathering crust, from a few metres up to 50 - 60 m thick, has been developed on the basement rocks.   The latter are overlain by the Arkalyk bauxite-bearing Group of Palaeocene-Eocene age, consisting in the lower part of sand-clay deposits, on which are located stony, friable, saccharoidal, and argillaceous bauxites.   The bauxites are replaced facies-wise by gibbsite--bearing kaolinitic clays.

The bauxite-bearing group contains a sequence of Palaeogene rocks, amongst which is the thin Chargai Group, consisting of coarse-grained quartzose sands and weakly cemented sandstones.   Above comes the Naurzum Group which consists of varicoloured finely-dispersed clays.   On them rests the Aral' Group, composed of compact gypsiferous montmorillonite clays.   The latter are overlain by the Zhilanda Group of reddish-brown loams with limestones and gypsum concretions.

The ore segregations and bauxite-bearing sectors of the Amangel'da region have been concentrated in five bauxite-bearing zones, near the contacts between the aluminosilicate (sand-clay) rocks of the Frasnian Stage and the carbonate rocks (limestones and dolomites of the Famennian.   Three bauxite--bearing zones (the Aktas, Lower Ashut, and Upper Ashut) extend submeridionally, one of them, the Arkalyk, has an oval shape, and the Ushtoba zone can be traced in a latitudinal direction.   Each zone forms an independent deposit, in addition to the Upper Ashut, consisting of two deposits, the Northern and Upper Ashut.

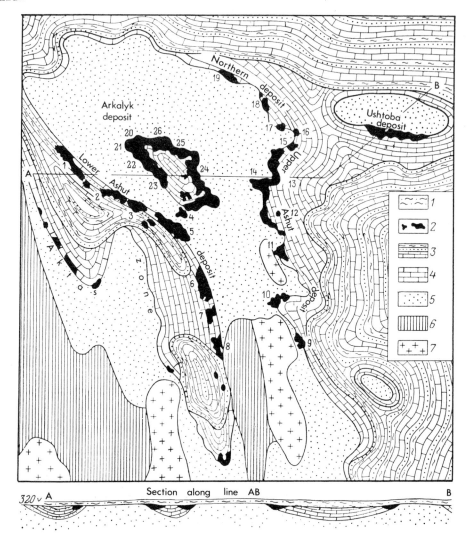

Fig. 128.   Geological-lithological map of the Palaeozoic basement of the Amangel'da group of bauxite deposits (After G. Kirpal')

1) sand-clay sediments; 2) bauxite ore segregations; 3) Lower Carboniferous limestones, argillites, siltstones, and sandstones; 4) Upper Devonian and Lower Carboniferous limestones, and dolomitized limestones; 5) Frasnian hydromica sand-clay shales and siltstones; 6) Precambrian quartzites, quartz-mica and mica-quartz graphitic schists, gneisses, and gneiss-granites; 7) granites.

Ore sectors (numbers on map): 1) West Arkalyk, 2) Geofizicheskii, 3) Ashut XI, 4) Ashut I, 5) Ashut V, 6) Ashut VI, 7) Ashut IX, 8) Ashut X, 9) Ashut XIV, 10) Ashut XIII, 11) Ashut XII, 12) Ashut III, 13) Ashut VIII, 14) Ashut VII, 15) Ashut IV, 16) Southeastern, 17) Southern, 18) Central, 19) Northwestern, 20) Arkalyk II, 21) Arkalyk I, 22) Arkalyk VI, 23) Arkalyk V, 24) Arkalyk VII, 25) Arkalyk IV, 26) Arkalyk III.

Most of the bauxite segregations of the deposits of the Amangel'da group belong to the contact-karst-basinal type, but some small segregations are typically karst in type.   The segregations usually consist of a few ore bodies, which lie horizontally or are weakly inclined and have a very uneven soil and top.

The bauxite segregations are confined to the marginal portions of pre-ore basins, formed in the areas of development of carbonate rocks.   In this respect, they lie on the slopes of depressions, restricted to the contact between the aluminosilicate and carbonate rocks (Fig. 129).   The lower horizons of some segregations in the deposit are arranged in linearly-extended valley--like karst and erosional-karst depressions along structures of the Palaeozoic basement.

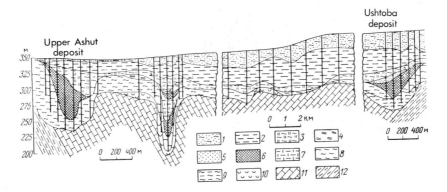

Fig. 129.   Geological section along the line BB (see Fig. 128) (After G. Kirpal').

1) Quaternary loams;   2) compact greenish-grey clays, Aral' Group;
3) mottled clay, Naurzum Group;   4) white, streaky clay, Naurzum
Group;   5) argillaceous clay, Chargai Group;   6) stony, friable,
and argillaceous bauxites and bauxitic clays (ore horizon);   7)
inter-ore, yellow, banded clays;   8) brown and yellow clays (sub--ore horizon);   9) dark-grey and black clays;   10) clays of weathering crust;   11) limestones;   12) sand-clay shales.

The relief of the pre-Cretaceous basement during the time of bauxite--formation was comparatively simple.   Here, we may recognize the Ulutau, Tasta, Arkalyk-Ashut, and Ustoba rises and the Akzhar-Ashut, Arkalyk, Upper Ashut, and Arkalyk-Tasta depressions.   The rises correspond to anticlinal structures, and the depressions, to synclines, consisting, as a rule, of carbonate and terrig-enous-carbonate rocks.   The relative increments in the relief vary from one metre up to several tens of metres.

During the time of bauxite-formation, continental water basins appar-ently existed in the depressed areas along the shores of which, being composed of aluminosilicate rocks, intense chemical weathering of the basement rocks took place.   On the slopes of most of the depressions, along the contact zones between the red sand-clay rocks of the Frasnian Stage and the carbonate rocks of the Famennian and Tournaisian, the formation of linearly-extended and equant karst funnels and depressions and their filling with bauxite-bearing sediments took place during the first phase of bauxite deposition.   During the phase of maximum bauxite deposition, the formation of bauxites spread on to the slopes of the depressions, consisting of aluminosilicate rocks.

The bauxites and genetically associated refractory clays occur at depths of 0 up to 100 m.   They are arranged in the form of a continuous chain of specific ore segregations, separated by sectors of low-quality rocks.   The bauxite segregations of varying dimensions are horizontal or gently-sloping lenses of irregular wavy outline in plan.

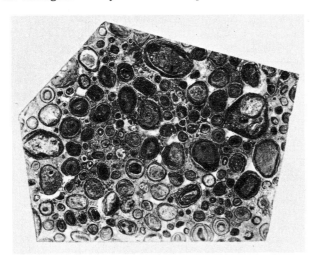

Fig. 130.
Stony bauxite with coarse-
-pisolitic structure,
Arkalyk sector.

The composition of the bauxites and the country rocks for all the deposits is comparatively uniform.   The bauxites usually consist of stony, friable, saccharoidal, and argillaceous varieties, and the refractory clays, consist of gibbsite-bearing kaolinitic varieties.   The mottled clays surrounding the bauxites also have a koalinitic composition.   Between the different varieties of bauxites, refractory clays, and the surrounding rocks, there are gradual transitions (Lisitsyna & Pastukhova, 1963).

The degree of participation of litholigical varieties of bauxites in the various deposits is variable.   The Arkalyk deposit consists mainly of stony (46.6%) and friable (29.1%) bauxites (Fig. 130).   In the Northern deposit, there is almost equal participation of stony (31-0%), friable (23.6%), and argillaceous (27.0%) bauxites.   The Upper Ashut deposit consists of saccharoidal (32.6%), argillaceous (26.2%), and friable (22.2%) bauxites, and stony types are of subordinate importance (11.0%).   In the Lower Ashut deposit, stony bauxites comprise 34.4%, friable bauxites 28.4%, saccharoidal and argillaceous types, 17.4%.

The principal ore-forming minerals of the bauxites are: gibbsite, hematite, and kaolinite.   Their content varies according to the variety and quality of the ores.   The minor minerals include: halloysite, quartz, goethite, rutile, gypsum, calcite, and others.

The chemical composition of the bauxites of the Amangel'da group is characterized by a significant variation in the components (in wt %): Al_2O_3, 28 - 60 (on average 46.42 - 57.11); SiO_2, 1.8 - 2 (on average 9.57 - 14.54); Fe_2O_3, 0.5 - 30 (on average 11.2 - 14.56).

In accordance with the requirements (the production of alumina, heavy metallurgy, abrazive industry), the bauxite ores are divided into the following grades or types: 1) hydrochemical (Bayer), 2) clinkering, 3) metallurgical, and 4) abrasive.

The Naurzum Deposit

The Naurzum deposit is located in the central part of the Turgai downwarp.   Rocks of the pre-Cretaceous deformed basement and horizontally overlying Cretaceous and Cainozoic rocks are involved in the structure of the deposit. The pre-Cretaceous pre-ore basement consists of rocks of the Lower Carboniferous, Lower and Middle Triassic, and Middle Jurassic stages (Fig. 131).

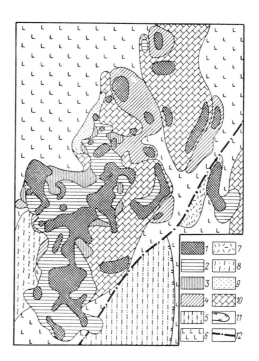

Fig. 131.
Geological-lithological map of the pre-Cretaceous basement and the bauxite-bearing sediments of the Naurzum deposit (After A. Yevlamp'ev & G. Kirpal')

1) Albian-Cenomanian bauxites and bauxitic rocks; 2) allites and bauxitic clays of the same age; 3) mottled and sandy clays (inter--ore horizon) of the same age; 4) kaolinitic clays of the weathering crust; 5) Middle Jurassic silts, argillites, sandstones, and brown coals; 6) basalts and dolerites; 7) quartz porphyries; 8) Lower--Middle Triassic siltstones, argill-ites, and sandstones; 9) upper Palaeozoic sandstones, siltstones, argillites, and conglomerates; 10) limestones with seams of sandstones, siltstones, and argillites; 12) tectonic faults

The Lower Carboniferous formations comprise an undifferentiated sequence of early Tournasian and Viséan age.   They are developed mainly in the central portion of the deposit and most of the bauxite segregations are confined to the belt of their development.   They consist mainly of limestones, and less frequently, sandstones, siltstones, argillites, and tuffites, and sometimes frequent interstratifications of these rocks.

Rocks of the Lower and Middle Triassic sequence (Tura Group) are extremely widely developed in the deposit and outside it. They consist of an eruptive-sedimentary sequence, among which eruptive rocks are of predominant importance and sedimentary formations are markedly inferior. The eruptive rocks consist mainly of basalts and to a lesser degree, dolerites and quartz porphyries. The sedimentary rocks consist of siltstones, argillites, sandstones, and shales.

Jurassic sediments have been developed mainly in the southern part of the Naurzum deposit, where they fill the Dokuchaevo coal-bearing depression. They consist of fine-grained polymict sandstones with seams of argillites and siltstones with a large quantity of coalified wood and brown coal.

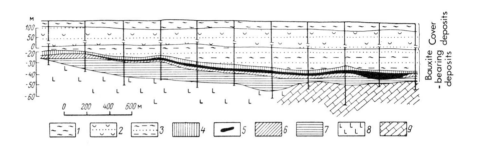

Fig. 132. Geological section through the Naurzum deposit (After G. Kirpal').

1) sandy loams, loams, and sands; 2) interstratification of gaize, sandstones, and sands (Tasaran Group); 3-7) Albian-Cenomanian bauxite--bearing sediments: 3) interlayering of lignitic and kaolinitic clays with quartz sands, 4) allites and kaolinitic clays with siderite; 5) stony, friable, and argillaceous bauxites; 6) bauxite-like clays and sandstones of brick-red colour, 7) interlayering of sandstones, siltstones, and argillites; 8) Lower and Middle Triassic basalts; 9) lower Viséan limestones.

The rocks of the Palaeozoic basement are almost everywhere overlain by formations of the weathering crust, the thickness of which reaches 50 m. A particularly thick weathering crust has been developed on the basalts and the Lower Carboniferous sediments; the Jurassic sediments have been relatively weakly affected by weathering. The upper zones of the weathering crust consist of mottled clays of montmorillonitic and kaolinitic composition.

The rocks of the weathering crust are overlain by continental bauxite--bearing deposits of Albian and Cenomanian age (Fig. 132). Their total thickness is 30 - 80 m. The bauxite-bearing sediments are overlain by marine deposits of the Maastrichtian Stage, developed in the southern part of the deposit and consisting of calcareous glauconite-quartz sandstones; their thickness is on average 10 m. Above come deposits of the Middle and Upper Eocene (Tasaran Group), distributed very widely over the whole area of the deposit. With a marked break, they cover all the above-mentioned sediments. They consist of cherty and sandy gaize, gaize clays, and glauconite-quartz sands and sandstones; their thickness does not exceed 100 m.

Upper Eocene - Lower Oligocene laminated clays (Chegan Group) are known only in the southern portion of the deposit, and in other places they have been removed by the Quaternary erosion; their thickness is 20 m.

The overlying Quaternary formations are developed everywhere, since the Naurzum deposit is located within the ancient Ubagan-Turgai trough, which has been filled with Quaternary lacustrine-alluvial sediments.

The Naurzum deposit is confined to the Kushmurum graben-syncline. The bauxite-bearing sediments within the deposit fill a gentle and shallow depression in the pre-Cretaceous basement, termed the Naurzum basin, with its floor at datum levels of -90 to +10 m. They consist of sand-clay sediments and to a lesser degree, bauxites, allites, and bauxitic clays.

In contrast to the other bauxite-bearing regions of the Turgai down-warp, the bauxites of the Naurzum deposit are not confined to karst depressions, and in the vertical sequence they occur in the middle portion of the bauxite--bearing deposits, and are associated with areas of continuous development of the Albian-Cenomanian continental bauxite-bearing sediments. In the sequence of Albian-Cenomanian bauxite-bearing sediments, four horizons are recognized: lower bauxitic, inter-ore, upper bauxitic, and lignite-bearing (Yevlamp'ev, 1965).

The lower bauxitic horizon has an insignificant area of distribution and is confined to two small depressions, located in Lower Carboniferous lime-stones. It has been found in several drill-holes and consists of stony, friable, and argillaceous bauxites, allites, and bauxitic clays, occurring in the lower portion of the horizon, the thickness of which does not exceed 13 m.

The inter-ore horizon consists of polymict sandstones and siltstones, and in rare cases, argillites, mutually interstratified; the thickness of the horizon does not exceed 30 m.

The upper bauxitic horizon is the principal one, and it is compara-tively widely distributed and maintains its thickness and quality. It occurs at a depth of 134 - 196 m, on average 160 m. This horizon consists of stony, friable, and argillaceous varieties of bauxites, and also allites, and bauxitic, mottled, and kaolinitic clays. The stony varieties predominate amongst the bauxites; argillaceous and friable bauxites occur rarely. The allites, as a rule, occur at the base of the horizon, and below the bauxites, and have a brick-red colour. Grey bauxites are developed in the upper part of the bauxite-ore horizon. They have apparently been formed as a result of intense sideritization of red bauxites and allites.

The bauxites, allites, and mottled clays of the upper bauxite-ore horizon, in contrast to the lower horizon, contain a significant quantity of sandy material. The thickness of the horizon varies from 1 to 15 m.

The supra-ore (lignite-bearing) horizon occurs at depths of 120 - - 180 m, and consists of quartzose sands and dark-grey clays, which are often interstratified and replace each other in facies. The sands are usually quartzose with a small trace of feldspar and fragments of carbonized wood. The clays are sandy, and sometimes bedded, kaolinitic in composition, with fragments of carbonized wood; the horizon is 80 m thick.

In time of formation, the lower bauxite-bearing and the inter-ore horizons may be assigned to the Albian, and the upper bauxite-bearing and the supra-ore (lignite-bearing) horizons, to the Cenomanian.

The Naurzum deposit consists of one principal bauxite-ore segregation, which comprises ten ore bodies, of varying size, irregular shape, complex outline, and different commercial value.   Ore body 1 in size and reserves is the largest; the predominant amount of bauxite reserves is concentrated in it.   Besides the principal segregation, two ore bodies, intersected by isolated drill-holes, have been discovered in the eastern portion of the deposit.

The bauxitic rocks of the Naurzum deposit are characterized by an increased amount of iron (22.5%), titanium oxide (4.8%), carbon dioxide (2.08%), and a relatively decreased amount of alumina (39.98%).   The increased amount of iron and titanium is explained by the fact that the source of the alumina came from the weathering crust on basalts.

The mineral composition of the bauxites is comparatively simple.   The principal ore-forming minerals are gibbsite, and also iron hydroxides and titanium oxide.   Spectral analyses of the bauxites and the surrounding rocks has demonstrated an increased amount, as compared with the clarke values, of vanadium, chromium, nickel, cobalt, zirconium, molybdenum, zinc, scandium, strontium, and niobium.

THE PRINCIPAL TYPES OF NON-BAUXITE ALUMINIUM RAW MATERIALS

Besides the bauxites, aluminous raw material for the production of alumina and aluminium on a commercial scale is obtained from the nepheline ores of the Kiya-Shaltyr and Tezhsar deposits, the nepheline concentrates of the apatite-nepheline ores of the Kola Peninsula, and the alunite ores of the Zaglik deposit.   Promising ores for the extraction of silicon-aluminium melts and aluminium are the high-alumina kyanite schists of the Keiv Group, located on the Kola Peninsula, and the kaolin clays of the Angren and other deposits.

DEPOSITS OF NEPHELINE ORES

Nepheline ores are the second in commercial importance after the bauxites in the nature of aluminium raw materials.   The commercial value of the nepheline ores is determined by the presence of nepheline, which contains up to 35.7% of alumina.   The urtites are richest in nepheline (75 - 85%).   These are high-grade nepheline ores, not requiring enrichment.   The nepheline syenites, which contain more than 5 - 7% of iron oxide, must be subjected to preliminary enrichment.

The nepheline syenites, including the theralite-syenites and apatite-nepheline ores are less rich in nepheline.   Moreover, they are an extremely valuable source of high-grade nepheline concentrates, which are obtained from the tailings of enriched apatite-nepheline ores.   The nephelines, in chemical and mineral composition, are a complex raw material.   From the nepheline, during extraction of limestone, we may obtain alumina, sodium carbonate, potash, and cement.

The treatment of the nepheline ores has been achieved by Soviet industry through the use of nepheline concentrates, obtained from the tailings of enriched apatite-nepheline ores of the Khibiny deposit. In addition, the nepheline ores of the Kiya-Shaltyr and to a lesser degree the Tezhsar deposits, have been used on a commercial scale.

Promising deposits for commercial use are those of Goryachegorsk, Andryushkina Creek and Tyulyul', Medvedkina, Kurgusul', and other deposits of nepheline ores of Eastern Siberia. Nepheline ores of similar quality are those of the Kubasadyr deposit, located in Central Kazakhstan, and the Turpi deposit in Tadzhikistan.

The Kiya-Shaltyr Deposit

The deposit is located in the northern part of the Kuznets Alatau. Its geological construction involves sedimentary, sedimentary-volcanogenic, and metamorphic rocks of different age (from Proterozoic to Devonian), and also intrusive rocks mainly of basic and alkaline composition.

The deposit, according to N. Burukhin, is confined to a massif of alkaline gabbroic rocks, cutting the volcanogenic-sedimentary rocks of the Aldan Stage and the limestones of the Usa Group of the Lower Cambrian (Fig. 133). The intrusive massif has a stock-like shape, extending meridionally for 3 km, with a width of 2 km. The northern portion of the massif consists mainly of trachytoid sectors of K-feldspathized gabbros, and the southern portion consists of hypidiomorphogranular gabbros, nephelinized in varying degree, and also urtites. The massif is surrounded by a zone of metasomatic hornfelses and skarns, from tens of metres up to 300 m wide.

Fig. 133
Diagrammatic geological map of the Kiya-Shaltyr deposit of nepheline ores (After A. Mostovsky).

1) schists, porphyrites, and their tuffs; 2) rhythmically-bedded limestones; 3) tuffites, tuffs, and limestones; 4) organogenic limestones; 5) urtites; 6) porphyritic ijolite-urtites and ijolites; 7) calcite-garnet ijolite-urtites and ijolites; 8) plagioclase-bearing urtites; 9) gabbro-pyroxenites and pyroxenites; 10) gabbros; 11) tectonic faults.

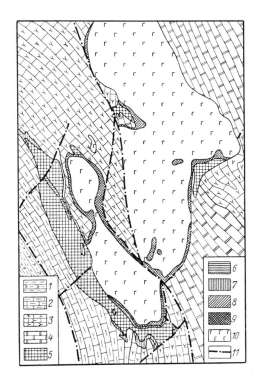

The ore body, consisting of urtite, occurs in the southwestern portion of the massif, along the contact between hypidiomorphogranular gabbros and carbonate rocks. In plan, it is horseshoe-shaped, and at depth, wedge-shaped.

The urtites are characterized by maintenance of mineral-petrographical composition, the almost complete absence of xenoliths of the country rocks, and little jointing. Dykes and schlieren bodies of ijolites, ijolite-porphyries, and diabase porphyrites occur in small quantity, and their thicknesses do not exceed 4 - 5 m (commonly 0.5 - 1.5 m); they comprise about 6% of the total volume of the urtite body.

The principal minerals of the urtites are nepheline (85% of the ore mass, with variations from 74 to 94%) and a titaniferous augite; less common are a titaniferous aegirine-augite, an alkaline hornblende, lepidomelane, apatite, titanomagnetite, pyrite, and pyrrhotite. The chemical composition of the ores is relatively constant (in wt %): Al_2O_3, 24 - 29 (on average 27.75); SiO_2, 39.0 - 0.42 (on average 40.44); average amounts in the ores, Fe_2O, 4.96; and $Na_2O + K_2O$, 13.26.

The conditions for calculating the reserves are as follows: a limiting content of alumina of not less than 24%, lime and soda in total not less than 10%, iron oxides not more than 7%, and silica not more than 43%.

The minimum amount of alumina is 26% in the block in question with a minimum amount of lime and soda of 11.5% in all, and a maximum average amount of iron oxide of 6%, and silica, 42%.

The nepheline ores of the Kiya-Shaltyr deposit are employed without preliminary enrichment for conversion to alumina in the Acha alumina combine.

The Goryachegorsk Deposit

The Goryachegorsk deposit is located in the area where the Kuznets Alatau merges with the Chulym-Yenisei basin, and is characterized by the development of rocks of two structural stages: the lower, consisting of metamorphosed and deformed carbonate sediments of the Upper Proterozoic, and the upper, consisting of interstratified sedimentary and volcanogenic members of Devonian formations. The intrusive rocks of the region belong in age to four complexes, and in this respect, the Goryachegorsk (Lower - Middle Devonian) complex consists of nepheline syenites, alkaline gabbroids, and alkaline syenites.

The deposit, as pointed out by V. Mekhalev, belongs to the massifs of nepheline-bearing rocks of complex petrographical composition, forming the crest and northwestern slopes of Mt Goryachei. The massif has a stock-like shape and occurs mainly in Lower Devonian basaltic porphyrites. The contacts between the massif and the surrounding rocks are steep, vertical in the southern portion of the member, and in the remaining places, dipping at 60 - 80° to the west. Its lower boundary has not been defined; the deepest drill-holes, taken to 470 - 540 m from the surface, did not pass out of the nepheline-bearing rocks.

The structure of the massif is zoned: three zones have been recognized from west to east, elongated meridionally, consisting of nepheline-bearing rocks of the alkaline gabbroid and nepheline syenite groups. The western portion of the massif, where mainly theralites, leucocratic theralites, feldspar ijolites, and syenite-ijolites are distributed, is distinguished by the absence of clear contacts and the presence of an interlayering of the rock types mentioned. In the central part, consisting of feldspar urtites and urtite-syenites, and in the eastern part, where nepheline syenites predominate, the contacts between the rocks are clearer. All the rocks are cut by dykes of syenite-porphyrites, aplites, bereshites, pegmatites, and alkaline lamprophyres; the strike of the dykes is usually southeasterly, and the dip steep to the southwest.

The ores of the Goryachegorsk deposit comprise all the above-listed varieties of rocks; in this respect, the difference between the quality and non-quality ores depends on the ratio of the principal rock-forming minerals, nepheline and feldspars: nepheline predominates in the quality ores, and it always exceeds 40% in amount, whereas feldspars predominate in the non-quality ores.

Allowing for the mineral and chemical composition, three types of ores may be recognized: 1) nepheline gabbroids (leucocratic theralites); 2) nepheline monzonites (theralite-syenites); and 3) nepheline syenites (feldspar urtites).

The conditions for calculating the reserves are as follows: 1) a minimum commercial alumina content (it is marginal in the sample): a) for the theralite-syenites and leucocratic theralites 21%, with a silica content of not more than 44% (silica module 0.48); 5) for the feldspar urtites 22%, with a silica content of not more than 49.0% (silica module 0.45); with a greater silica content, the amount of alumina must satisfy the above-indicated modules; 2) a minimum thickness of the ore interval, included in the scope of the calculation of the reserves, of 10 m; 3) a maximum thickness of the seams of non--quality rocks and non-nepheline dykes, included in the scope of the calculation of the reserves, of 10 and 5 m respectively (and with the impossibility of spatially separating the dykes, their inclusion in the scope of the calculation without limits on the thickness); 4) a separate calculation of the reserves of the individual types of ores (theralite-syenites, leucocratic theralites, and feldspar urtites).

The nepheline ores of the Goryachegorsk deposit are suggested for use in the conversion into alumina with preliminary enrichment. They serve as an additional raw-material basis for the Acha alumina combine.

The Nepheline Concentrates

At the present time, in order to obtain alumina in the Volkhov aluminium plant and the Pikalevsk alumina combine, nepheline concentrates have been used in significant quantities; these concentrates are the products of enrichment of the apatite-nepheline ores.

More than half of the reserves of the nepheline ores are included in the complex apatite-nepheline ores of the deposits of the Murmansk district, being the principal source for obtaining an apatite concentrate; the nepheline concentrate is obtained as a by-product, from portion of the extracted and treated ores.

The apatite-nepheline ores contain only 10 - 14% alumina, and the nepheline concentrate contains 29% alumina. Of the total reserves of the apatite-nepheline ores, 30 - 35% may be provisionally assigned to the proportion

of aluminium raw material. Of the apatite-nepheline ores of the Khibiny deposit,
0.7 tonne of nepheline may be obtained from each tonne of apatite concentrate.

There are at present no officially enforced technical conditions on
the nepheline ore and the nepheline concentrate as a raw material for the production of alumina.   In assessing the nepheline-bearing rocks as an aluminous
raw material, the alkali module is of significant importance (the molecular
ratio of the total of alkalies, potassium and sodium, to aluminium).   The best-
-quality rocks must have this module close to unity.   Account is also taken of
the silica module, which must not exceed 3.3 - 3.4.   With an iron-oxide content
of more than 7%, the nepheline ore must be enriched (and the iron removed).

DEPOSITS OF ALUNITE ORES

The alunite ores are a complex raw material, since during their treatment we may obtain, in addition to alumina, potash fertilizers and sulphuric
acid.   The alunite ores are formed under various geological conditions, although
the commercial deposits are associated with young volcanism.   In this respect,
the alunite ores are formed as a result of the action of volcanogenic sulphurous
gases and solutions, enriched in sulphuric acid, on the country rocks.

The largest deposit in the USSR is the Zaglik occurrence in the Azerbaidzhan SSR, the Aktash, Gusshai, Kattaisai, Urgaz, and Ravatsai deposits in
the Uzbek SSR, the Began'kov in the western part of the Ukraine, and the Iska
deposit in the Soviet Far East.

The Zaglik Deposit

The Zaglik deposit is located in the Dashkesan region of the Azerbaidzhan SSR.   The region of the deposit, according to R. Kofman and K. Aliev,
represents a synclinal fold with gentle dips, and consists of Jurassic tuffogenic
rocks, which have undergone metasomatism in the area of the deposit, under the
influence of hydrothermal solutions, and have formed alunite ores (Fig. 134).

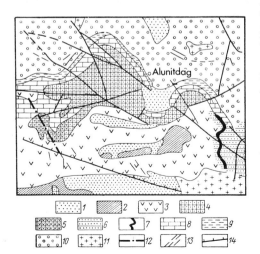

Fig. 134.
Geological map of the Zaglik-Dashkesan
group of alunite deposits (After M.-A.
Kashkai).

1) alluvial, eluvial, and deluvial Quaternary formations;   2) Lower Cretaceous-Upper
Jurassic diabase porphyrites;   3-9) Kimmeridgian upper group of tuffs and tuffites,
4) alunitized, pyrophyllitized, and kaolin-
itized sequence,   5) tuffites, underlying
and overlying the alunite sequence,   6)
hornfelses after Oxfordian-Kimmeridgian
and in part, Middle Jurassic rocks,   7)
skarn-iron-ore segregation,   8) Lusitanian
limestones,   9) Callovian argillites and
sandstones with seams of marls;   10) Middle
Jurassic volcanogenic clastic rocks of
porphyrite composition;   11) gabbroids and
granitoids;   12) tectonic faults;   13) dykes
of vein rocks of basic composition;   14)
axis of Dashkesan syncline.

The Zaglik deposit is a homoclinally arranged sequence of hydrothermally altered volcanic rocks with dips of 10 - 15° to 190 - 210° (SW). The marker horizon is the limestones of Lusitanian age (upper Oxfordian - lower Kimmeridgian), up to 175 m thick. The limestones are conformably overlain by Kimmeridgian tuffogenic rocks, consisting of tuffs, tuffites, tuff-sandstones, and tuff-breccias. These tuffogenic rocks have undergone alunitization, and the degree of alunitization varies throughout the entire thickness.

In this connexion, alternation and interbedding of the alunitized and non-ore layers has been observed. Change in the degree of alunitization has also been recorded along the strike. The thickness of the non-ore layers varies from a few metres up to 10 - 20 m.

The main ore body has a layer-like form, conformable with the general attitude of the sequence. In the southeastern sector, it splits into two members, and the upper member rapidly thins out. The lower member has the greater area of distribution. Its extent along strike is from 1.5 to 2.6 km.

For the calculation of the reserves of the alunite ores, conditions have been laid down that require a marginal amount of alunite of 25%, a minimum commercial content per block of 48%, and a maximum amount of non-alunite alumina of 4.9%. The minimum thickness of the ore layers, included in the calculation of the reserves, is 2 m; the maximum thickness of the stripped rocks to useful material per block is 5 : 1; the marginal content of alunite for calculating suitable reserves is 10%. In order to calculate the amount of alunite in the samples on the basis of sulphur trioxide, a coefficient of 2.55 has been accepted.

The ores consist mainly of alunite and quartz, and clay minerals are present in small amount (about 5). The alunite of the deposit is characterized by a K-Na composition with a predominance of K_2O over Na_2O. The amount of alunite is uneven and varies, on the basis of individual intervals of sampling, from 10 to 80%, but is most commonly 40 - 60%, comprising on average per block, 51 - 57%, and per locality, 53%.

The average amount of Na_2O per block is 1.3 - 1.8%, K_2O, 3.4 - 4.2%, and the amount of non-alunite alumina does not exceed 2.6%.

DEPOSITS OF KYANITE ORES

Soviet specialists have developed an electrothermic method of obtaining aluminium and its alloys from high-alumina rocks such as kyanite, sillimanite, and andalusite schists. High-alumina schists are known in many regions of the Soviet Union. The largest reserves of kyanite schists have been investigated on the Kola Peninsula, where they may be extracted by open-cut methods. The deposits of sillimanite schists occur in the Irkutsk district (Kitoi), and in the Buryat ASSR (Kyakhta), and kyanite-sillimanite placers have been discovered in the Ukraine, etc.

The deposits of kyanite, sillimanite, and andalusite have been formed as a result of intense metamorphism of aluminosilicate rocks and, in the first instance, clay shales. The association between these deposits and the zones of intense metamorphism is dependent on their restriction to ancient metamorphic complexes. Examples are the metamorphic complexes of the Kola

Peninsula, the Patom plateau, and the Eastern Sayana. Kyanite and sillimanite schists when used, must be subjected to enrichment with the object of obtaining pure mineral concentrates, suitable for electrochemical treatment.

The VAMI and Hydrotsement institutes have set up requirements for concentrates of high-alumina ores of kyanite, sillimanite, and andalusite composition, which may be treated by the electrochemical method for obtaining aluminium and its alloys. The composition of the concentrates must meet the following requirements: an alumina content of not less than 57%, silica not more than 37.5%, iron oxide not more than 1.5%, titanium oxide not more than 0.5%, total soda and potash not more than 0.7%, and total lime and magnesia not more than 0.6%.

The Keiv Group of Deposits

The Keiv group includes the following deposits of kyanite ores: the Vol'gel'urta, Tyashi-Manyuk, Chervurta, Bezymyannoe, Bol'shoi Rov, Nussa, Novaya Shuururta, etc. (Fig. 135). The last is the largest and possesses the largest amount of kyanite (40.9%) with a small content of deleterious additives.

The deposits are located in the central part of the Kola Peninsula, on the watershed ridge, extending northeastwards for more than 200 km; the ridge consists of individual high points, termed keivs. The width of the metamorphic rocks of the Keiv Group, surrounding the kyanite-ore deposits, varies from a few hundreds of metres on the flanks up to 10 - 15 km in its central part.

The Keiv ridge, according to A. Karyakin, consists of biotite and biotite-garnet gneisses, mica, staurolite, and kyanite schists, quartzites, and metamorphosed carbonate rocks, comprising the Keiv Group of Late Archaean age. The rocks of the Keiv Group occur in the form of a complex synclinorium, which

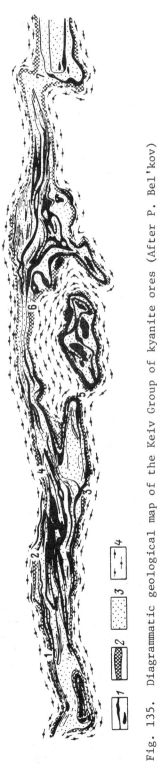

Fig. 135. Diagrammatic geological map of the Keiv Group of kyanite ores (After P. Bel'kov)

1) amphibolites and schists of the Keiv Group; 2) productive kyanite schists; 3) non-productive schists; 4) biotite, garnet-biotite, and amphibole gneisses. Deposits of kyanite ores (figures on map): *1)* Vol'gel'urta, *2)* Tyashi-Manyuk, *3)* Chervurta, *4)* Bezymyannoe, *5)* Bol'shoi Rov, *6)* Novaya Shuururta, *7)* Nussa.

has an asymmetrical structure and a southeasterly strike.   The synclinorium
consists of several synclinal structures.   The formation of the folds was
accompanied by injection of anorthosites and porphyritic granites, dykes and
sills of amphibolites, and also alkaline granites.

The Keiv deposits of kyanite ores are associated with kyanite, quartz-
-kyanite, and staurolite-kyanite schists, which form the productive horizon, up
to 80 - 150 m thick, traceable over a distance of 140 km.   The deposits represent individual sectors of the belt of kyanite schists.

Three principal types of kyanite ores are recognized according to
the morphological features of the kyanite in the deposits: fibrous-acicular,
paramorphic, and concretionary.   Porphyroblastic and concretionary-paramorphic
types of ores are of minor importance.   The best ores are the coarse-concre-
tionary kyanite schists, which on enrichment provide high-quality concentrates
with a content of 57% alumina, and a minimum amount of deleterious components.

According to the amount of kyanite in the schists, we recognize rich
ores with a content of this mineral above 40%, medium ores, 30 - 40%, and lean
ores, with less than 20%.

Temporary conditions for the largest and higher-quality deposits of
Novaya Shuururta are as follows: a marginal content of kyanite of 25%, a
minimum commercial content of kyanite (per block) of 30%, a minimum commercial
thickness of the ore bodies of 5 m, and a maximum thickness of the seams of
non-quality ores and overburden, included in the commercial scope, of 5 m.
Ores of the paramorphic type have been reckoned separately and are assigned
to the non-commercial types.

The principal ore-forming minerals are kyanite, quartz, staurolite,
muscovite, graphite, and carbonaceous matter, with rutile and plagioclase
sometimes present.   The crystalline schists in some cases contain 10 - 40%
of sillimanite, and rarely garnet.

The Keiv deposits of kyanite ores have vast reserves, which in scale
have no equal in the world and exceed all other deposits taken together.

KAOLINS AND HIGH-ALUMINA CLAYS

The question of using kaolin and high-alumina clays as a raw material
for the production of aluminium has long been studied, especially in countries
with a vigorous aluminium industry, but having no bauxite deposits or possessing
limited resources of bauxite ores.   Such countries include the USA, Canada,
West Germany, Austria, Norway, Sweden, and a number of others.

The kaolin and high-alumina clays are regarded as potential sources
for the production of aluminium.   In recent times, experiments have been con-
ducted in a number of countries on the most economical methods of extracting
alumina from them.   In the USA, eight companies have constructed experimental
plants for the treatment of non-bauxite raw materials.   On a commercial scale,
alumina has been obtained from clays, treated with hydrochloric acid;   the
nitric acid method is also widely used.   During the treatment of 1000 tonnes
of clay per day, the cost of 1 tonne of alumina is 60.6 dollars.   The cost
per tonne of alumina on the world market is about 65 dollars.   In Georgia

(USA) in 1966, a plant was constructed for the treatment of clays to produce 180,000 tonnes of alumina per year.

In the Soviet Union, deposits of kaolins and high-alumina clays are widely distributed. However, up till now they have not been used for conversion to alumina. In this respect, there are not specific requirements for these kinds of raw materials. It is evident that the clay raw material for conversion to alumina must contain not less than 30 - 32% of aluminium oxide with a minimum amount of iron oxide and silica. An ideal composition in this respect is that of a pure kaolinite. In addition to the pure kaolinite clays, kaolinite concentrates, obtained during the mechanical enrichment of clays, may also be used in the aluminium industry.

Recently in the USSR, kaolinite concentrates of the Prosyanovsk deposit have been employed for the production of silicon-aluminium alloys, and with subsequent addition of aluminium, silumin, analogous to the synthetic material, has been obtained.

The Angren deposit of brown coal and kaolins is promising for the production of alumina and aluminium on a huge commercial scale.

The Angren Deposit

The Angren deposit is located 110 km southeast of Tashkent in the valley of the River Angren. It is restricted to a gentle synclinal fold, bounded on the northeast and the west by the foothills of the Kuraminsk and Chatkal ranges.

The productive sequences of sand-clay kaolin rocks and coals of Jurassic age, with a total thickness of 130 - 180 m, occurs in a weathering crust of Paleozoic rocks (primary kaolins). They have been divided into two groups, the lower coal-bearing unit, and the upper kaolin group. The secondary supra-coal kaolins, investigated as an alumina raw material, comprise a thick sequence of interbedded finely-dispersed clays, sandy clays, and kaolinized sandstones and siltstones, overlain by deposits of Cretaceous, Tertiary, and Quaternary age, from 20 to 200 m thick, on average 90 m.

The productive sequence of secondary kaolins includes mottled kaolins, up to 60 m thick, and on average amount 20 m thick (the kaolin group) and grey kaolins, 20 - 30 m thick (the supra-coal subgroup of the coal-bearing group). The clay material of the kaolinite rocks consists of kaolinite, and a sandy and powdery portion of quartz grains. The presence of hydromicas, feldspars, calcite, and heavy minerals, has been established.

The non-enriched secondary kaolins contain (in wt %): Al_2O_3, 23 - 24; SiO_2, 60; Fe_2O_3, 1.72 - 3.79; $K_2O + Na_2O$, 1.5 - 3.0. The kaolinite concentrate, obtained for commercial use during the production of alumina, may have the following composition (in wt %): Al_2O_3, 33.0; SiO_2, 47.5; CaO, 1.2; MgO, 0.25; Na_2O, 0.13; K_2O, 1.7; Fe_2O_3, 1.7. It is planned to use the concentrates by sintering with limestone, which will ensure the production of non-alkaline self--dispersed clinkers. At present, technological plant investigations are being carried out on the treatment of the Angren kaolins for conversion to alumina. The use of other deposits of kaolins depends on the results obtained during technological experiments on the kaolins of the Angren deposit.

REFERENCES

ABRAMOV V.P. (АБРАМОВ В.П.), 1970: Тиман -- новый бокситоносный бассейн (Timan, a new bauxite-bearing basin). *Razv. Okhr. Nedr*, No.2, pp.3-7.

AGEENKO N.F. (АГЕЕНКО Н.Ф.), 1970: Новые данные по Обуховскому месторождению бокситов и перспективы геосинклинальных месторождений Салаира (New data on the Obukhovo bauxite deposit and prospects of the geosynclinal deposits of Salair) *in* "Новые данные по геологии и полезным ископаемым Западной Сибири" (*'New Data on the Geology and Mineral Deposits of Western Siberia'*), Vol.5, pp.19-26.   Tomsk Univ. Press.

ARKHANGEL'SKY A.D. (АРХАНГЕЛЬСКИЙ А.Д.), 1954: Типы бокситов СССР и их генезис (Types of bauxites of the USSR and their origin), *in* "Избранные труды" (*'Published Works'*), Vol.2, pp.509-622.   Akad. Nauk SSSR Press, Moscow.

BARDOSSY G. (БАРДОШИ Д.), 1957: Геология бокситовых месторождений Венгрии (The geology of the bauxite deposits of Hungary). *Izv. Akad. Nauk SSSR, ser. geol.*, No.9, pp.3-18.

BASS Yu.B., RYABCHUN V.K., SLAVUTSKY M.V. & SHALYT E.S, (БАСС Ю.Б., РЯБЧУН В.К., СЛАВУТСКИЙ М.В., ШАЛЫТ Е.С.), 1971: Бокситы платформенной части Украинской ССР (южная провинция) (Bauxites of the platform part of the Ukrainian SSR (southern province)), *in* "Платформенные бокситы СССР" (*'The Platform Bauxites of the USSR'*), pp.93-128.   Nauka Press, Moscow.

BOL'SHUN G.A. (БОЛЬШУН Г.А.), 1970: Условия образования бокситов Урала (The conditions for the formation of bauxites in the Urals). *Sov. Geol.*, No.9, pp.74-85.

BUSHINSKY G.I. (БУШИНСКИЙ Г.И.), 1962: Роль карста в образовании и распространении бокситовых месторождений (The role of karst in the formation and distribution of bauxite deposits). *Byull. Mosk. Obshch. Ispyt. Prir., 37,* No.6, pp.143-144.

BUSHINSKY G.I. (БУШИНСКИЙ Г.И.), 1971: Геология бокситов (*The Geology of Bauxites*) Nedra Press, Moscow.

GLADKOVSKY A.K. & SHAROVA A.K. (ГЛАДКОВСКИЙ А.К., ШАРОВА А.К.), 1951: Бокситы Урала (*The Bauxites of the Urals*).   Gosgeoltekhizdat, Moscow.

GOLOVENOK V.K. & PUSHKIN G.Yu. (ГОЛОВЕНОК В.К., ПУШКИН Г.Ю.), 1964: О находке докембрийских бокситов (на Патомском нагорье в Сибири) (The discovery of Precambrian bauxites (on the Patom Plateau in Siberia)), *Litol. polezn. Iskop.*, No.1, pp.114-116.

GORETSKY Yu.K. (ГОРЕЦКИЙ Ю.К.), 1960: Закономерности размещения и условия образования основных типов бокситовых месторождений (Distribution patterns and conditions of formation of the principal types of bauxite deposits). *Trudj vses. nauchno-issled. Inst. miner. Syr'ya, 5* (n.s.).

GRIGOR'EV V.N. (ГРИГОРЬЕВ В.Н.), 1968: Генезис верхнебаширских бокситовых пород Средней Азии (The origin of the upper Bashkirian bauxite rocks of Middle Asia), *Litol. polezn. Iskop.*, No.1, pp.44-55.

GUTKIN E.S. & RODCHENKO Yu.M. (ГУТКИН Е.С., РОДЧЕНКО Ю.М.), 1965: Тектоника Северо-Уральского бассейна и её связь с бокситовым оруденением (The tectonics of the North Urals basin and its association with bauxite mineralization). *Izv. Akad. Nauk SSSR, ser. geol.*, No.2, pp.56-66.

IL'INA N.S. (ИЛЬИНА Н.С.), 1958: Геология и генезис боксонских бокситов в Восточных Саянах (The geology and origin of the Bokson bauxites in the Eastern Sayany) *in* "Бокситы, их минералогия и генезис" (*'Bauxites, Their Mineralogy and Origin'*), pp.267-281.   Moscow.

(Instructions for Applying the Classification of Reserves to Bauxite Deposits) Инструкция по применению классификации запасов к месторождениям бокситов.   GKZ.   Gosgeoltekhizdat, Moscow (1962).

KIRPAL' G.R. (КИРПАЛЬ Г.Р.), 1964: Эпохи бокситонакопления в меловое и палеогеновое время на территории Тургайского прогиба (Epochs of bauxite deposition during Cretaceous and Palaeogene time in the area of the Turgai downwarp).   *Geologiya rudn. Mestorozh.*, *6*, No.6, pp.110-122.

KIRPAL' G.R. (КИРПАЛЬ Г.Р.), 1966: Закономерности размещения и условия образования тургайских бокситов (Distribution patterns in and conditions of formation of the Turgai bauxites), *in* "Генезис бокситов" (*'The Origin of Bauxites'*), pp.193-199.   Nauka Press, Moscow.

KIRPAL' G.R. (КИРПАЛЬ Г.Р.), 1971*a*: Состояние минерально-сырьевой базы алюминиевой промышленности и задачи по дальнейшему её расширению (The state of the mineral raw-material basis of the aluminium industry and the objects of further expansion).   *Trudy sib. nauchno-issled. Inst. Geol. Geofiz. miner. Syr'ya, 126*, pp.4-18.

KIRPAL' G.R. (КИРПАЛЬ Г.Р.), 1971*b*: Оценка бокситоносности и дальнейшее направление поисково-разведочных работ на бокситы в Сибири и на Дальнем Востоке (An assessment of the bauxite content and the future direction of exploratory-research work on bauxites in Siberia and the Far East). *Trudy sib. nauchno-issled. Inst. Geol. Geofiz. miner. Syr'ya, 121*, pp.4-20.

KIRPAL' G.R. (КИРПАЛЬ Г.Р.), 1972: Перспективы поисков бокситов на Русской платформе (Prospects for finding bauxites on the Russian Platform). *Sov. Geol.*, No.8, pp.36-50.

KIRPAL' G.R. & KHATSKEVICH V.A. (КИРПАЛЬ Г.Р., ХАЦКЕВИЧ В.А.), 1971: Континентальные отложения меловой системы (Continental deposits of the Cretaceous System) *in* "Геология СССР. Т.34 -- Тургайский прогиб" (*'Geology of the USSR. Vol.34. -- Turgai Downwarp'*), pp.238-265.   Nedra Press, Moscow.

KISELEV L.I. (КИСЕЛЕВ Л.И.), 1963: О возрасте древней коры выветривания в Мугоджарах (The age of the ancient weathering crust in the Mugodzhary). *Vestn. Akad. Nauk Kaz. SSR, 3*, pp.70-75.

KISELEV L.I. (КИСЕЛЕВ Л.И.), 1970: Триасовая система.   Древняя кора выветривания.   Мугоджары.   (The Triassic System.   The Ancient Weathering Crust.   Mugodzhary) *in* "Геология СССР" (*'The Geology of the USSR'*), Vol.21, pp.345-350.   Nedra Press, Moscow.

KLEKL' V.N. (КЛЕКЛЬ В.Н.), 1969: Древние коры выветривания КМА и перспективы поисковых работ на бокситы (The ancient weathering crusts of the KMA and the prospects for finding bauxites in them).   *Litol. polezn. Iskop.*, No.5, pp.5-16.

KONNOV L.P.   (КОННОВ Л.П.), 1972:   Геология и генизис бокситов Средней Азии *(The Geology and Origin of the Bauxites of Middle Asia)*.   Nedra Press, Moscow.

KRIVTSOV A.I.   (КРИВЦОВ А.И.), 1968-1969:   Мезозойские и кайнозойские бокситы СССР, их генезис и промышленное значение *(The Mesozoic and Cainozoic Bauxites of the USSR, Their Origin and Commercial Value)*, Vols 1 and 2.   Nedra Press, Moscow.

LISITSYNA N.A. & PASTUKHOVA M.V. (ЛИСИЦЫНА Н.А., ПАСТУХОВА М.В.), 1963: Структурные типы мезо-кайнозойских бокситов Казахстана и Западной Сибири (Structural types of Mesozoic-Cainozoic bauxites of Kazakhstan and Western Siberia).   *Trudy geol. Inst. Mosk., 95.*

NIKITINA A.P., VITOVSKAYA I.V. & NIKITIN K.K.   (НИКИТИНА А.П., ВИТОВСКАЯ И.В., НИКИТИН К.К.), 1971:   Минералого-геохимические закономерности формирования профилей и полезных ископаемых коры выветривания; некоторые вопросы методики их изучения *(Mineralogical-Geochemical Patterns in the Formation of Profiles and Mineral Deposits of the Weathering Crust; Some Problems in Methods of Studying Them)*.   Nauka Press, Moscow.

PEIVE A.V. (ПЕЙВЕ А.В.), 1947:   Тектоника Североуральского бокситового пояса (The tectonics of the North Urals bauxite belt), *in*   "Материалы по познанию геологического строения СССР" *('Data on the Geological Structure of the USSR')*, n.s., вұp. *4* (8), 204 pp.

PEL'TEK E.N.   (ПЕЛЬТЕК Е.Н.), 1971:   Месторождения бокситов Енисейского кряжа и Сибирской платформы (Ангаро-Енисейская провинция) (The bauxite deposits of the Yenisei Ridge and the Siberian Platform (the Angara-Yenisei Province)), *in*   "Платформенные бокситы СССР" *('The Platform Bauxites of the USSR')*, pp.221-262.   Nauka Press, Moscow.

RODCHENKO Yu.M. (РОДЧЕНКО Ю.М.), 1964:   Некоторые особенности геологического строения и достоверность результатов разведки североуральских бокситовых месторождений (Some features of the geological structure and confirmation of the results of exploration of the Northern Urals bauxite deposits).   *Trudy Inst. Geol. ural'. Fil., 64,* pp.177-223.

SAPOZHNIKOV D.G.   (САПОЖНИКОВ Д.Г.), 1971*a*:   Основные бокситоносные провинции СССР   (The principal bauxite provinces of the USSR), *in* "Платформенные бокситы СССР" *('The Platform Bauxites of the USSR')*, pp.5-21. Nauka Press, Moscow.

SAPOZHNIKOV D.G.   (САПОЖНИКОВ Д.Г.), 1971*b*:   Типы платформенных бокситов СССР, их особенности и условия образования   (The types of platform bauxites of the USSR, their features and conditions of formation), *in* "Платформенные бокситы СССР" *('The Platform Bauxites of the USSR')*, pp.320-351.   Nauka Press, Moscow.

STRAKHOV N.M. (СТРАХОВ Н.М.), 1947:   Железорудные фации и их аналоги в истории Земли   (The iron-ore facies and their equivalents during the history of the Earth).   *Trudy Inst. geol. Nauk Mosk., 73.*

TENYAKOV V.A. (ТЕНЯКОВ В.А.), 1971:   О принципиальной аналогии в источнике и способе формирования вещества бокситов в платформенных и геосинклинальных областях (A fundamental equivalent in the source and method of formation of the material of bauxites in platform and geosynclinal districts). *Dokl. Akad. Nauk SSSR, 198,* pp.198-201.

TENYAKOV V.A. & AKAEMOV S.T. (ТЕНЯКОВ В.А., АКАЕМОВ С.Т.), 1972: "Латеритные коры выветривания" Гвинеи. Некоторые принципальные вопросы геологии и возраста ('The laterite weathering crusts' of Guinea. Some fundamental problems of their geology and age). *Dokl. Akad. Nauk SSSR*, *202*, pp.1159-1161.

TENYAKOV V.A., ELDIN M.G., SAPRYKINA N.V. & VINOKUROV Ts.K. (ТЕНЯКОВ В.А., ЭЛДИН М.Г., САПРЫКИНА Н.В., ВИНОКУРОВ Ц.К.), 1972: О поглощенном комплексе осадочных бокситов (в связи с проблемой реконструкции палеофациальных условий их накопления) (The absorption complex of bauxites (in connexion with the problem of reconstructing the paleofacies conditions of their deposition)). *Dokl. Akad. Nauk SSSR*, *202*, pp.1423-1426.

TYURIN B.A. (ТЮРИН Б.А.), 1958: Месторождения гиббситовых бокситов Амангельдинского бокситорудного района в Центральном Казахстане (The deposits of gibbsite bauxites of the Amangel'da bauxite-bearing region in Central Kazakhstan), *in* "Бокситы, их минералогия и генезис" (*'Bauxites, Their Mineralogy and Origin*), pp.416-430. Akad. Nauk SSSR Press, Moscow.

YERSHOV A.D. (Ed.) (ЕРШОВ А.Д. (глав. ред.), 1962: Требования промышленности к качества минерального сырья. Алюминий. (Quality requirements of industry for mineral raw-materials. Aluminium). *Vses. nauchno--issled. Inst. miner. Syr'ya*, vyp. *35*.